T0299726

CHEMISTRY AND TECHNOLOGY OF PLANT SUBSTANCES

Chemical and Biochemical Aspects

CHEMISTRY AND TECHNOLOGY OF PLANT SUBSTANCES

Chemical and Biochemical Aspects

Edited by
Alexander V. Kutchin, DSc
Lyudmila N. Shishkina, DSc
Larissa I. Weisfeld, PhD

Reviewers and Advisory Board Members
Gennady E. Zaikov, DSc,
Ilya N. Kurochkin, DSc,
Alexander N. Goloshchapov, PhD

AAP | APPLE ACADEMIC PRESS

Apple Academic Press Inc.
3333 Mistwell Crescent
Oakville, ON L6L 0A2 Canada

Apple Academic Press Inc.
9 Spinnaker Way
Waretown, NJ 08758 USA

Library and Archives Canada Cataloguing in Publication

Chemistry and technology of plant substances : chemical and biochemical aspects / edited by Alexandr V. Kutchin, DSc, Lyudmila N. Shishkina, DSc, Larissa I. Weisfeld, PhD ; reviewers and advisory board members, Gennady E. Zaikov, DSc, Igor N. Kurochkin, DSc, Alexander N. Goloshchapov, PhD.

Includes bibliographical references and index.
Issued in print and electronic formats.
ISBN 978-1-77188-560-7 (hardcover).--ISBN 978-1-315-20746-9 (PDF)

1. Botanical chemistry. I. Kutchin, Alexandr V., editor II. Shishkina, Ludmila N., editor III. Weisfeld, Larissa I., editor

QK861.C54 2017   572'.2   C2017-902312-8   C2017-902313-6

Library of Congress Cataloging-in-Publication Data

Names: Kutchin, Alexandr V., editor. | Shishkina, Ludmila N., editor. | Weisfeld, Larissa I., editor.
Title: Chemistry and technology of plant substances : chemical and biochemical aspects / editors, Alexandr V. Kutchin, Lyudmila N. Shishkina, Larissa I. Weisfeld.
Description: Toronto; New Jersey : Apple Academic Press, 2017. | Includes bibliographical references and index.
Identifiers: LCCN 2017015048 (print) | LCCN 2017015859 (ebook) | ISBN 9781315207469 (ebook) | ISBN 9781771885607 (hardcover : alk. paper)
Subjects: | MESH: Plants--chemistry | Plant Physiological Phenomena | Plant Extracts | Phytochemicals
Classification: LCC QK861 (ebook) | LCC QK861 (print) | NLM QK 861 | DDC 572/.2--dc23
LC record available at https://lccn.loc.gov/2017015048

Apple Academic Press also publishes its books in a variety of electronic formats. Some content that ap-pears in print may not be available in electronic format. For information about Apple Academic Press products, visit our website at **www.appleacademicpress.com** and the CRC Press website at **www.crc-press.com**

ABOUT THE EDITORS

Alexander V. Kutchin, DSc

Alexander V. Kutchin, DSc in chemistry, is a Professor of Organic Chemistry and a corresponding member of the Russian Academy of Sciences as well as Director of the Institute of Chemistry of Komi Scientific Center of Ural Branch of Russian Academy of Sciences. He is a leading expert in the field of organic and organometallic synthesis, chemistry, and technology of natural compounds. Dr. Kutchin is the founder of the recognized scientific school of "Scientific bases of chemistry and technologies complex processing of vegetal raw material," established in 1994. He has received several honors and awards, including the Nesmeyanov RAS Prize; the Academician I. Ya. Postovskiy Prize; the Order of "For Merit for the Motherland"; the Order of Friendship (Russian Federation); the Sorokin Award, issued by the Komi Republic Government, and the Komi Republic Government Award for Contribution to the Field of Science Research (Russian Federation). He has published over 700 scientific papers, including many articles and patents. He is a member of the editorial boards of the journals *Russian Chemical Bulletin* and *Chemistry of Plant Raw Material*. His field of interest includes organic synthesis, asymmetric synthesis, chemistry of natural compounds, medicinal chemistry, and biotechnology.

Lyudmila N. Shishkina, DSc

Lyudmila N. Shishkina, DSc in chemistry, is a Professor of Biophysics and Head of the Department of Physicochemical Problems in Radiobiology and Ecology at the Emanuel Institute of Biochemical Physics of the Russian Academy of Sciences, Moscow, Russia. She is also a member of the Scientific Counsel of the Russian Academy of Sciences on Radiobiology. She is laureate of the medal "In Memory of Academician N. M. Emanuel" for achievements in chemical and biochemical physics. She has given several invited lectures and oral presentations at different international conferences. Her work includes over 200 research publications in different leading Russian and international journals and textbooks, two monographs, and several author certifications. Her fields of interest include biochemical physics, chemical kinetics, regulation of oxidation processes in systems of different complexity, radiobiology, and physico-chemical ecology.

Larissa I. Weisfeld, PhD

Larissa I. Weisfeld, PhD, is a senior researcher at the Emanuel Institute of Biochemical Physics, Russian Academy of Sciences, Moscow, Russia; a member of the All-Russia Vavilov Society of Geneticists and Breeders; the author of about 300 publications in scientific journals as well as several patents and conference proceedings; and the co-author of publications on three new cultivars of winter wheat. Her main field of interest concerns basic problems of chemical mutagenesis, mutational selection, and the mechanism of action of para-amonobensoic acid. She has worked as a scientific editor in the publishing house Nauka (Moscow) and in the jour-nals *Genetics* and *Ontogenesis.* She has co-edited several books with Apple Academic Press, including *Ecological Consequences of Increasing Crop Productivity: Plant Breeding and Biotic Diversity; Biological Systems, Biodiversity, and Stability of Plant Communities; Temperate Crop Science and Breeding: Ecological and Genetic Study;* and *Heavy Metals and Other Pollutants in the Environment: Biological Aspects.*

CONTENTS

LIST OF CONTRIBUTORS

Ekaterina S. Alinkina, PhD
Emanuel Institute of Biochemical Physics of Russian Academy of Sciences, Laboratory of Physical Chemical Regulation of Biological Systems Researcher, 4, Kosygin St., Moscow 119334, Russia. E-mail: tmish@rambler.ru

Ekaterina S. Ayskhanova, PhD
Associate Professor, Chechen State University, Department of Criminal Law, Chechen State University, 17, Dudayev Blvd, Grozny, Chechen Republic 364060, Russia. E-mail: gishkaieva.66@mail.ru

Maksim P. Bei, PhD
Institute of Chemistry of New Materials of the National Academy of Science of Belarus, the Laboratory of Forest-and Petrochemical Products, Researcher, 36, Skoriny St, Minsk 220141, Republic of Belarus. E-mail: beymaksim@gmail.com, aspirin55@yandex.ru

Sarra A. Bekuzarova, DSc
Honored the Inventor of the Russian Federation, Professor, Gorsky State Agrarian University, d. 37, Kirov St., Vladikavkaz, Republic of North Ossetia Alania, 362040, Russia. E-mail: bekos37@mail.ru

Eugene Iv. Boreko, DSc
Associated Professor, Republican Research and Practical Center for Epidemiology and Microbiology, Laboratory of Influenza and Influenza-like diseases, Chief Research Scientist, 23 Filimonov St., Minsk 220114, Republic of Belarus. E-mail: bei@mail.by

Elena B. Burlakova, DSc
Professor, State Prize winner, Deputy Director, Head of the Laboratory of the Emanuel Institute of Biochemical Physics pf RAS, Moscow, Russia

Danil D. Efryushin
Polzunov Altai State Technical University, Post-graduate Student, 46, Lenin Pr., Barnaul 656038, Russia. E-mail: vadandral@mail.ru

Vladimir F. Eremin, DSc
Associated Professor, the Republican Research and Practical Center for Epidemiology and Microbiology, Laboratory of HIV and Opportunistic Infections Diagnosis Laboratory of HIV and Opportunistic Infections Diagnosis, Head of Department, 23 Filimonov St., Minsk 220114, Republic of Belarus. E-mail: veremin@mail.ru

Valeriy N. Erokhin, PhD
Senior Scientist of Biochemical Physics, Emanuel Institute of Biochemical Physics of Russian Academy of Sciences, Laboratory of Quantitative Oncology, Senior Scientist, 4, Kosygin St., Moscow 119334, Russia. E-mail: <valery@sky.chph.ras.ru>

Lujdmila D. Fatkullina, PhD
Emanuel Institute of Biochemical Physics, of Russian Academy of Sciences, Laboratory of Physicochemical Regulation of Biological Systems, Senior Scientist, 4, Kosygin St., Moscow 119334, Russia

Natalia V. Galaiko, PhD

Institute of Technical Chemistry of Ural Branch of Russian Academy of Sciences, Laboratory of Biologically Active Compounds, Researcher Assistant, 3, Akad. Korolev St., Perm 614013, Russia. E-mail: natagalaiko@gmail.com

Sergey N. Gaydamaka, PhD

M.V. Lomonosov Moscow State University, Department of Chemistry, Researcher, Leninskye gory, 1/11b, Moscow 119991, Russia. E-mail: s.gaidamaka@gmail.com

Marina A. Gladchenko, PhD

M.V. Lomonosov Moscow State University, Department of Chemistry, Senior Scientist, Leninskye gory, 1/11b, Moscow 119991, Russia. E-mail: gladmarina@yandex.ru

Marina N. Gorbunova, PhD

Institute of Technical Chemistry of Ural Branch of Russian Academy of Sciences, Laboratory of Biologically Active Compounds, Senior Scientist, 3, Akad. Korolev St., Perm 614013, Russia. E-mail: mngorb@yandex.ru

Victoria V. Grishko, PhD

Associated Professor, Institute of Technical Chemistry of Ural Branch of Russian Academy of Sciences, Head of Laboratory of Biologically Active Compounds, 3, Akad. Korolev St., Perm 614013, Russia. E-mail: grishvic@gmail.com

Farid M. Gumerov, DSc

Professor, Kazan State Technology University, Head of Department Theoretical Basics of Heat, 68, K. Marks St., Kazan 420015, Russia. E-mail: gum@kstu.ru

Ekaterina V. Igosheva

Institute of Technical Chemistry of Ural Branch of Russian Academy of Sciences, Laboratory of Biologically Active Compounds, Research Assistant, 3, Akad. Korolev St., Perm 614013, Russia. E-mail: ekig00@mail.ru

Gleb S. Ivanov, PhD

Saint Petersburg State Institute of Technology (Technical University), Laboratory of Molecular Pharmacology, Senior Scientist, 26, Moskovsky Pr., Saint Petersburg 190013, Russia. E-mail: glebsi@hotmail.com

Evgeniy S. Izmest'ev, PhD

Institute of Chemistry of Komi of Science Center of Ural Brunch of Russian Academy of Sciences, Researcher, 48, Pervomaiskaya St., Syktyvkar 167982, Russia. E-mail: evgeniyizmestev@rambler.ru

Johnny G. Kachmazov, PhD

Senior Lecturer, the Department of A.A. Tibilov South Ossetian State University, 8, Moscow St., Tskhinval 100001, Republic of South Ossetia. E-mail: yogu@mail.ru

Tatiana V. Khurshkainen, PhD

Associated Professor, Institute of Chemistry of Komi of Science Center of Ural Brunch of Russian Academy of Sciences, Senior Scientist, 48, Pervomaiskaya St., Syktyvkar 167982, Russia. E-mail: hurshkainen@chemi.komisc.ru

Nona L. Komissarova, PhD

Emanuel Institute of Biochemical Physics of Russian Academy of Sciences, Senior Scientist, 4, Kosygin St., Moscow 119334, Russia. E-mail: komissarova@polymer.chph.ras.ru

Vadim V. Konshin, DSc

Associated Professor, Polzunov Altai State Technical University, Head the (Sub)Department, 46, Lenin Pr., Barnaul 656038, Russia. E-mail: vadandral@mail.ru

Anastasiya V. Konysheva (Pereslavtseva)
Institute of Technical Chemistry of Ural Branch of Russian Academy of Sciences, Laboratory of Biologically Active Compounds, Junior Researcher, 3, Akad. Korolev St., Perm 614013, Russia. E-mail: Pereslavceva89@mail.ru

Vyacheslav B. Kovalev, PhD
Associated Professor, Astrakhan State University, Associated Professor of Organic, Inorganic & Pharmaceutical Chemistry Department, 1, Shaumyan sq., Astrakhan 414000, Russia. E-mail: chemkovalevne@mail.ru

Elena V. Koverzanova
Emanuel Institute of Biochemical Physics of Russian Academy of Sciences, Senior Scientist, 4, Kosygin St., Moscow 119334, Russia. E-mail: koverlena@list.ru

Gulnaz F. Krainova, PhD
Institute of Technical Chemistry of Ural Branch of Russian Academy of Sciences, Laboratory of Biologically Active Compounds, Researcher Assistant, 3, Akad. Korolev St., Perm 614013, Russia. E-mail: Gulja.perm@mail.ru

Alexander V. Kutchin, DSc
Professor, correspondent member of RAS, Institute of Chemistry of Komi of Science Center of Ural Brunch of Russian Academy of Sciences, Director, 48, Pervomaiskaya St., Syktyvkar 167982, Russia. E-mail: kutchin-av@chemi.komisc.ru

Yuliya K. Lukanina, PhD
Emanuel Institute of Biochemical Physics of Russian Academy of Sciences, Moscow, Laboratory of Physical Chemistry of the Synthetic and Natural Polymer Compositions, Researcher, 4, Kosygin St., Moscow 119334, Russia. E-mail: julialkn@gmail.ru

Kseniya M. Marakulina
Emanuel Institute of Biochemical Physics of Russian Academy of Sciences, Moscow, Laboratory of Physicochemical Problems in Radiobiology and Ecology, Engineer-Researcher, 4, Kosygin St. 119334, Moscow, Russia. E-mail: maksmy@mail.ru

Lidiya I. Mazaletskaya, PhD
Emanuel Institute of Biochemical Physics of Russian Academy of Sciences, Laboratory of Oxidation of Organic Substances, Leading Scientist 119334, 4, Kosygin st., Moscow 119334, Russia. E-mail: lim@sky.chph.Russin Acadtmi of Siences.ru

Tamara A. Misharina, DSc
Emanuel Institute of Biochemical Physics of Russian Academy of Sciences, Head of Laboratory of Flavor Chemistry, 4, Kosygin St., Moscow 119334, Russia. E-mail: tmish@rambler.ru

Valentina P. Murygina, PhD
M.V. Lomonosov Moscow State University, Department of Chemistry, Senior Scientist, 1/11b, Moscow 119991, Russia. E-mail: vp_murygina@mail.ru

Alexey V. Nazarov
Institute of Ofnical Chemistry, Ural Branch of Russian Academy of Sciences, Post-graduate student, Laboratory of Biologically Active Compounds, Post-graduate student, 3, Akad. Korolev St., Perm 614013, Russia. E-mail: Nazarov_AV@mail.ru

Svyatoslav B. Nosachev, PhD
Astrakhan State University, Associated Professor, Associated Professor of Organic, Inorganic & Pharmaceutical Chemistry Department, 1, Shaumyan Sq., Astrakhan 414000, Russia. E-mail: sbn86chem@yandex.ru

Daria S. Novikova
Saint Petersburg State Institute of Technology (Technical University), Laboratory of Molecular Pharmacology, Senior Scientist, 26, Moskovsky Pr., Saint Petersburg 190013, Russia. E-mail: dc.novikova@gmail.com

Victoriya S. Pekhtereva
Institute for Nature Management of National Academy of Science of Belarus, Researcher, 10, Francis Skoriny St., Minsk 220114, Republic of Belarus. E-mail: nature@ecology.basnet.by

Svetlana V. Pestova
Institute of Chemistry of Komi of Science Center of Ural Brunch of Russian Academy of Sciences, Junior Researcher, 28, Syktyvkar, Kommunisticheskaya St. 167000, Russia. E-mail: pestova-sv@chemi.komisc.ru

Maria V. Pitelina, PhD
Astrakhan State University, Associated Professor, 1, Shaumyan sq., Astrakhan 414000, Russia. E-mail: petelka-m@mail.ru

Irina G. Plashchina, PhD
Emanuel Institute of Biochemical Physics of Russian Academy of Sciences, Moscow, Head of Laboratory of Physicochemical Modification of Biopolymers, 4, Kosygin St., Moscow 119334, Russia, +7(495)9397171. E-mail: igplashchina@sky.chph.Russin Acadtmi of Siences.ru

Tamara I. Ponomareva
Federal Center for Integrated Arctic Research of Russian Academy of Science, Junior Researcher, Northern Dvina Embankment, 23, Arkhangelsk 163000, Russia. E-mail: ponomtamara@gmail.com

Andrey V. Protopopov, PhD
Polzunov Altai State Technical University, Associated Professor, 46, Lenin Pr., Barnaul 656038, Russia. E-mail: a_protopopov@mail.ru

Svetlana A. Rubtsova, DSc
Institute of Chemistry of Komi of Science Center of Ural Brunch of Russian Academy of Sciences, Researcher, 48, Pervomaiskaya St., Syktyvkar 167982, Russia. E-mail: rubtsova-sa@chemi.komisc.ru

Svtelana B. Selyanina, PhD
Associated Professor, Federal Center for Integrated Arctic Research of Russian Academy of Science, Senior Scientist, 23, Arkhangelsk, Northern Dvina Embankment 163000, Russia. E-mail: smssb@yandex.ru

Ekaterina V. Shchepetova, PhD
Associated Professor, Astrakhan State University, 1, Shaumyan Sq., Astrakhan 414000, Russia. E-mail: org@asu.edu.ru

Nataliya I. Sheludchenko
Emanuel Institute of Biochemical Physics of Russian Academy of Sciences, Moscow, Laboratory of Physicochemical Problems in Radiobiology and Ecology, Researcher, 4, Kosygin St., Moscow 119334, Russia. E-mail: shishkina@sky.chph.ras.ru

Oksana G. Shevchenko, PhD
Institute of Chemistry of Komi of Science Center of Ural Brunch of Russian Academy of Sciences, Senior Scientist, 28, Kommunisticheskaya St., Syktyvkar 167982, Russia.

Lyudmila N. Shishkina, DSc
Professor, Emanuel Institute of Biochemical Physics of Russian Academy of Sciences, Moscow, Laboratory of Physicochemical Problems in Radiobiology and Ecology, Head of Laboratory, 4, Kosygin St., Moscow 119334, Russia. E-mail: shishkina@sky.chph. <shishkina@sky.chph.ras.ru>

Tamara V. Sokolova, PhD
Institute for Nature Management of National Academy of Science of Belarus, Senior Scientist, 10, Francis Skoriny St., Minsk 220114, Republic of Belarus. E-mail: nature@ecology.basnet.by

Victor P. Strigutskiy, PhD
Institute for Nature Management of National Academy of Science of Belarus, Senior Scientist, 10, Francis Skoriny St., Minsk 220114, Republic of Belarus. E-mail: nature@ecology.basnet.by

Irina A. Tolmacheva, PhD
Institute of Technical Chemistry of Ural Branch of Russian Academy of Sciences, Laboratory of Biologically Active Compounds, Senior Scientist, 3, Akad. Korolev St., Perm 614013, Russia. E-mail: tolmair@gmail.com

Alexey R. Tomson, PhD
Institute for Nature Management of National Academy of Science of Belarus, Head of Department, 10, Francis Skoriny St., Minsk 220114, Republic of Belarus. E-mail: nature@ecology.basnet.by

Mikhail A. Torlopov, PhD
Associated Professor, Institute of Chemistry of Komi of Science Center of Ural Brunch of Russian Academy of Sciences, laboratory of chemistry of plant polymers, Senior Scientist, 48 Pervomaiskaya St., Syktyvkar 167982, Russia. E-mail: torlopov-ma@chemi.komisc.ru

Marina V. Trufanova, PhD
Federal Center for Integrated Arctic Research of Russian Academy of Science, Senior Scientist, Northern Dvina Embankment, 23, Arkhangelsk 163000, Russia. E-mail: mtrufanova@yandex.ru

Alexander R. Tsyganov, DSc
Academic of National Academy of Belarus, Institute for Nature Management of National Academy of Science of Belarus, Project Leader, 10, F. Skoriny, Minsk 220114, Republic of Belarus. E-mail: nature@ecology.basnet.by

Alexey G. Tyrkov, DSc
Professor, Astrakhan State University, Head of Chemistry Department, 1, Shaumyan Sq., Astrakhan 414000, Russia. E-mail: tyrkov@rambler.ru

Elena V. Udoratina, PhD
Associated Professor, Institute of Chemistry of Komi of Science Center of Ural Brunch of Russian Academy of Sciences, Laboratory of Chemistry of Plant Polymers. Head of Department, 48 Pervomaiskaya St., Syktyvkar 167982, Russia. E-mail: udoratina-ev@chemi.komisc.ru

Sergey D. Varfolomeev, DSc
Professor, Emanuel Institute of Biochemical Physics of Russian Academy of Sciences, Corresponding Member of Russian Academy of Sciences, Scientific Chief of Institute, 4, Kosygin St., Moscow 119334, Russia. E-mail: sdfarv@skychph.ras.ru; M.V. Lomonosov Moscow State University, Department of Chemistry, 1, Leninsky gory, Moscow 119991, Russia

Anatolyi V. Velikorodov, DSc
Professor, Astrakhan State University, Head of Organic, Inorganic & Pharmaceutical Chemistry Department, 1, Shaumyan sq., Astrakhan 414000, Russia. E-mail: avelikorodov@mail.ru

Violetta B. Volieva, PhD
Emanuel Institute of Biochemical Physics of the Russian Academy of Sciences, Head of Laboratory, 4, Kosygin St., Moscow 119334, Russia. E-mail: komissarova@polymer.chph.ras.ru

Larissa I. Weisfeld, PhD
N. M. Emanuel Institute of Biochemical Physics RAS, d. 4, Kosygin Street, Moscow, 119334, Russia, E-mail: liv11@yandex.ru

Anatolij P. Yuvchenko, PhD
Institute of Chemistry of New Materials of the National Academy of Science of Belarus, the Laboratory of Forest- and Petrochemical Products, Deputy Director, 36, Skoriny St, Minsk 220141, Republic of Belarus. E-mail: mixa@ichnm.basnet.by

Irina V. Zhigacheva, DSc
Emanuel Institute of Biochemical Physics of Russian Academy of Sciences, Laboratory of Physical-Chemical Principles of Regulation of the Biological Systems, Leading Scientist, 4, Kosygin St. 119334, Russia. E-mail: zhigacheva@mail.ru

LIST OF ABBREVIATIONS

A549	lung carcinoma tumor cells
A_{tt}	indicated value of SHF-channel attenuator
AA	acetylated lignin
A/A_0	amplitude ratio of signal, shot at 50–0.1 mW
AAPH	2,2-azobis(amidinopropane)dihydrochloride
ABA	abscisic acid
Ac_2O	acetic anhydride, or ethanoic anhydride
ADP	adenosine diphosphate
AE	antiradical efficiency
AFB	ash-free biomass of the substances
AGU	anhydro-D-glucose units
AICAR	5-aminoimidazole-4-carboxamide-1-β-D-ribofuranoside
Al_2O_3	aluminum oxide
AMP	adenosine 5′-monophosphate
AMPK	AMP-activated protein kinase
AOX	alternative oxidase
ATP	adenosine triphosphate
ATPase	adenosintriphosphatasa
BAP	cytokinin 6-benzylaminopurine
BS	brassinosteroids
C	liquid crystal
CaMKKβ	Ca^{2+}/calmodulin-dependent protein kinase β
CF	cellulose formate
CFs	cellulose formates
CCO	cytochrome oxidase
CD	conversion degree
CHMS	chloromethylsilatrane
CHNS	elemental analyzer (Vario Micro Cube "Elementar")
C–H	carbon–hydrogen bond
$(C_2H_5)_2O$	diethyl ether, also known as ethoxyethane, ethyl ether, sulfuric ether, or simply ether

CH_2	methylene group
$CHCl_3$	chloroform
CH_2Cl_2	bichloromethane or methylene chloride
$(C_2H_5)_3N$, Et_3N	triethylamine
C_2H_5OH	ethanol
CH_3COCl	acetyl chloride
$CH_3CONHNH_2$	acetic acid hydrazide
$(CH_3)_2CO$	acetone
CH_3COOH	acetic acid or ethanoic acid
CH_3COONa	sodium acetate or the sodium salt of acetic acid
CH_3I	methyl iodide or iodomethane
CH_3MgI	Grignard reagent
$C_1–H$	carbon atom at position 1–hydrogen bonds
$(C_2H_5)_2O$	diethyl ether, also known as ethoxyethane, ethyl ether, sulfuric ether, or simply ether
$C_3–O$	carbon atom at position 3–oxygen bonds
C_6	carbon atom at position 6 of anhydroglucose unit of cellulose
^{13}C NMR	carbon-13 nuclear magnetic resonance
C–O	carbon–oxygen bonds
COD	chemical oxygen demanded rate of content of organic substances
$(COCl)_2$	oxalyl chloride
$C_6–O–H$	carbon atom at position 6–oxygen–hydrogen bonds
COO^-	carboxyl ion
$C_6–OSO_3$	sulfone groups substitution at carbon atom at position 6
COSY	two-dimensional mononuclear correlation spectroscopy
CPA	citraconopimaric acid
CrO_3	chromium oxide (VI)
Cr_2O_3	chromium oxide (III)
CS	cellulose sulfate
CSPs	cold shock proteins
Cs_2CO_3	cesium carbonate
DCC	N,N'-icyclohexylcarbodiimide

DEPT	distortionless enhancement by polarization transfer
DMAP	4-(dimethylamino)pyridine
DMF	dimethylformamide or the amide of formic acid
DP	degree of polymerization
DPPH	2.2-diphenyl-1-picrylhydrazyl
DS	degree of substitution
DS_{HCO}	degree of formyl groups substitution
DTA	differential termical analysis
DTH	plaque forming cells
EB	epibrassinolide
24-EB	24-epibrassinolide
EC	concentration of the extract
EC_{50}	dose inhibiting virus cytopathogenic activity by 50%
EGCG	epigallocatechin-3-gallate
EPR	electron paramagnetic resonance
ETC	electron transport chain
EtOH	ethanol
f	the stoichiometric coefficient of the inhibition
FCCP	carbonylcyanide-p-trifluoromethoxyphenylhydrazone
FIBAN K-4	polypropylene-grafted-acrylic acid
FFA	free fatty acids
fk_7	the inhibitory efficiency
FRET	fluorescence resonance energy transfer
FTIR	Fourier transform infrared spectroscopy
G	gauss, this value is measured in units of magnetic field
GLUT4	glucose transporter
GM	germatrane
ΔH	quantity that characterizes the width of the EPR-signal
HCOOH	formic acid
HIV-1	human immunodeficiency virus type 1
HL	hydrolytic lignin
HMBC	heteronuclear multiple-bond correlation spectroscopy

^1H NMR	proton nuclear magnetic resonance
HPA	heteropolyacid
HSPs	heat shock proteins
HSQC	heteronuclear single-quantum correlation spectroscopy
H_2SO_4	sulfuric acid
$H_3PMo_{12}O_{40}$	phosphomolybdic acid
$H_3PW_{12}O_{40}$	phosphotungstic acid
$H_4SiMo_{12}O_{42}$	siliconmolybdic acid
I	the concentration of free radicals
I_2	iodine
IAA	indoleacetic acid
i-$C_5H_{11}ONO$	isoamyl nitrite, ester isoamyl alcohol, and nitrous acid
IF IL-4	interleukin 4 is a cytokine that induces differentiation of naive helper T cells (Th0 cells) to Th2 cells N-α produced by monocytes/macrophages, lymphoblastoid cells, fibroblasts, and a number of different cell types following induction by viruses, nucleic acids, glucocorticoid hormones, and low-molecular weight substances
IFN, IFN-α, IFN-γ	interferones
IL, IL-2, IL-4	interleukins
InH	antioxidant and/or antioxidants
IR	infra-red spectroscopy
k_7	the rate constant of the interaction of antioxidants with peroxy radicals
ISQ	innovative single quad
KBr	potassium bromide
K_2CO_3	potassium carbonate
KF	potassium fluoride
KOH	potassium hydroxide, or caustic potash
kV	kilovolt
L	lecithin
LKB1	liver kinase B1
LP	lipid peroxidation
LPO	lipid peroxidation
mABA	m-amino benzoic acid

% mass	weight percent
MCC	microcrystalline cellulose
m-CPBA	3-chlorperoxybenzoic acid
MeOH	methyl alcohol
MHz	megahertz
mitoK$^+_{ATP}$	mitochondrial ATP making-sensitive potassium
MK-l	ignin acylated by myristic acid
mL	milliliter
mol%	mole percent
MPA	maleopimaric acid
Ms	methanesulfon
MS	melanoma tumor cells
MTC	maximum of studied compound doses not cause tissue morphological changes
NaBH$_4$	sodium tetrahydridoborate
NADH-	nicotinamide adenine dinucleotide
NADP (NAD(P)H)	nicotinamide adenine dinucleotide phosphate
Na–SC–Na	cellulose sulfate sodium salt
NH$_2$NH$_2\cdot$H$_2$O	hydrazine hydrate
NH$_2$OH$\cdot$HCl	hydroxylamine hydrochloride
NH$_2$SO$_3$H	aminosulfonic acid
NH$_2$SO$_3$H–DMF	system of aminosulfonic acid–dimethylformamide
NMR	nuclear magnetic resonance
NOESY	nuclear Overhauser effect spectroscopy
NtAOX	hormone induces concentration-dependent gene expression
oABA	o-amino benzoic acid
[O]–m-CPBA	$meta$-chloroperoxybenzoic acid
OS	organic substances
pABA	p-amino benzoic acid
PC	phosphatidylcholine
PC	powdered cellulose
PCD	programed cell death
PEK	porcine embryo kidney cells
PGR	plant growth regulator
PFC	plaque-forming cells
PK	lignin acylated by palmitic acid
PL	phospholipids
PmitoK$^+_{ATP}$	ATP-sensitive potassium channel

PNA	phosphotungstic acid
ppm	parts per million
PTA	phosphotungstic acid
PTP	permeability transition pore
pUCP	plant uncoupling protein
pUCPs	unsaturated fatty acids
Py, C_5H_5N	pyridine
RD	rhabdomyosarcoma tumor cells
RMA	rosin-maleic anhydride adduct
R_1NHNH_2	hydrazine derivatives
ROOH	hydroperoxides
ROS	reactive oxygen species
RRR	growth regulators and plant development
$R=SO_3Na$, CHO, H	substituted groups sodium sulfone, formil, and hydrogen groups
$R'=SO_3Na$, H	substituted groups sodium sulfone and hydrogen groups
SA	salicylic acid, 2-hydroxybenzoic acid
SCF	supercritical fluid
SEM	scanning electron microscopy
SHAM	salicylhydroxamic acid
SL	sulphated lignin
S=O	double bond oxygen and sulfur atoms
$SOCl_2$	thionyl chloride
SOD	superoxide dismutase
TAK1	transforming growth factor β-activated kinase-1
TBA	2-thiobarbituric acid
TBAI	tetra-n-butylammonium iodide
TBA-RS	thiobarbituric acid reactive substances
TBHP	*tert*-butyl hydroperoxide
t-BuOH	*tert*-butyl alcohol or 2-methylpropan-2-ol
t-BuOK	potassium *tert*-butoxide
TCA	citric acid cycle
TEC_{50}	reduction of 50% of radical concentration
THF	tetrahydrofuran
TI	ratio between MTC and MAC (minimum active concentration) of the preparations
TLC	thin layer chromatography

Ts	*p*-toluenesulfonyl
TsCl	4-toluenesulfonyl chloride (*p*-toluenesulfonyl chloride)
TVD	thermal vacuum deposition
UASB	upflow anaerobic sludge blanket
UCPs	uncoupling proteins
UCP	uncoupling protein
VFAs	volatile fatty acids
VO(acac)$_2$	vanadyl acetylacetonate
VSV	vesicular stomatitis virus
XRD	X-ray diffraction
XSA	X-ray crystal structure analysis
YC	Yuanhuacine
ν (Greek letter nu)	the fluctuations of the valences of the functional groups

FROM FUNDAMENTAL SCIENCE TO NEW TECHNOLOGIES

Wood chemistry, initially defined as the basis for the manufacturing technology of rosin and turpentine and charcoal burning, currently represents itself as a science that combines many disciplines. The study of the chemical composition of wood and all morphological components of wood, development of methods for isolating individual chemical compounds, synthesis on their basis of new derivatives, as well as identification of the biological and pharmacological activity are the subject of wood chemistry in the modern sense. This circumstance requires the cooperation of researchers in different scientific disciplines: chemists, biologists, physiologists, ecologists, and physicians.

Natural renewability, as well as a unique structure and biological activity of the components of plant raw materials, make them an inexhaustible source for obtaining medicinal preparations, valuable technical products, and intermediates for organic synthesis. Integrated and rational approach to the use of all substances contained in natural raw materials—processing of plant raw materials—dramatically increases the yield of beneficial products. Biologically active compounds of natural origin represent themselves as knowledge-intensive and competitive products that are in demand in the world market, and the tendency towards their use is constantly growing. The implementation of this perspective first of all requires research into the chemical composition of plant raw materials, estimation of the ratio of its various components, specification of their biological activity, development of the methods for chemical transformations of natural compounds, as well as generalization and systematization of the available information.

The presented studies prove that the chemistry of natural compounds firmly occupies the leading positions in the world and domestic science. The range of research papers presented in the book is quite wide: from fundamental and applied problems of wood chemistry and organic synthesis to biological activity of natural compounds.

At the initiative of the Institute of Chemistry of the Scientific Center of Ural Branch of the Russian Academy of Sciences for many years, many conferences have been conducted, showcasing the achievements in the field of chemistry and technology of plant substances.

Corresponding Member, Russian Academy of Science
—Alexander V. Kutchin

PREFACE

A widespread use of antioxidants, among which phenolic compounds are dominant, is associated with their ability to participate in the regulation of oxidative processes in food, industrial and cosmetic oils and fats, as well as polymeric materials and various biological systems. A relatively high toxicity of synthetic compounds for humans and an active participation of natural antioxidants in the side reactions, which significantly reduces their inhibiting efficiency, make it necessary to expand the search for new promising medications suitable for practical use.

Over the past years, researchers have focused their efforts on the development of methods for synthesis and modification of plant raw materials to create semisynthetic antioxidants having less toxicity as compared to synthetic compounds. The unique structure and biological activity of the components of plant raw materials are widely used by synthetic chemists. What is more, being natural renewable resources, they represent an inexhaustible source of obtaining new substances and materials with valuable practical properties and a wide range of applications, both as pharmaceuticals, valuable industrial products, and intermediates, for organic synthesis.

The book is intended for a wide circle of specialists in the field of organic, physical, biological, and medicinal chemistry, chemical processing of plant raw materials, and physicochemical biology.

—**Lyudmila N. Shishkina**

INTRODUCTION

We live in an amazing time when an inquiring human eye is capable of crossing the border of the XX-th and XXI-st centuries and throwing a look into the third millennium. Science and technology have reached such a level at which the forecast for the development of human society in the long term becomes a real and scientifically justified.

—Academician N.M. Emanuel

The book "*Chemistry and Technology of Plant Substances: Chemical and Biochemical Aspects*" is devoted to the methods for synthesis of new compounds on the basis of plant raw materials, modification of their components, and research into physicochemical properties and biological activity of the obtained compounds. It covers a wide range of problems associated with the use of the components of plants to produce new substances with a wide variety of purposes. The scope of the problems under discussion has led to the situation when none of the authors can be a universal expert, so the book is a collective monograph written by the experts from the leading scientific research institutions of different regions in Russia and Belarus. As a result, their collective contribution to the solution of the discussed problems is much more effective than if the book was written by one author.

Biologically active compounds of natural origin represent themselves as actively demanded products, and the tendency toward their use is constantly growing due to a relatively low toxicity of the compounds, synthesized on the basis of the components of natural raw materials as compared to synthetic analogues. Coniferous trees, along with turpentine and rosin, are the richest source of various chemical compounds. Low molecular weight components extracted from plant raw materials are widely used in fine organic synthesis, representing the basis for chemical transformations. High molecular weight polysaccharides of coniferous wood greenery, water-soluble pectic substances, and hemicelluloses possess a set of valuable physicochemical properties and potential physiological activity. Polysaccharides of plant origin, first of all cellulose, represent themselves an available and biocompatible polymer that could serve as the basis for creating new anticoagulant medicines.

This book is represented by 15 chapters and divided into two sections. The first section "Components of Plant Origin: Synthesis, Modification, and Properties" is represented by eight chapters. Three chapters describe the methods for synthesis of new compounds based on low molecular weight plant components, as well as give insight into their properties and transformations. They are: *Synthesis and Transformations of 2,3-Secotriterpene Derivatives of Betulin* by Irina A. Tolmacheva, Natalia V. Galaiko, Ekaterina V. Igosheva, Anastasiya V. Konysheva, Alexey V. Nazarov, Gulnaz F. Krainova, Marina N. Gorbunova, Eugene I. Boreko, Vladimir F. Eremin, and Viktoriya V. Grishko (Chapter 1); *Synthesis and Membrane-Protective Properties of Sulfur-Containing Derivatives of Monoterpenoides with Monosaccharide Fragments* by Alexander V. Kutchin, Svetlana V. Pestova, Evgeny S. Izmest'ev, Oksana G. Shevchenko, and Svetlana A. Rubtsova (Chapter 2); *The Synthesis and Properties of New Oxygen- and Nitrogen-Containing Terpene Acid Derivatives* by Maksim P. Bei, and Anatolij P. Yuvchenko, performed by team of the authors from scientific and research institutions in Perm, Syktyvkar and Minsk (Chapter 3).

Four chapters of this section are devoted to the methods for transformation of high molecular weight components of plant origin with the aim of obtaining a variety of useful compounds. These works include: *Structural and Chemical Modification of Cellulose in Phosphotungstic Acid–Formic Acid System and Sulfation Prepared Derivatives* by Elena V. Udoratina and Michael A. Torlopov (Chapter 4); *Biocatalytic Conversion of Lignocellulose Materials to Fatty Acids and Ethanol with Subsequent Esterification* by Sergey D. Varfolomeev, Marina A. Gladchenko, Sergey N. Gaydamaka, Valentina P. Murygina, Violetta B. Volieva, Nona L. Komissarova, Farid M. Gumerov, Rustem A. Usmanov, and Elena V. Koversanova (Chapter 6); *Mechanism of Ammonia Immobilization by Peat and Obtaining of Peat-Based Sorbent* by Alexander R. Tsyganov, Aleksey Em. Tomson, Victor P. Strigutskiy, Victoriya S. Pehtereva, Tamara V. Sokolova, Svetlana B. Selyanina, Marina V. Trufanova, and Tamara Ig. Ponomareva (Chapter 7), and *Problem of Modification of Technical Lignins Using Acylation Method* by Andrey V. Protopopov, Danil D. Efryushin, and Vadim V. Konshin (Chapter 8) and carried out by the authors from Moscow, Syktyvkar, Astrakhan, Kazan, Minsk, Arkhangelsk, and Barnaul. The study *Effect of*

Complexation with Phospholipids and Polarity of Medium on the Reactivity of Phenolic Antioxidants, performed in IBCP RAS (Moscow) by Lyudmila N. Shishkina, Lidiya I. Mazaletskaya, Kseniya M. Marakulina, Yuliya K. Lukanina, Irina G. Plashchina, and Nataliya I. Sheludchenko (Chapter 5), presents the findings of research into the impact exerted by the structure of semisynthetic phenolic antioxidants and the properties of the medium on their inhibiting efficiency.

The second section "Biological Activity of Plant Substances" includes seven chapters performed by researchers from the leading scientific and research institutions of St. Petersburg, Moscow, Syktyvkar and Vladikavkaz. Four chapters are devoted to the detailed analysis of biological and pharmacological activity of the compounds of plant origin when administered to animals as well as to prevention of oxidative stress in plants. These research works include: *Compounds of Plant Origin as AMP-Activated Protein Kinase Activators* by Daria S. Novikova, Gleb S. Ivanov, Alexander V. Garabadzhiu, and Viacheslav G. Tribulovich (Chapter 9); *Effects of Low Doses of Savory Essential Oil Dietary Supplementation on Lifetime and the Fatty Acid Composition of the Ageing Mice Tissues* by Tamara A. Misharina, Valery N. Yerokhin, and Lujdmila D. Fatkullina (Chapter 10); *Plant Growth and Development Regulators and their Effect on the Functional State of Mitochondria* by Irina V. Zhigacheva and Elena B. Burlakova (Chapter 12), and *Antiradical Properties of Essential Oils and Extracts from Spices* by Tamara A. Misharina and Ekaterina S. Alinkina (Chapter 14). The studies *Technology for Obtaining of Biopreparations and Investigation of their Effectiveness* by Tatyana V. Khurshkainen and Alexander V. Kutchin (Chapter 11) and *Amaranth—Bioindicator of Toxic Soil* by Sarra A. Bekuzarova, Johnny G. Kachmazov, and Ekaterina S. Ayskhanova (Chapter 13) analyze the effectiveness of using biological products in agriculture. The research work *The Chemical Composition of Essential Oils from Wild-Growing and Introduced Plants of the Astrakhan Region* by Anatoly V. Velikorodov, Vyacheslav B. Kovalev, Svyatoslav B. Nosachev, Alexey G. Tyrkov, Maria V. Pitelina, and Ekaterina V. Shchepetova (Chapter 15) is devoted to the differences in the composition of essential oils of various plants growing in the Astrakhan region.

Methodological approaches to the synthesis of new compounds based on the components of plant raw materials, methods of their modification, and modern ideas about the mechanisms of their activity in complex systems presented in the research studies will undoubtedly be of interest not only to the specialists in the field of fundamental sciences, but also to the young researchers and engineers occupied in recycling crop residues.

—**Alexander V. Kutchin**
—**Lyudmila N. Shishkina**
—**Larissa I. Weisfeld**

PART I

Components of Plant Origins: Synthesis, Modification, and Properties

CHAPTER 1

SYNTHESIS AND TRANSFORMATIONS OF 2,3-SECOTRITERPENE DERIVATIVES OF BETULIN

IRINA A. TOLMACHEVA[1*], NATALIA V. GALAIKO[1],
EKATERINA V. IGOSHEVA[1], ANASTASIYA V. KONYSHEVA[1],
ALEXEY V. NAZAROV[1], GULNAZ F. KRAINOVA[1],
MARINA N. GORBUNOVA[1], EUGENE I. BOREKO[2],
VLADIMIR F. EREMIN[2], and VICTORIA V. GRISHKO[1]

[1]*Institute of Technical Chemistry, Ural Branch, Russian Academy of Sciences, 3, Academician Korolev St., Perm 614013, Russia*

[2]*The Republican Research and Practical Center for Epidemiology and Microbiology, 23, Filimonov St., Minsk 220114, Republic of Belarus*

*Corresponding author. E-mail: tolmair@gmail.com

CONTENTS

ABSTRACT

The review is devoted to the results of our studies on the synthesis and modification of 2,3-secotriterpene compounds. The original 2,3-secotriterpenoids were synthesized by Beckmann fragmentation of the A-ring of available derivatives of betulin (betulonic acid, its methyl ester and allobetulone). Directed chemical modification of the initial 2,3-secotriterpenoids has resulted in a wide spectrum of mono- and bi-functionalization products as well as products of heterocyclization and recyclization of the A-ring. Compounds with high antiviral and cytotoxic activity among the obtained products were found.

1.1   INTRODUCTION

Pentacyclic triterpenoids of natural and semisynthetic origin are characterized by their high biological activity as a result of the fact that these compounds can be obtained from readily available sources. Typical representatives of this group of triterpenoids are betulin, a major component of birch bark Betula, and its derivatives betulinic and betulonic acids. These compounds are perspective candidates for synthesis of therapeutically active agents [1–6]. The majority of the known reactions of betulin, and its derivatives are carried out at C-3, C-28, and C-30 carbon atoms which occur without affecting the carbon skeleton. Examples of rearrangement of the carbon skeleton of lupane triterpenoids are less common, in particular the A-ring fragmentation with the A-secoderivatives formation by chemical [7–10], photochemical [11, 12], and microbiological [13] transformations. The production of various ring-A-secoderivatives which are characterized by different biological activity [8, 14, 15] and modification of secoderivatives via the introduction of various pharmacophoric groups usually results in the enhancement or variation of biological activity of the obtained compounds [2, 16].

The goal of the present review was to generalize the obtained results of own researches on the synthesis and chemical modification of 2,3-secotriterpene compounds that resulted to a wide spectrum of linear and cyclic heteroatom 2,3-secotriterpene derivatives. Compounds featured by anticancer and antiviral action have been revealed among the produced 2,3-seco- and A-pentacyclic derivatives.

1.2 SYNTHESIS OF THE BASIC 2,3-SECOTRITERPENOIDS

According to the published data, the most common strategy for the synthesis of A-secotriterpenoids used an approach based on the Beckmann rearrangement, such as oxidative cleavage of relevant diosfenols or the Beckmann fragmentation of ketoximes [17]. We have studied the Beckmann fragmentation of betulonic acid (1), its methyl ester (2), and allobetulone (3), which are obtained from the readily available lupane triterpenoid betulin. α-Hydroxyoxime (7–9) were converted into the corresponding 2,3-secoaldehydonitriles (10–12) by the action p-toluenesulfonic acid chloride. 2,3-Secotriterpene hydroxynitriles (13–15) and carboxynitriles (16–18) were obtained by reduction or oxidation of the C-3-aldehyde group of the aldehydonitriles (10–12). Compounds (17, 18) were converted into the corresponding methyl esters (19) and (20) under classical alkylation conditions by methyliodide [18–20] (Fig. 1.1).

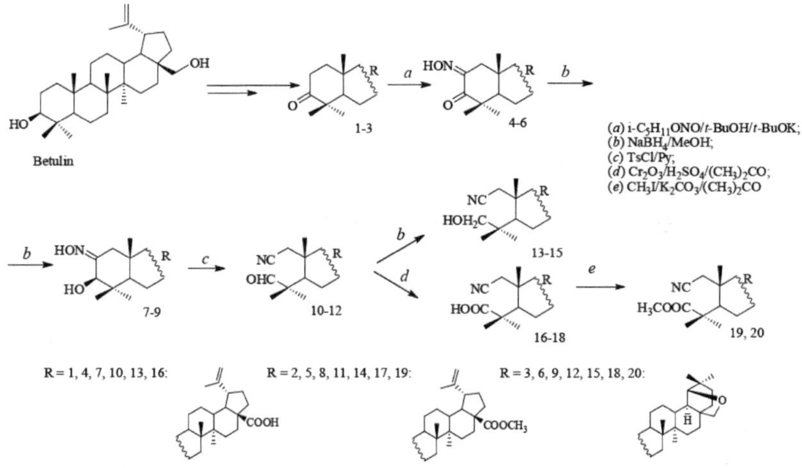

FIGURE 1.1 Scheme of synthesis of the basic 2,3-secotriterpenoids (10–20) [18–20].

Structures of 2-hydroximino-3-oxo derivatives (4–6) with anticonfiguration of the oxime group were additionally confirmed by X-ray structural analysis (XSA) for 18αH-oleanane ketoxime (6) [21] (Fig. 1.2).

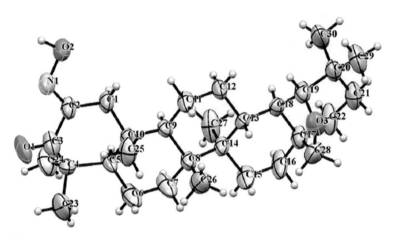

FIGURE 1.2   Molecular structure of ketoxime (6) with representation of the atoms by thermal vibration ellipsoids of 50% probability [21].

It was found that (10) and (16) were capable of suppressing reproduction of Herpes virus type 1 (EC_{50} = 1.9 and 21.3 μM; ratio MTC/EC_{50} = 14.0 and 9.7, respectively). Moreover, (16) was slightly active against flu A virus (MTC/EC_{50} = 1.6), whereas (10) exhibited moderate antiviral activity (EC_{50} = 7.7 μM, MTC/EC_{50} = 3.5) [19].

1.3   SYNTHESIS OF A-PENTACYCLIC TRITERPENE α,β-ALKENENITRILES

It is known that 2,3-secotriterpenoids are key intermediates in the synthesis of five-membered ring A derivatives. For example, Dieckmann condensation of 2,3-secodicarboxylic acid methyl esters gives five-membered ring A 1-methoxycarbonyl-2-oxo-*A*-norderivatives which subsequent transformations lead to cyclopentanone derivatives [22, 23] or epiceanothic acid [24]. Given that, 2,3-secotriterpene aldehydonitriles (10–12) are highly electrophilic aldehyde group and C-1-activated methylene protons of the C-2 of the nitrile group, we have investigated the possibility of nitrile anion intramolecular cyclizations of these compounds to form the five-membered ring A structures—α,β-alkenenitriles (31–33), including Thorpe–Ziegler cyclization [25] of their dinitrile (24–26) and methylketone (38–39) derivatives [26, 27] (Fig. 1.3).

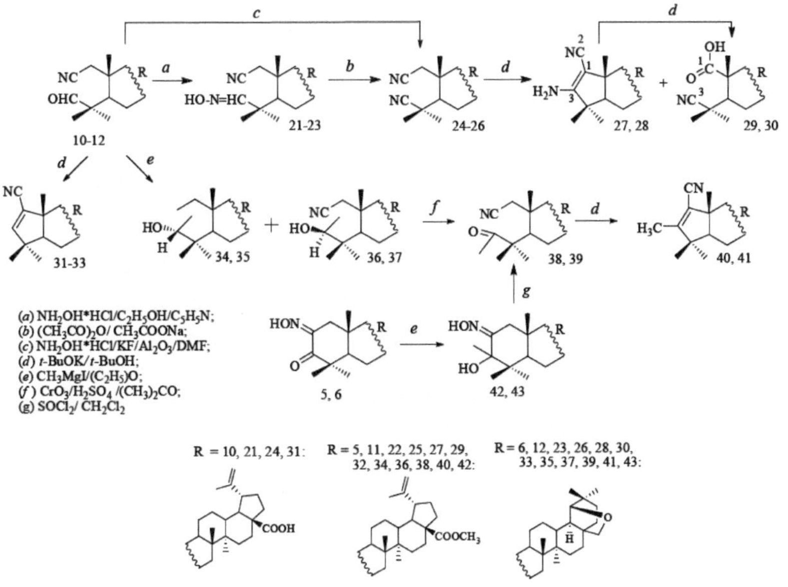

FIGURE 1.3 Scheme of synthesis of A-pentacyclic triterpene α,β-alkenenitriles (27), (28), (31–33), (40), and (41) [26, 27].

Dinitrile secointermediates (24–26) based on 2,3-secoaldehydenitriles were obtained using two approaches: (1) a two-step method involving the secoaldoxime intermediate formation and (2) one-pot method of direct dinitrile formation under heterogeneous catalysis on KF/Al$_2$O$_3$ [26]. A-seco-triterpenoids with a methylketone group (38), (39) were synthesized: (1) from epimeric 3-hydroxy-3-methyl-1-cyano-2,3-seco-triterpenoids (34–37) of the lupane and 19β,28-epoxy-18αH-oleanane types, which were formed by a Grignard reaction [27] and (2) by the Beckmann fragmentation of 3-methylsubstituted hydroxyoxime (42), (43).

The Thorpe–Ziegler intramolecular cyclizations of 2,3-secotriterpene derivatives (10–12), (25), (26) and (38), (39) were carried out using the t-BuOK/t-BuOH system. For dinitriles (25), (26), the formation of two types of products was observed after the reaction. Those were five-membered ring A enaminonitriles (27), (28) and 4-cyano-2-nor-1,3-secotriterpene acids (29), (30). Possible formation of 4-cyano-2-nor-1,3-secotriterpene acids (29) and (30) from the enaminonitriles (27), (28) was confirmed in the reaction between enaminonitrile (27) and t-BuOK/t-BuOH [26]. The

structure of 1,3-secoacid (30) and alkenenitrile (33) were confirmed by
XSA [26] (Figs. 1.4 and 1.5).

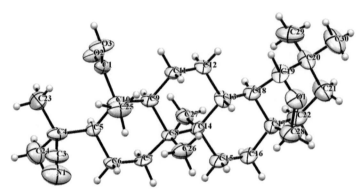

FIGURE 1.4   The structure of 4-cyano-2-nor-1,3-seco-acid (30) in thermal ellipsoids of
50% probability [26].

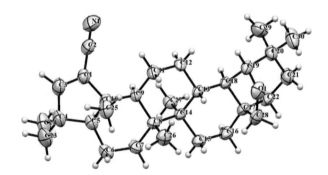

FIGURE 1.5 Structure of alkenenitrile (33) in thermal ellipsoids of 50% probability [26].

The structure of compound (35) and the relative configuration of asymmetric C-3 were confirmed by XSA [27] (Fig. 1.6).

According to XSA, methylketone (39) formed two conformational
isomers with (+)- and (−)-synclinal positioning of the methylketone relative to the C-5 H-bond of the 2,3-seco-triterpene backbone. Their existence was possible through mutual rotation of the substituents around the
C(4)–C(5) bond [27] (Fig. 1.7).

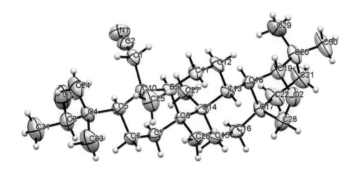

FIGURE 1.6   The structure of compound (35) in thermal ellipsoids of 50% probability [27].

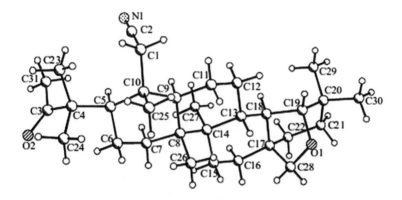

FIGURE 1.7   The structure of methylketon (39) from XSA [27].

Lupane-type triterpenoids exhibited a more pronounced antiviral effect against influenza virus A/FPV/Rostock/34 (H7N1). Ketoxime (4) and 2,3-secoaldoxime (22) demonstrated the highest levels of inhibitory activity and MTC_0/EC_{50} ratio (EC_{50} 12.9 and 22.7 µM, respectively). The antiviral effect of lupane 2,3-secoaldoxime (21) with a free C(28)-carboxyl group against influenza virus A remains at the level of the initial 2,3-seco-aldehydenitrile (10) [19]. Lupane derivatives with a free C-28-carboxyl group ketoxime (4) and 2,3-secoaldoxime (21) inhibited reproduction of the human immunodeficiency virus type 1 (HIV-1) most effectively (EC_{50} 0.06 µM) [26]. Their activity was probably due to the presence of an additional C-2 or C-3 oxime fragment in their structures. The key influence of oxime fragment on antiviral properties of triterpenoids is confirmed by the fact that the 2,3-secoaldoxime (21) antiHIV activity was about 250 times higher than that of its 2,3-tricarboxylic secoanalogue [8]. 18αH-Oleanane methylketone (39) exhibited the most effective inhibitory activity (EC_{50} 7.2 µg/mL). Although the activity of β-substituted lupane A-pentacyclic α,β-alkenenitrile (40) (EC_{50} 39.92 µg/mL) was about half that of α,β-alkenenitrile (32) (EC_{50} 50.46 µM), the range of its therapeutic index (TI) was 2–10 times greater than the analogous index for (32) and (39) [27].

1.4   SYNTHESIS, TRANSFORMATION AND BIOLOGICAL EVALUATION OF 2,3-SECOTRITERPENE ACETYLHYDRAZONES AND THEIR DERIVATIVES

Many nitrogen-containing derivatives of betulin, betulinic, and betulonic acids (amides, hydrazones, and hydrazides) effectively inhibit the replication of HIV-1, influenza, and herpes viruses [28–30]. At the same time, A-secotriterpenoids, active in regards to HIV-1 and the herpes simplex virus, are frequently considered as prospective antiviral agents [8, 19]. In continuation of research on chemical transformation of 2,3-seco-triterpenoids and in consideration of the reported high antiangiogenic and antiviral activity of betulonic acid hydrazones and hydrazides, we synthesized 2,3-secotriterpene acylhydrazones based on lupane and 19β,28-epoxy-18α-oleanane derivatives. Triterpene 2,3-seco-acylhydrazones (50–70) were prepared by reaction of acylated hydrazines with yields of 20–68% with 2,3-secoaldehydonitriles (11) or (12) in EtOH in the presence of catalytic amounts of glacial acetic acid [31, 32] (Fig. 1.8).

Antiviral properties of the synthesized hydrazones (50–70) had been studied using two models of the virus infection: (1) protection of a cell system against vesicular stomatitis virus (VSV) affection and (2) the suppression of the VSV reproduction in primarily infected cells. Among the obtained compounds, acetylhydrazones (44) and (57) were selected due to their high inhibitory activity against VSV (EC_{50} 0.85 and 0.21 µg/mL) and low toxicity at that the acetylhydrazone of 1-cyano-19β, 28-epoxy-2,3-seco-2-nor-18αH-olean-3-al (57) was combining prophylactic (EC_{50} 0.00016 µg/mL) and therapeutic (EC_{50} 0.21 µg/mL) activities [32]. The acetylhydrazone (57) shows dose-dependent (1.0–100.0 mg/kg) stimulating effect on formation of plaque-forming cells (PFC) under local/ systematic immunization conditions and suppressing effect on intensity of the delayed-type hypersensitivity reaction. At concentrations 30 nM and 40 μM, compound (57) has shown in vivo inhibitory properties on production of IL-2, IL-4, and IFN-γ, while not influencing synthesis of IFN-α [33].

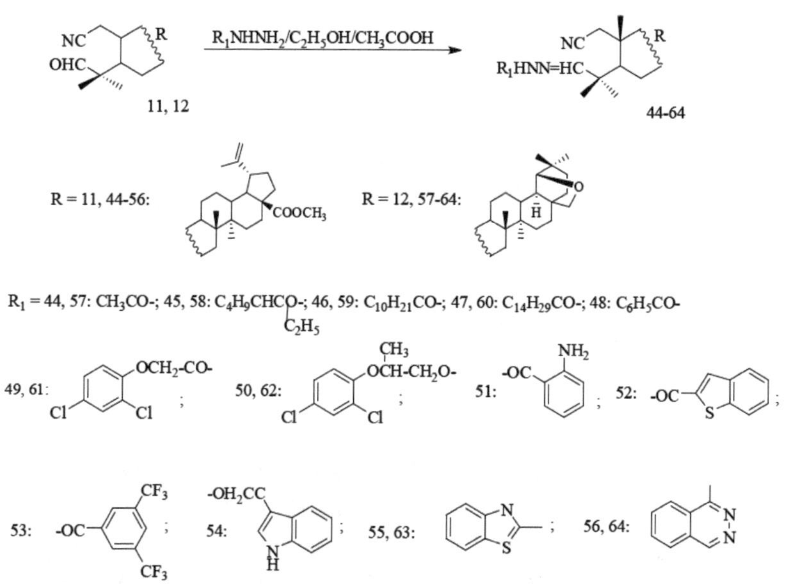

FIGURE 1.8   Synthesis of 2,3-secotriterpene acetylhydrazones (44–64) [31, 32].

We have conducted a detailed study on improving the antiviral activity of 2,3-secotriterpene acetylhydrazones (44), (57), and therefore, the conversion of these compounds has was carried out, and the novel 2,3-secotriterpene with acetylhydrazone fragments were synthesized.

18βH-Oleanane 2,3-seco-aldehydonitrile (66) There is correctly. Save (66)was synthesized from 3-oxo-18βH-glycyrrhetinic acid methyl ester [34] as described previously [18, 19] and was converted into acetylhydrazone (66) [35] (Fig. 1.9).

Lupane acetylhydrazone (44) analogues with residual ethyl ester of β-alanine (76) and 2-aminopropane (77) at C(28) were synthesized [35]. The process of compound (50) derivative formation with the C-28 amide fragment is shown in Figure 1.10.

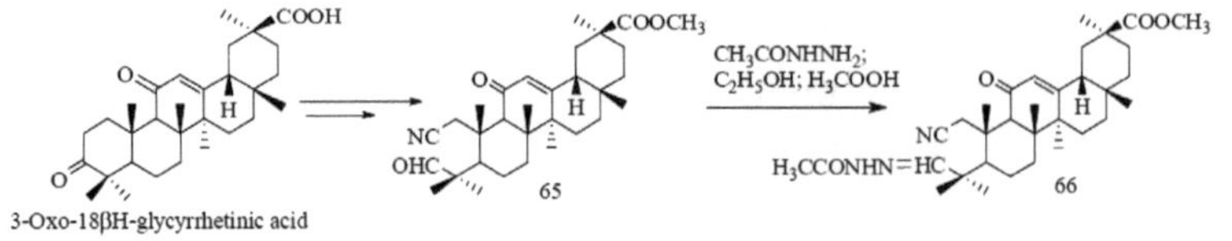

FIGURE 1.9   Synthesis of 18βH-oleanane acetylhydrazone (66) [35].

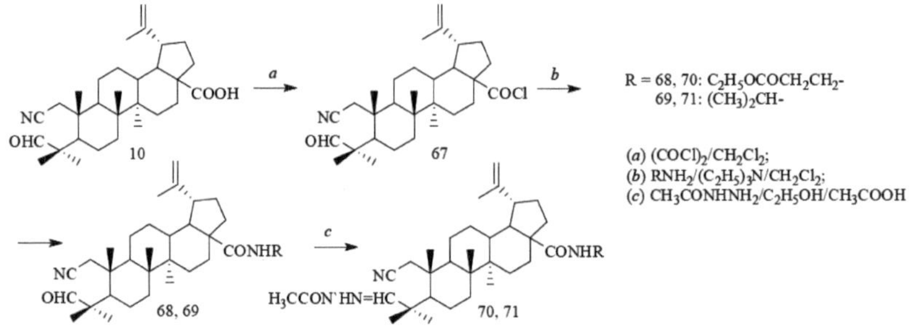

FIGURE 1.10   Synthesis of 2,3-secotriterpene Hydrazone (70), (71) [35].

A-Secotriterpene hydrazonohydrazides (76–81) were prepared from C-3-carboxy derivatives (17) and (18) [18] through the corresponding acid chlorides (72) and (73) that were used directly without purification in the hydrazinolysis reaction with subsequent acid-catalyzed condensation of the resulting 1-cyano-3-hydrazides (74) and (75) with aldehydes [36] (Fig. 1.11).

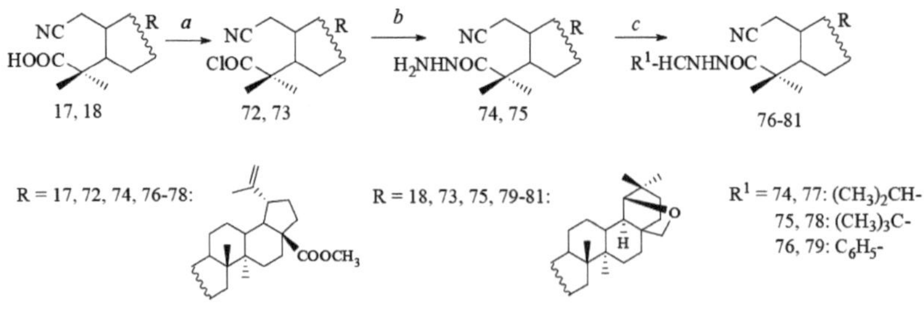

FIGURE 1.11   Synthesis of 2,3-secotriterpene hydrazonohydrazides (82–87) [36].

The synthesized compounds (76–81) did not exhibit preventative antiVSV activity [32].

Acetylhydrazones (44), (57), and (66) were used to obtain heterocyclic triterpene derivatives. Heating of compounds (44), (57), and (66) in acetic anhydride under reflux resulted in a mixture of (R)- and (S)-isomers (the ratio as 2:1) of 1,3,4-oxadiazolines (82–87) were easily separated by column chromatography. Also, N',N'-diacetyl derivatives (88–90) were detected in the reaction mixture and isolated as minor components. On the other hand, the refluxing of compounds (44), (57), and (66) with a threefold excess of acetic anhydride in pyridine in the presence of triethylamine lead to the formation of N',N'-diacetyl derivatives (88–90) [35] (Fig. 1.12).

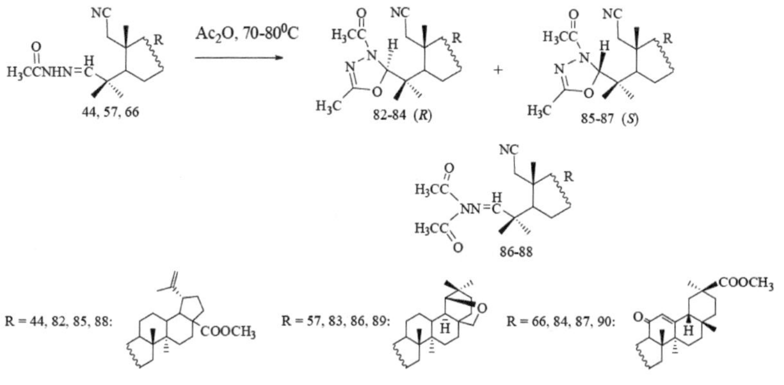

FIGURE 1.12 Synthesis of 2,3-secotriterpene 1,3,4-oxadiazolines (82–87) [35].

XSA of (85) proved its structure as 3(R)-(30-acetyl-50-methyl-20,30-dihydro-10,30,40-oxadiazolyl-20)-1-cyano-2,3-seco-2-nor-19β,28-epoxy-18αH-oleanane [35] (Fig. 1.13).

Studies have shown that replacement of the ester group at position C-28 of acetylhydrazone (44) by amide fragment or introduction of a heterocyclic fragment into acetylhydrazone (57) resulted in partial or complete loss of the antiviral activity. Only two of 2,3-secolupane derivatives (82) and (85) that were structurally different from (44) due to the presence of an oxadiazoline fragment in their molecules showed excellent EC$_{50}$ of 0.62 and 0.34 μM, respectively, exceeding antiVSV activity of known triterpene derivatives of cyclosolone [37] and betulinic acid [38]. However, derivatives (82) and (85) were fairly toxic to porcine embryo kidney cells and had a low selectivity index (TI: 3 and 5, respectively).

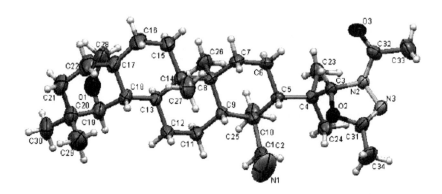

FIGURE 1.13   The structure of compound (85) in thermal ellipsoids of 50% probability [35].

Based on high cytotoxicity of lupane (*R*)- and (*S*)-1,3,4-oxadiazolines (82) and (85) in the antiviral assay, we evaluated the antiproliferative effects of compounds (82–87) on A-549, MS, and RD tumor cells. The studies also revealed a greater sensitivity of MS cells towards (*R*)-isomers (82) and (85), as indicated by their cytotoxicity indices (IC$_{50}$ 5.8 and 11.4 µM, respectively). Both isomers of the 18βH-oleanane type (84) and (87) exhibited weak or no activity against all cell lines tested. Thus, the configuration of the 1,3,4-oxadiazoline C-20 chiral center and the triterpene skeleton type appeared to be the most important determinants of cytotoxic properties of the tested compounds. Early proapoptotic effect of (*R*)-isomer (82) on RD tumor cells was confirmed by AnnexinV/Propidium Iodide analysis [35].

1.5   SYNTHESIS AND ANTIVIRAL ACTIVITY OF C-3 AND/OR C-28-SUBSTITUTED 2,3-SECOTRITERPENOIDS

Synthetic transformations of betulinic and betulonic acids relate mainly to modifications at the C-3 and C-28 atoms. A wide spectrum of C-3 and C-28 functionally substituted lupane derivatives has been obtained, and some of the compounds displayed high biological activity [2–4, 16]. The synthesis of super low-dosing antiviral agents Bevirimat [39] and Betulavir [40] can be regarded as the most striking achievements in this area. The analysis of basic regularities for displaying antiHIV properties made it possible to find the relationship between the antiviral activity and the nature of C-3 and C-28 substituent(s) of lupane acids, as well as the impact of the position of substituent on the molecular mechanism of antiretroviral activity [41, 42].

2,3-Secotriterpene acids attract special attention due to the possibility of both simultaneous and independent substitution of their C-3 and/or C-28 functional groups, which results in various derivatives differing in the substituent nature and position.

Lupane and oleanane mono- and dicarboxylic-acid derivatives (10), (16–18) [18, 19] were converted in situ by treatment with oxalyl chloride in CH_2Cl_2 into the corresponding acid chlorides, reaction of which with primary amines, natural and synthetic amino acids, and amino alcohols in the presence of Et_3N produced in 20–58% yields C-3 and/or C-28 mono- and diamides (91–137) [43–46] (Fig. 1.14).

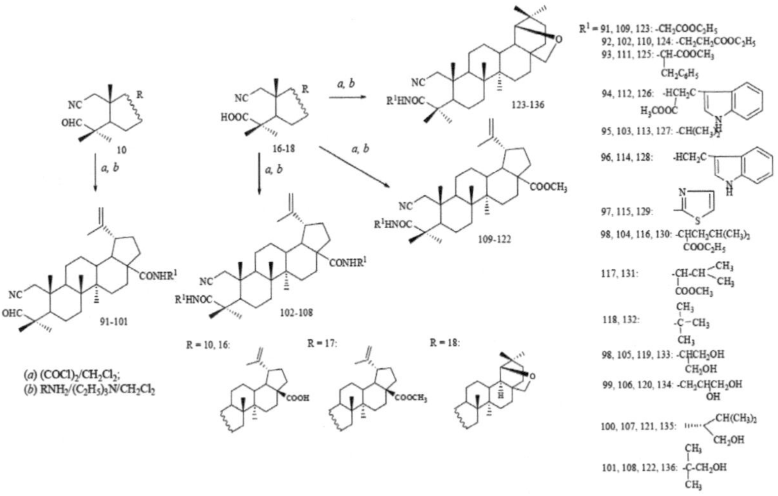

FIGURE 1.14 Synthesis of 2,3-secotriterpene C-3 and/or C-28 mono- and diamides (91–136) [43–46] .

Obtaining of bis-2,3-seco-derivatives (137–139) with lysine is caused by that unlike the amines used in the synthesis of the above-mentioned triterpene amides, the biogenic acid lysine contains two primary amino groups, which are equally involved in the formation of amide bonds [45] (Fig. 1.15).

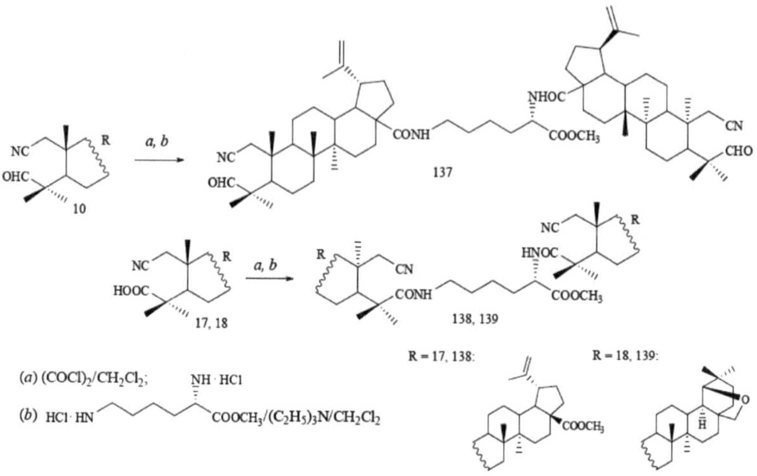

FIGURE 1.15   Synthesis of C3–C3' and C28–C28' biscondensed amides (137–139) [45].

Assessment of the antiviral activity of the 2,3-secoamides showed that introducing pharmacophore groups into the C-3 and/or C-28 position of the carbon skeleton of A-secotriterpenoids could produce compounds with high antiherpes activity. Thus, the mean-effective concentrations EC_{50} for C-3/C-28 mono- (92) and diamides (102) with ethyl-β-alaninate were 8.7 and 4.1 μM, respectively. The diamide conjugate (102) also exhibited antiHIV activity (EC_{50} 5.1 μM) [45]. The 2,3-seco-triterpene conjugates with ethyl-β-alaninate had higher MTC/EC_{50} ratios than the starting A-seco-acids. This indicated that they were promising as antiherpes agents. An analogous influence of β-alanine on increased antiviral activity was noted for betulinic acid [47]. Amides with amino alcohols (98–101), (105–108), (119–122), and (133–136) demonstrated in general a lower level of antiviral activity than starting 2,3-seco-triterpene acids (10) and (16) [19]. However, this group also contained the most active compound

(98) (EC$_{50}$ 5.7 μM). The MTC/EC$_{50}$ ratio of (98) also indicated that amidation of the C(28) carboxylic acid of (10) reduced the toxicity level of the starting aldehyde acid.

Examples of introducing dicarboxylic acids as pharmacophores capable of increasing antiviral activity of lupane derivatives were also published [48, 49]. Introducing the substituent into the C-3 position was more effective for the ester derivatives. Ester 2,3-secoderivatives (140–150) produced by reaction of 3-hydroxy derivatives (14) and (15) [20] with anhydrides of acetic, succinic, 2,2-dimethylsuccinic, 2,2-dimethylglutaric, and 3,3-dimethylglutaric acids in pyridine in the presence of 4-dimethylaminopyridine (DMAP) in 33–71% yields [46] (Fig. 1.16).

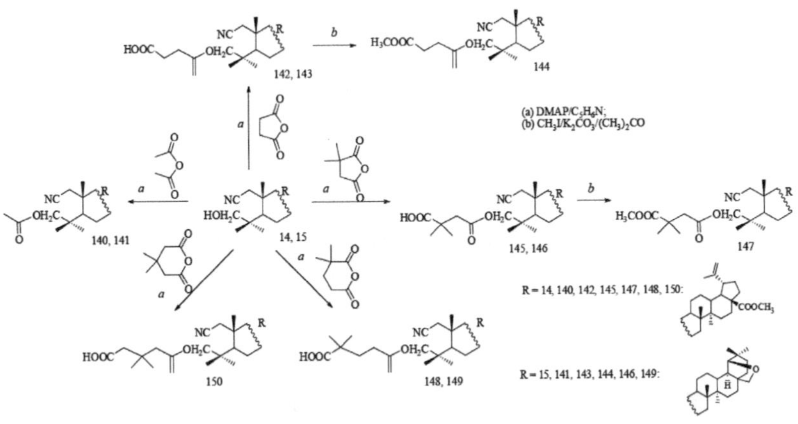

FIGURE 1.16   Synthesis of 2,3-secotriterpene esters (140–150) [46].

New C-3/C-28 mono- and diallylamides (151–153) of 2,3-secolupane and C-28 allylamides of 2-oxime derivatives (4) and (7) were synthesized [50] (Fig. 1.17).

Formation of A-secolupane C-3/C-28 mono- and diallylamides (151–153) was used as an example to demonstrate that a one-step synthesis of 2,3-secotriterpene amides was possible by fragmentation of triterpene α-hydroxy- and α-ketoximes by oxalylchloride or thionylchloride with subsequent amidation of the 2,3-secotriterpenoic acids formed in situ by allylamine. The proposed method enabled the yield of 2,3-secolupane C-3/C-28 diamides to be increased by five times compared with the previously described five-step method for preparing A-secotriterpene amides from betulonic acid.

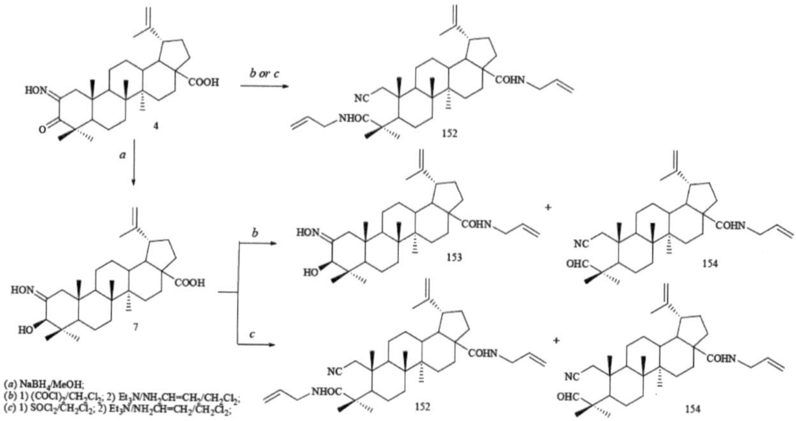

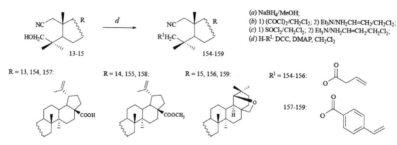

FIGURE 1.17 Synthesis of C-3/C-28 mono- and diallylamides (151–153) [50].

Ester 2,3-seco-derivatives (154–159) with a vinyl fragment are produced by reaction of 3-hydroxy derivatives (13–15) [20] with vinylacetic and vinylbenzoic acids in the presence of DMAP and *N,N'*-dicyclohexylcarbodiimide [51] (Fig. 1.18).

FIGURE 1.18 Synthesis of 2,3-secotriterpene esters (154–159) [51].

2,3-Secotriterpene C-3/C-28 mono- and amides (151–153) and C-28 ester (154–159) with the unsaturated fragments were interesting as potential biologically active agents and as constituents to solubilizing polymeric matrices.

1.6   SYNTHESIS OF TRITERPENE A-CONDENSED AZOLES

Many examples of the chemical transformation of triterpenoids give more active semisynthetic derivatives which indicate the promise for derivatives obtained by introducing a heterocyclic fragment condensed with ring A of the triterpene skeleton [52]. Lupane and 18αH-oleanane α-hydroximino ketones (5), (6), cyclic precursors of semisynthetic 2,3-pentacyclic triterpenoids, provide a platform for the annelation of triterpene skeleton with heterocyclic rings. Heating α-hydroximino ketones (5), (6) in refluxing pyridine with excess acetyl chloride leads to formation of three reaction products: N-acetyl-substituted enamines (160), (161), C(1)–C(2)-fused isoxazoles (162), (163), and C(2)–C(3)-fused oxazoles (164), (165) which the chemical yields 15–18%, 31–36%, and 24–26%, respectively [21] (Fig. 1.19). Cyclization of N-acetylhydrazones obtained in situ from α-hydroximino ketones (5), (6) gave C(2)–C(3)-fused 2'-N-acetyl-1,2,3-triazoles (166), (167) which was hydrolyzed in unsubstituted 1,2,3-triazoles (168), (169). The acylation of 18αH-oleanane 1,2,3-triazole (169) by acetyl chloride in pyridine leads to formation of a mixture of 1'-N- and 3'-N-acetyl derivatives (171) and (172) with virtually identical yields (48% and 47%, respectively). We should note that using hydrazine hydrate instead of acetylhydrazide in the preparation of the 1,2,3-triazoles from α-hydroximino ketones (5), (6) enhances the yields of 2'-N-acetyl-1,2,3-triazoles (166), (167) to 60%. Up to 10%, 1'-N-acetyl-1,2,3-triazoles (170), (171) were formed in this reaction [21] (see Fig. 1.19).

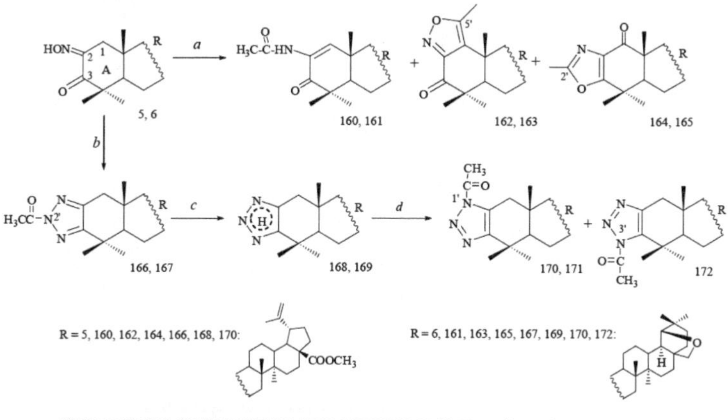

(a) CH$_3$COCl/C$_5$H$_5$N; (b) CH$_3$CONHNH$_2$/CH$_3$COOH; (c) KOH/C$_2$H$_5$OH; (d) CH$_3$COCl/C$_5$H$_5$N/ DMAP

FIGURE 1.19   Synthesis of triterpene azoles (160–172) [21] .

The structures of compounds (160) and (161) were conclusively confirmed by XSA in the case of enamine (160) [21] (Fig. 1.20).

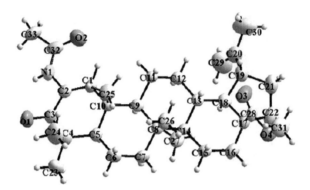

FIGURE 1.20 Molecular structure of *N*-acetyl-substituted enamines (160) with representation of the atoms by thermal vibration ellipsoids of 50% probability [21].

The identification of these products as triterpene C(1)–C(2)-fused isoxazoles (162), (163) and C(2)–C(3)-fused oxazoles (164), (165) was confirmed by XSA for oxazole (164) [21] (Fig. 1.21).

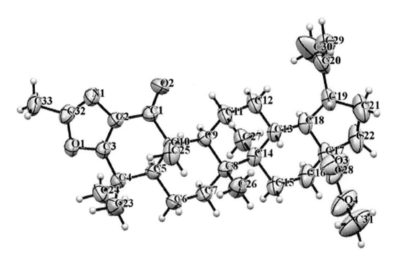

FIGURE 1.21 Molecular structure of oxazole (164) with representation of the atoms by thermal vibration ellipsoids of 50% probability [21].

In cytotoxic studies with triterpene, heterocyclic derivatives were found to have unsubstituted 1,2,3-triazole (168)-exhibited higher cytotoxic activity (IC_{50} 5.6–16.8 µM) relative to tumor cells of RD TE32, lung A549, and MS [21].

1.7 CONCLUSIONS

Betulin from the birch bark as a promising domestic raw material to prepare A-secotriterpenoids can be regarded. The approach, developed by us, to the synthesis of A-secotriterpenoids includes the Beckmann fragmentation of α-hydroxyoxymes of betulin derivatives with the formation of 2,3-secotriterpene aldehydonitriles. Oxidation–reduction transformations and subsequent targeted modification of 2,3-secoaldehydonitriles in C-1, C-3, and C-28-carbon centers resulted in a wide spectrum of linear and cyclic heteroatom 2,3-secotriterpene derivatives including amide and ester mono- and diconjugates, functionalized hydrazones, hydrazides, aldoximes, dinitriles, and substituted heterocyclic derivatives. Methods for synthesis of A-pentacyclic α,β-alkenenitriles on the basis of 2,3-secotriterpenoids have been designed.

The conducted investigations have been resulted in the experimentally proven prospection for the 2,3-secotriterpenoids under elaboration with a different (as compared with the known A-secotriterpenoids) combination of functional groups. Promising compounds with anticancer and antiviral action (antiHIV, in particular) have been revealed among the produced 2,3-seco- and A-pentacyclic derivatives, including compounds combining inhibiting action against several types of virus.

ACKNOWLEDGMENTS

This work is supported by the Russian Foundation for Basic Research (Grants No. 13-03-00629a, 14-03-00256a, 14-03-96007-ural-a, 15-03-01701a) for the MK-5386.2016.3. The reported study was partially supported by the Government of Perm Krai, research project No. C-26/056.

KEYWORDS

- triterpenoids
- Beckmann fragmentation
- A-secotriterpenoids
- heterocyclization
- hydrazones
- amides
- esters

REFERENCES

1. Cichewicz, R. H.; Kouzi, S. A. Chemistry, Biological Activity, and Chemotherapeutic Potential of Betulinic Acid for the Prevention and Treatment of Cancer and HIV Infection. *Med. Res. Rev.* **2004**, *24*(1), 90–114.
2. Tolstikova, T. G.; Sorokina, I. V.; Tolstikov, G. A.: Tolstikov, A. G.; Flekhter, O. B. Biological Activity and Pharmacological Prospects of Lupine Terpenoids: Natural Lupane Derivatives. *Russ. J. Bioorg. Chem.* **2006**, *32*(3), 261–276 (in Russian).
3. Krasutsky, P. A. Birch Bark Research and Development. *Nat. Prod. Rep.* **2006**, *23*, 919–942.
4. Alakurtti, S.; Makela, T.; Koskimies, S.; Yli-Kauhaluoma, J. Pharmacological Properties of The Ubiquitous Natural Product Betulin. *Eur. J. Pharm. Sci.* **2006**, *29*(1), 1–13.
5. Fulda, S.. Betulinic Acid for Cancer Treatment and Prevention. *Int. J. Mol. Sci.* **2008**, *9*(6), 1096–1107.
6. Laszczyk, M. N.. Pentacyclic Triterpenes of the Lupane, Oleanane and Ursane Group as Tools in Cancer Therapy. *Planta Med.* **2009**, *75*(15), 1549–1560.
7. Kazakova, O. B.; Khusnutdinova, E. F.; Kukovinets, O. S.; Zvereva, T. I.; Tolstikov, G. A.. Effective Synthesis of 2,3-Seco-2,3-Dicarboxyplatanic Acid. *Chem. Nat. Compd.* **2010**, *46*, 393–396 (in Russian).
8. Wei, Y.; Ma, C.-M.; Hattori, M. Synthesis and Evaluation of A-Seco Type Triterpenoids for Anti-HIV-1 Protease Activity. *Eur. J. Med. Chem.* **2009**, *44*(10), 4112–4120.
9. Urban, M.; Sarek, J.; Klinot, J.; Korinkova, G.; Hajduch, M.. Synthesis of A-Secoderivatives of Betulinic Acid With Cytotoxic Activity. *J. Nat. Prod.* **2004**, *67*(7), 1100–1105.
10. Kazakova, O. B.; Medvedeva, N. I.; Tolstikov, G. A.; Kukovinets, O. S.; Yamansarov, E. Y.; Spirikhin, L. V.; Gubaidullin, A. T. Synthesis of Terminal Acetylenes Using POCl3 in Pyridine as Applied to Natural Triterpenoids. *Mendeleev Commun.* **2010**, *20*(4), 234–236 (in Russian).

11. Deng, Y.; Snyder, J. K. Preparation of a 24-Nor-1,4-Dien-3-One Triterpene Derivative from Betulin: A New Route to 24-Nortriterpene Analogues. *J. Org. Chem.* **2002**, *67*(9), 2864–2873.

12. Simoneit, B. R. T.; Xu, Y.; Neto, R. R.; Cloutier, J. B.; Jaffe, R. Photochemical Alteration of 3-Oxygenated Triterpenoids: Implications for the Origin of 3,4-Seco-Triterpenoids in Sediments. *Chemosphere*, **2009**, *74*(4), 543–550.

13. Akihisa, T.; Takamine, Y.; Yoshizumi, K.; Tokuda, H.; Kimura, Y.; Ukiya, M.; Nakahara, T.; Yokochi, T.; Ichiishi, E.; Nishino, H. Microbial Transformations of Two Lupane-Type Triterpenes and Anti-Tumor-Promoting Effects of the Transformation Products. *J. Nat. Prod.* **2002**, *65*(3), 278–282.

14. Maitraie, D.; Hung, C.-F.; Tu, H.-Y.; Liou, Y.-T.; Wei, B.-L.; Yang, S.-C.; Wang, J.-P.; Lin, C.-N. Synthesis, Anti-Inflammatory, and Antioxidant Activities of 18β-Glycyrrhetinic Acid Derivatives as Chemical Mediators and Xanthine Oxidase Inhibitors. *Bioorg. Med. Chem.* *17*(7), **2009**, 2785–2792.

15. Flekhter, O. B.; Medvedeva, N. I.; Tolstikov, G. A.; Savinova, O. V.; Boreko, E. I.; Dolgushin, F. M. Betulonic Amides Modified at Cycle a by Amino Acids: Synthesis and Inhibition of Flu A Virus Reproduction. *Russ. J. Bioorg. Chem.* **2009**, *35*(1), 118–122 (in Russian).

16. Tolstikov, G. A.; Flekhter, O. B.; Shultz, E. E.; Baltina, L. A.; Tolstikov, A. G. Betulin And Its Derivatives Chemistry And Biological Activity. *Chem. Sustainable Dev.* 2005, *13*, 1–29 (in Russian).

17. Shernyukov, A. V.; Salakhutdinov, N. F.; Tolstikov G. A. Methods of the Synthesis of A-Seco Derivatives of Pentacyclic Triterpenoids. *Russ. Chem. Bull.* **2013**, *62*(4), 878–895 (in Russian).

18. Tolmacheva, I. A.; Nazarov, A. V.; Maiorova, O. A.; Grishko, V. V. Synthesis of Lupane and 19β,28-Epoxy-18a-Oleane 2,3-Seco-Derivatives based on Betulin. *Chem. Nat. Compd.* **2008**, *44*(5), 606–611 (in Russian).

19. Tolmacheva, I. A.; Grishko, V. V.; Boreko, E. I.; Savinova, O. V.; Pavlova, N. I. Synthesis and Antiviral Activity of 2,3-Seco-Derivatives of Betulonic Acid. *Chem. Nat. Compd.* **2009**, *45*(5), 673–676 (in Russian).

20. Anikina, L. V.; Tolmacheva, I. A.; Vikharev, Yu. B.; Grishko, V. V. The Immunotropic Activity of Lupane and Oleanane 2,3-*Seco*-Triterpenoids. *Russ. J. Bioorg. Chem.* **2010**, *36*(2), 240–244 (in Russian).

21. Galaiko, N. V.; Nazarov, A. V.; Tolmacheva, I. A.; Slepukhin, P. A.; Vikharev, Yu. B.; Maiorova, O. A.; Grishko, V. V. Synthesis of Triterpene A-Condensed Azoles. *Chem. Het. Comp.* **2014**, *50*(1), 65–75 (in Russian).

22. Konoike, T.; Takahashi, K.; Kitaura, Y.; Kanda, Y. Synthesis of [2-13C]-Oleanolic Acid and [2-13C]-Myricerone, *Tetrahedron*, **1999**, *55*, 14901–14914.

23. Khudobko, M. V.; Mikhailova, L. R.; Baltina, Jr. L. A.; Spirikhin, L. V.; Baltina, L. A.. Synthesis f 2,11-Dioxo-Norolean A(1)-12,18(19)-Dien-30-Oic Acid. *Chem. Nat. Compd.* **2011**, *47*(1), 76–78 (in Russian).

24. Zhang, P.; Xu, L.; Qian, K.; Liu, J.; Zhang, L.; Lee, K.-H.; Sun, H. Efficient Synthesis and Biological Evaluation of Epiceanothic Acid and Related Compounds. *Bioorg. Med. Chem. Lett.* **2011**, *21*(1), 338–341.

25. Fleming, F. F.; Shook, B. C. Nitrile Anion Cyclizations. *Tetrahedron*, **2002**, *58*, 1–23.

26. Grishko, V. V.; Galaiko, N. V.; Tolmacheva, I. A.; Kucherov, I. I.; Eremin, V. F.; Boreko, E. I.; Savinova, O. V.; Slepukhin, P. A. Functionalization, Cyclization and Antiviral Activity of A-Secotriterpenoids. *Eur. J. Med. Chem.* **2014**, *83*, 601–608.

27. Pereslavtseva, A. V.; Tolmacheva, I. A.; Slepukhin, P. A.; El'tsov, O. S.; Kucherov, I. I.; Eremin, V. F.; Grishko, V. V.. Synthesis of A-Pentacyclic Triterpene α,β-Alkenenitriles. *Chem. Nat. Compd.* **2014**, *49*(6), 1059–1066 (in Russian).

28. Muherjee, R.; Jaggi, M.; Rajendran, P.; Siddiqui, M. J. A.; Srivastava, S. K.; Vardhan, A.; Burman, A. Betulinic Acid And Its Derivatives As Anti-Angiogenic Agents. *Bioorg. Med. Chem. Lett.* **2004**, *14*(9), 2181–2184.

29. Mukherjee, R.; Jaggi, M.; Rajendran, P.; Srivastava, S. K.; Siddiqui, M. J. A.; Vardhan, A.; Burman, A. Synthesis of 3-O-Acyl/3-Benzylidene/3-Hydrazone/3-Hydrazine/17-Carboxyacryloyl Ester Derivatives of Betulinic Acid as Anti-Angiogenic Agents. *Bioorg. Med. Chem. Lett.* **2004**, *14*(12), 3169–3172.

30. Flekhter, O. B.; Boreko, E. I.; Nigmatullina, L. R.; Pavlova, N. I.; Nikolaeva, S. N.; Savinova, O. V.; Eremin, V. F.; Baltina, L. A.; Galin, F. Z.; Tolstikov, G. A.. Synthesis and Antiviral Activity of Hydrazides and Substituted Benzalhydrazides of Betulinic Acid and Its Derivatives. *Russ. J. Bioorg. Chem.* **2003**, *29*(3), 296-301 (in Russian).

31. Tolmacheva, I. A.; Galaiko, N. V.; Grishko, V. V.. Synthesis of Acylhydrazones from Lupane and 19b,28-Epoxy-18a-Oleanane 2,3-Seco-Aldehydonitriles. *Chem. Nat. Compd.* **2010**, *46*, 39–43 (in Russian).

32. Galayko, N. V.; Tolmacheva, I. A.; Grishko, V. V.; Volkova, L. V.; Perevozchikova, E. N.; Pestereva, S. A. Antiviral Activity of 2,3-Secotriterpenic Hydrazones of the Lupane and 19β,28-Epoxy-18a-Oleanane Types. *Russ. J. Bioorg. Chem.* **2010**, *36*(4), 516–521 (in Russian).

33. Gein, S. V.; Grishko, V. V.; Baeva, T. A.; Tolmacheva, I. A. Immunoregulatory In Vitro/In Vivo Effects of 2,3-Secotriterpene Acetylhydrazone. *Int. J. Pharm.* **2013**, *9*, 74–79.

34. Tolmacheva, I. A.; Galaiko, N. V.; Grishko, V. V. Synthesis of 1-Cyano-2,3-Secoderivatives of Glycyrrhetic Acid. *Chem. Nat. Compd.* **2011**, *47*(2), 246–249 (in Russian).

35. Grishko, V. V.; Tolmacheva, I. A.; Galaiko, N. V.; Pereslavceva, A. V.; Anikina, L. V.; Volkova, L. V.; Bachmetyev, B. A.; Slepukhin, P. A. Synthesis, Transformation and Biological Evaluation of 2,3-Secotriterpene Acetylhydrazones and Their Derivatives. *Eur. J. Med. Chem.* **2013**, *68*, 203–211.

36. Galaiko, N. V.; Tolmacheva, I. A.; Volkova, L. V.; Grishko, V. V. Synthesis of 2,3-Secotriterpene Hydrazonohydrazides of the Lupane and 19β,28-Epoxy-18α-oleanane Types. *Chem. Nat. Compd.* **2011**, *48*(1), 72–74 (in Russian).

37. Dargan, D. J.; Gait, C. B.; Subak-Sharpe, J. H. Effect of Cicloxolone Sodium on the Replication of Esicular Stomatitis Virus in BSC-1 Cells. *J. Gen. Virol.* **1992**, *73*(2), 397–406.

38. Kaminska, T.; Kaczor, J.; Rzeski, W.; Wejksza, K.; Kandefer-Szerszen, M.; Witek, M. A. A Comparison of the Antiviral Activity of Three Triterpenoids Isolated from *Betula alba* Bark. *Ann. Univ. Mariae Curie-Sklodowska, Sect. C. Biol.* **2004**, *LIX*, 1–7.

39. Kashiwada, Y.; Hashimoto, F.; Cosentino, L. M.; Chen, C. H.; Garrett, P. E.; Lee, K. H. *J. Med. Chem.* **1996**, *39*(5), 1016–1017.

40. Stonik, V. A.; Tolstikov, G. A. Natural Compounds and the Creation of Domestic Drugs. *Vestnik Ross. Akad. Nauk* **2008**, *78*(8), 675–687 (in Russian).
41. Aiken, Ch.; Chen, Ch. H. Betulinic Acid Derivates as HIV-1 Antivirals. *Trends Mol. Med.* **2005**, *11*, 31–36.
42. Lee, K.-H. Discovery and Development of Natural Product-Derived Chemotherapeutic Agents Based on a Medicinal Chemistry Approach. *J. Nat. Prod.* **2010**, *73*, 500–516.
43. Tolmacheva, I. A.; Igosheva, E. V., Grishko, V. V.; Zhukova, O. S.; Gerasimova, G. K. The Synthesis Of Triterpenic Amides On The Basis Of 2,3-*Seco*-1-Cyano-19β,28-Epoxy-18α-Oleane-3-Oic Acid. *Russ. J. Bioorg. Chem.* **2010**, *36*(3), 377–382 (in Russian).
44. Tolmacheva, I. A.; Igosheva, E. V.; Vikharev, Yu. B.; Grishko, V. V.; Savinova, O. V.; *Boreko, E. I.; Eremin, V. Ph.* Synthesis And Biological Activity of Amides Of 28-Methoxy-28-Oxo-1-Cyano-2,3-*Seco*-Lup-20(29)-En-3-Oic Acid. *Chem. Nat. Compd.* **2012**, *48*(3), 426–431 (in Russian).
45. Igosheva, E. V.; Tolmacheva, I. A., Vikharev, Yu. B.; Grishko, V. V.; Savinova, O. V.; *Boreko, E. I.; Eremin, V. Ph.* Synthesis And Biological Activity of Mono- and Diamides Of 2,3-Secotriterpene Acids. *Russ. J. Bioorg. Chem.* **2013**, *39*(2), 186–193 (in Russian).
46. Tolmacheva, I. A.; Igosheva, E. V.; Savinova, O. V., Boreko, E. I.; Grishko, V. V. Synthesis and Antiviral Activity of C-3(C-28)-Substituted 2,3-Seco-Triterpenoids. *Chem. Nat. Compd.* **2014**, *49*(6), 1050–1058.
47. Soler, F.; Poujade, C.; Evers, M.; Carry, J.-C.; Henin, Y.; Bousseau, A.; Huet, T.; Pauwels, R.; De Clercq, E.; Mayaux, J.-F.; Le Pecq, J.-B.; Dereu, N. Betulinic Acid Derivatives: A New Class of Specific Inhibitors of Human Immunodeficiency Virus Type 1 Entry. *J. Med. Chem.* **1996**, 39, 1069–1083.
48. Sun, I.-C.; Wang, H.-K.; Kashiwada, Y.; Shen, J.-K.; Cosentino, L. M.; Chen, C.-H.; Yang, L.-M.; Lee, K.-H. Synthesis and Structure-Activity Relationships of Betulin Derivatives as Anty-HIV Agents. *J. Med. Chem.* **1998**, *41*, 4648–4657.
49. Qian, K.; Kuo, R.-Y.; Chen, C.-H.; Huang, L.; Morris-Natschke, S. L.; Lee, K.-H. Anti-AIDS Agents 81. Design, Synthesis, And Structure–Activity Relationship Study Of Betulinic Acid And Moronic Acid Derivatives As Potent HIV Maturation Inhibitors. *J. Med. Chem.* **2010**, *53*, 3133–3141.
50. Krainova, G. F.; Tolmacheva, I. A.; Gorbunova, M. N.; Grishko V. V. Synthesis of Lupane and A-Secolupane Allylamides. *Chem. Nat. Compd.* **2013**, *49*(2), 281–285 (in Russian).
51. Krainova, G. F.; Tolmacheva, I. A.; El'tsov, O. S.; Gorbunova, M. N.; Grishko, V. V. Synthesis of Triterpene Vinyl Esters Containing of the Lupane and A-Secolupane Types. *Chem. Nat. Compd.,* **2016,** *52*(2), 256–261 (in Russian).
52. Kvasnica, M.; Urban, M.; Dickinson, N. J.; Sarek, J. Pentacyclic Triterpenoids with Nitrogen- and Sulfur-Containing Heterocycles: Synthesis and Medicinal Significance. *Nat. Prod. Rep.* **2015,** *32*(9), 1303–1330.

CHAPTER 2

SYNTHESIS AND MEMBRANE-PROTECTIVE PROPERTIES OF SULFUR-CONTAINING DERIVATIVES OF MONOTERPENOIDES WITH MONOSACCHARIDE FRAGMENTS

ALEXANDER V. KUTCHIN[1], SVETLANA V. PESTOVA[1*], EVGENY S. IZMEST'EV[1], OKSANA G. SHEVCHENKO[2], and SVETLANA A. RUBTSOVA[1]

[1]*Institute of Chemistry of Komi Scientific Centre of Ural Branch of the Russian Academy of Sciences, 48, Pervomaiskaya St., Syktyvkar 167982, Komi Republik, Russia*

[2]*Institute of Biology of Komi Scientific Centre of Ural Branch of the Russian Academy of Sciences, 28, Kommunisticheskaya St., Syktyvkar 167982, Komi Republik, Russia*

Corresponding author. E-mail: pestova-sv@chemi.komisc.ru

CONTENTS

ABSTRACT

Synthesis of sulfides proceeding from neomenthanethiol and diacetone-galacto-, diacetonefructopyranose, monoacetoneglucofuranose with yields up to 98% was carried out. Under oxidation of sulfides with diacetone-protected carbohydrate moieties, new diastereomeric sulfoxides were synthesized. As oxidants, m-CPBA, CHP/VO(acac)$_2$, and TBHP/VO(acac)$_2$ systems were used. Based on 6-thiodiacetonegalactopyranose, 1-thiodiacetonefructopyranose, and terpenic thiols: neomentanethiol, isobornanethiol, $trans$-verbenethiol, myrthenethiol, and cis-myrtanethiol both symmetrical disulfides with terpene and carbohydrate moieties of up to 41% and 13%, respectively, and asymmetrical disulfides containing simultaneously terpene and monosaccharide fragments in an amount of 51–90% by weight of the reaction products were prepared. To increase the water solubility of sulfides and sulfoxides obtained, the removal of the isopropylidene protection was performed. The estimation of membrane-protective and antioxidant properties of thioglycosides with monoterpene and monosaccharide fragments on the basis of their ability to inhibit H_2O_2-induced hemolysis of erythrocytes and to break the accumulation of secondary products of lipid peroxidation (LPO) was carried out.

2.1 INTRODUCTION

Natural and semisynthetic thioglycosides possess a wide range of biological activity. Among low toxicity thioglycosides [1], the compounds with antiviral, antitumor [2, 3], anticoagulant, and anti-inflammatory activity [4] are known.

Due to a large number of free hydroxyl groups, the natural monosaccharides represent hydrophilic compounds in most cases. When a biologically active monosaccharide fragment is introduced into the structure of the lipophilic fragment, such as a terpene, this makes it possible to produce the entire conjugate being soluble in water, which facilitates its further application in pharmacology [5, 6].

The presence of a sulfur atom, which is able to be oxidized, in the structure, can impart antioxidant properties caused antiperoxide activity of the sulfur-containing groups [7, 8]. The features of a chemical structure of sulfur-containing compounds synthesized based on natural monoterpenoids also suggest the availability of these properties [9], which is confirmed by our studies of membrane-protective and antioxidant activity of disulfides bearing monosaccharide and monoterpene fragments [10].

2.2 MATERIALS AND METHODOLOGY

For the evaluation of toxicity, antioxidant and membrane-protective activities of the compounds synthesized, a 0.5% (v/v) suspension of laboratory mice erythrocytes in the phosphate-buffered saline (pH 7.4) was used. Ethanol was utilized as a solvent. The toxicity (in vitro) was assessed according to their capacity to induce hemolysis. Ethanol solutions were added to the erythrocyte suspension which was incubated at 37 °C for 5 h in a thermostatic Biosan ES20 shaker (Latvia). The control samples contained an appropriate volume of ethanol. Membrane-protective and antioxidant activities were assessed based on the degree of inhibition of hemolysis induced and retardation of accumulation of LPO secondary products. For this purpose, 30 min after the inclusion to the erythrocyte suspension of the tested compounds, the hemolysis was initiated by H_2O_2 (0.9 µM) or 2,2-azobis(amidinopropane)dihydrochloride (AAPH, 5 mmol) solution. The reaction mixture was incubated in a thermo-stated shaker under slow stirring at 37°C for 5 h. An aliquot was taken every 1 h, centrifuged for 5 min at 1600 g, and the hemolysis degree was measured at λ 524 nm on a ThermoSpectromic Genesys 20 spectrophotometer (the USA) based on the hemoglobin amount in the supernatant [11]. The hemolysis degree was calculated relative to the total hemolysis of the sample [12, 13]. The content of LPO secondary products reacting with 2-thiobarbituric acid (TBA-RS) was determined spectrophotometrically [14]. Each experiment was repeated three to five times. The statistical treatment and diagram plotting were performed using the Microsoft Office Excel 2007 program package.

2.3 RESULTS AND DISCUSSION

In this paper, we have studied membrane-protective and antioxidant activity of sulfur-containing glycosides earlier synthesized [15, 16] (Fig. 2.1 [15, 16]). Sulfides I, II and sulfoxides III, IV were obtained from (1S,2S,5R)-2-isopropyl-5-methylcyclohexanethiol (neomenthanethiol) and D-galactose [15], as well as D-glucose and D-fructose [16].

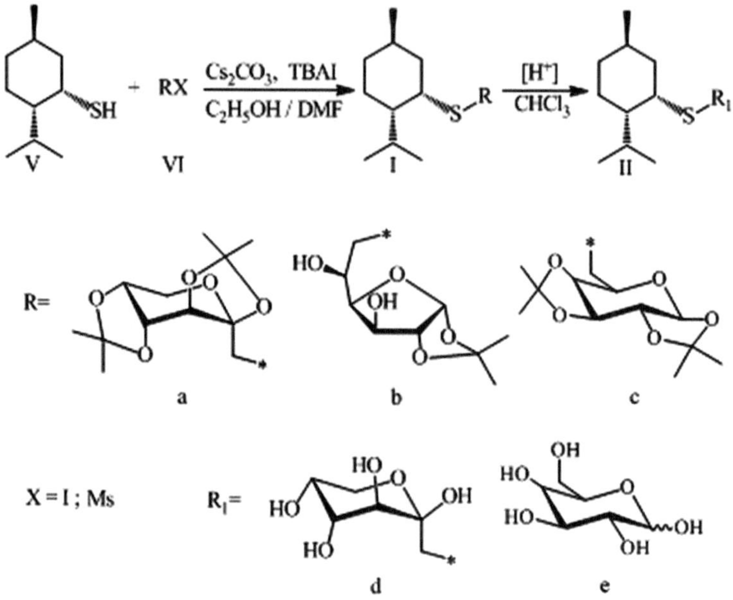

FIGURE 2.1 Synthesis of thioglycosides.

Synthesis of the sulfides I was carried out in refluxing ethanol with Cs_2CO_3 (see Fig. 2.1 [15, 16]). The yield of sulfides with isopropylidene-protected carbohydrate moieties reached 95%. To obtain the compound V, a well-known method based on the substitution of *p*-toluenesulfonyl group in the corresponding tosylate to thioacetyl fragment followed by reduction of the resulting thioacetate to thiol was used. The compounds VI were prepared by reacting of galacto- and fructopyranose acetonides with iodine in the presence of benzimidazole (BzIm) and Ph_3P, as well as by reacting of glucofuranose-1,2-monoacetonide with methanesulfonyl chloride.

For obtaining of the sulfoxides III, oxidation of the sulfides I by *meta*-chloroperoxybenzoic acid (*m*-CPBA) and by the systems *tert*-butyl hydroperoxide (TBHP)–vanadyl acetylacetonate (VO(acac)$_2$), cumene hydroperoxide (CHP)/VO(acac)$_2$ (Fig. 2.2 [15, 16]) was carried out. The sulfoxides were obtained as diastereomeric mixtures, which were separated by chromatographic methods. Using as an oxidant TBHP/VO(acac)$_2$, the sulfoxides were formed with the highest yield. The diastereoselectivity of sulfide oxidation was not higher than 37%. Formation of the sulfoxides with fructopyranose fragment, when CHP/VO(acac)$_2$ was used, proceeded nondiastereoselectively (*de* 0).

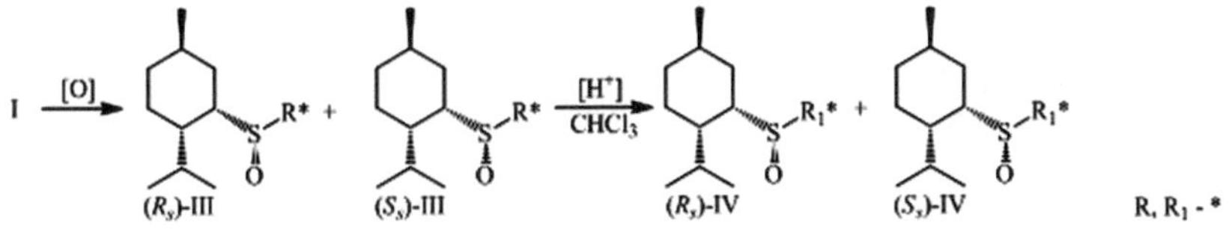

FIGURE 2.2 Thioglycoside oxidation.

Note: [*O*]: *m*-CPBA—*meta*-chloroperoxybenzoic acid; TBHP—*tert*-butyl hydroperoxide; CHP—cumene hydroperoxide; *—see radical in Figure 2.1.

To increase water solubility of the obtained sulfides Ia,c and sulfoxides (R_s)-IIIc and (S_s)-IIIc, removal of the isopropylidene protection was performed by trifluoroacetic acid in chloroform. The resulting sulfides IId,e and sulfoxides (R_s)-IVe, (S_s)-IVe were synthesized as anomeric mixtures. Removing of isopropylidene protection from the sulfide and sulfoxides with monoacetoneglucofuranose fragment and the sulfoxides having diacetonefructopyranose moiety failed. All attempts to remove the acetonide protection from these compounds led to a complete resinification of reaction mixture. In all cases, yield of the synthesized sulfides reached 98%, except the sulfide with the fructose moiety with free OH groups; its yield was 62%.

One of the most suitable and affordable objects for estimation of membrane-protective and antioxidant activity of new compounds is the nonnuclear red blood cells of mammals. Screening of biological activity of the synthesized compounds was performed in vitro using red blood cells [12, 17–20]. As the toxicity of sulfides and sulfoxides is able to limit their further application, preliminary hemolytic activity of the compounds must be assessed [11, 13].

The studies found that sulfides I(a,c) and IIe in the concentration 100 μM were highly toxic against the erythrocytes of mammals blood (Fig. 2.3), which did not allow to estimate their membrane-protective and antioxidant activities in the present model system. However, their sulfoxides III and IV did not possess pronounced hemolytic activity.

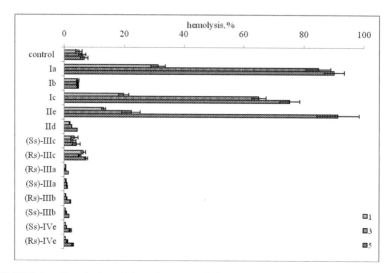

FIGURE 2.3   Hemolytic activity of compounds in concentration 100 µM.
Note: 1, 3, and 5 h of incubation.

Comparison of the results shown in Figure 2.4a and b exhibits that in decreasing order of membrane-protective activity, the test compounds can be arranged in the following order: IIg = (S_s)-IIIa > (S_s)-IVe > (R_s)-IVe = (R_s)-IIIa > (R_s)-IIIc > (S_s)-IIIc. The statistically significant membrane-protective activity in H_2O_2-induced hemolysis of erythrocytes for the compounds (R_s)-IIIb, (Ss)-IIIb and Ib has not been revealed.

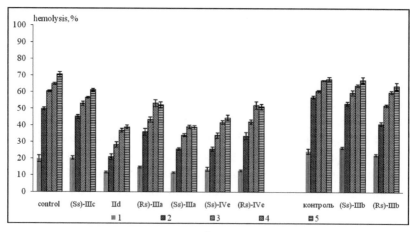

a

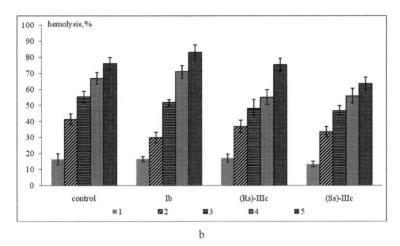

b

FIGURE 2.4   Parameters of H_2O_2-induced hemolysis of erythrocytes in presence of test compounds (100 µM). (a) and (b) Dynamics of hemolysis (1–5 h of incubation).

Generally, the action mechanism of the test conjugates can be reduced to a simple oxidation, where sulfides act as "traps" of free radicals in the initiation stage of peroxide decomposition:

$$ROOH + R_1SR_2 \rightarrow ROH + R_1S(O)R_2$$
$$ROOH + R_1S(O)R_2 \rightarrow ROH + R_1S(O)_2R_2,$$

where R_1 – terpene, and R_2 – monosaccharide moiety.

Sulfides deactivate hydroperoxides in the system, by the way of stoichiometric interaction with them, to obtain a sulfoxide and an alcohol. This reaction, however, is too slow to prevent a significant accumulation of hydroperoxide under autooxidation and, therefore, gives only a small contribution to the overall antioxidant effect of sulfides [21]. Sulfoxides, which are formed at a later stage of oxidation, act as inhibitors of ionic hydroperoxide decomposition. Therefore, they make the most significant contribution to the total antioxidant activity of sulfides [22].

The authors of the paper [23] proposed a mechanism according to which sulfoxides act by formation of molecular complexes with hydroperoxides, thereby preventing the disintegration of the lasts to free radicals. The sulfoxide activity largely depends on their thermal stability, but no explanation of this fact is given. This conclusion is refuted in [21], where there is a proof of the sulfoxide antioxidant effect, which is derived from their ability to form sulfenic acids inhibiting the hydroperoxide decomposition. Presumably, the given process of forming of unstable sulfenic acids can be described by the following reactions:

$$\text{Oxidant inducer} \rightarrow RO_2\bullet \text{ (initiation)}$$
$$RO_2\bullet + RH \rightarrow ROOH + R\bullet$$
$$R\bullet + O_2 \rightarrow RO_2\bullet \text{ (fast)} \rightarrow \text{propagation}$$
$$2RO_2\bullet \rightarrow \text{molecular products (termination)}$$
$$R_1SOR_2 \rightarrow R_1SOH + R_2 - \text{(thermal decomposition of sulfoxide)}$$
$$2R_1SOH \rightarrow R_1SOSR_1 + H_2O$$
$$R_1SOH + nRO_2\bullet \rightarrow \text{molecular products (inhibition)}$$
$$R_1SOH \rightarrow \text{molecular products (decomposition of sulfenic acid)}$$
$$(R_1 - \text{Terp-}, R_2 - \text{monosaccharide moiety})$$

In our study, the sulfoxides containing diisopropylidene galacto- (c, e) and fructopyranose (a, d) fragments are the most active against the hydroxyl radical. This can be probably explained by the fact that sulfoxides are able to have an instantaneous antioxidant effect, while the starting sulfides to exhibit antioxidant properties have to be prereacted with a small amount of oxygen [24]. Most of the conjugates (IId, (S_s)-IVe, (R_s)-IVe) having free hydroxyl groups in the monosaccharide fragments demonstrated high membrane-protective activity. Exceptions are the investigated sulfides and sulfoxides with glucopyranose moiety (((R_s)-IIIb, (S_s)-IIIb, Ib). They contain two free hydroxyl groups in their structure, but none of them has shown high activity in this model system.

The compounds exhibited the highest membrane-protective activity not only protected cells from hemolysis, but also significantly inhibited the accumulation of lipid peroxidation secondary products (TBA-RS) in a suspension of erythrocytes (Fig. 2.5), what confirms the presence of antioxidant activity with the test compounds in this model system. If we consider the dynamics of cell death in the acute oxidative stress conditions, we could conclude that, in general, the investigated group of compounds is active only in the early stages and low-active in the final process stage.

The activity of the given sulfur-containing compounds is due to their action inhibiting the peroxide decomposition, but they are not able to except the total process. This phenomenon can be eliminated by an internal synergism when besides sulfur-containing groups possessing antiperoxidant activity, the fragment with already available antiradical activity is introduced into the structure of a biologically active compound [25].

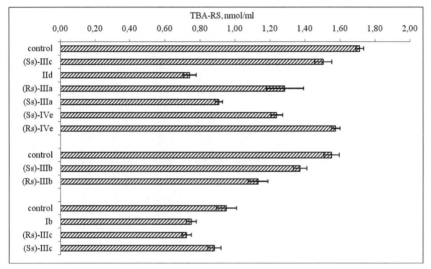

FIGURE 2.5 Parameters of H_2O_2-induced hemolysis of erythrocytes in the presence of test compounds (100 µM): the content of LPO secondary products.

Further, estimation of the antioxidant and membrane-protective activity of our compounds obtained was carried out using AAPH (2,2'-*azo*bis-(amidinopropane) dihydrochloride) as a source of free radicals, whose decomposition occurs as follows:

$$R–N = N–R \rightarrow 2R\bullet + N_2, \text{ where } R = –C(CH_3)_2C(=NH)NH_2$$

$$R\bullet + O_2 \rightarrow ROO\bullet$$

In the aqueous phase under physiological temperatures, the peroxide radicals ROO˙ [26] not being able unlike H_2O_2 to penetrate into the cell are formed from AAPH with constant velocity and exerted influence upon the erythrocyte on the outer side of membrane [17].

Apparently, from the data presented in parts a and b of Figure 2.6, considerable membrane-protective and antioxidant activities under the conditions of oxidative stress induced by peroxide radical are typical only for the sulfoxides (R_s)-IIIa, (R_s)-IIIb, (S_s)-IIIc, (S_s)-IVe, and (R_s)-IVe. Membrane-protective activity of the compounds (S_s)-IIIb, IId, and (S_s)-IIIa was very weak or fully not observed in the experiment. As in the previous experiment, the biological activity of the conjugates was dependent on the structure of a sulfur-containing group.

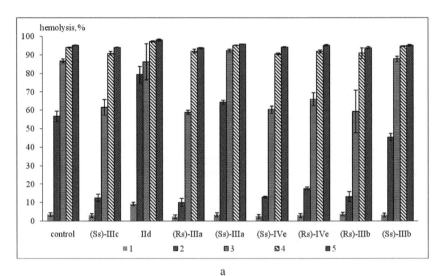

a

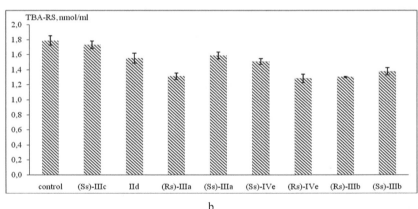

b

FIGURE 2.6 Parameters of AAPH-induced hemolysis of erythrocytes in the presence of test compounds (100 μM): (a) dynamics of hemolysis (1–5 h of incubation) and (b) the content of LP secondary products.

Moreover, we carried out the synthesis of disulfides bearing terpene and/or carbohydrate moieties in a variety of combinations [10] under oxidation of the corresponding thiol compounds (1:1) by iodine at room temperature (Fig. 2.7[10]). The starting terpene materials were used, such as neomenthanethiol VIIA, isobornanethiol VIIB, *cis*-myrtanethiol VIIC, myrtenethiol VIID, and *trans*-verbenethiol VIIE, as well as 1-thiodiace-tonefructopyranose VIIIa and 6-thiodiacetonegalactopyranose VIIIc as carbohydrate substances.

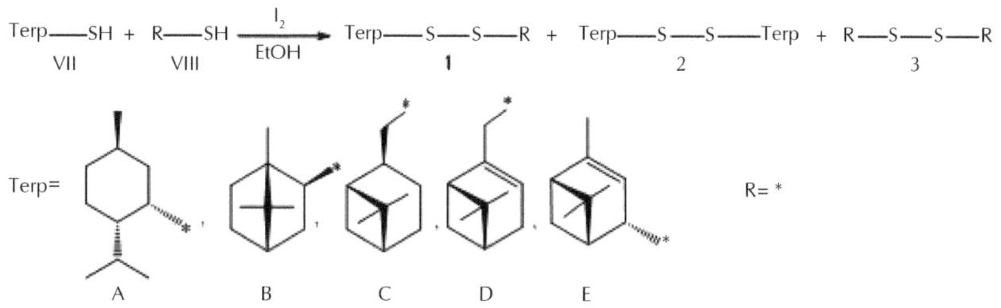

FIGURE 2.7   Synthesis of disulfides.

*Denotes monosaccharide fragment (R) and its notation are shown in Figure 2.1.

We found that the main products of oxidation of the presented terpene and carbohydrate thiol mixtures by iodine were unsymmetrical disulfides 1. The maximum amount of them was formed under the oxidation of a mixture of diacetonefructopyranosyl thiol VIIIa and isobornanethiol VIIB (up to 90% of the total number of products), the smallest one (51%) was in the case when neomentanethiol VIIA as a thiol was used. The terpene-substituted symmetrical disulfides 2 were formed in amounts of 7–41%, and disulfides with monosaccharide fragments were yielded 3–13%, what probably indicates the low speed of their formation.

As for mono sulfur compounds, a screening of the bioactivity, such as membrane protective and antioxidant, was performed for disulfides obtained in the concentration 10 μM. Concentration was decreased 10 times as compared with the previous experiment to reduce the hemolytic activity of the conjugates.

Among all the conjugates researched, the hemolytic activity was found only in unsymmetrical disulfides containing neomenthyl (1Ac, 1Aa) or isobornyl (1Ba) fragments. The addition of these compounds to an eryth-rocyte suspension in the concentration 10 μM caused 2.1–3.4 times excess of a spontaneous hemolysis level.

Most of the tested compounds showed an ability to inhibit the death of erythrocytes under the acute stress conditions induced by the addition of hydrogen peroxide. The membrane-protective activity of compounds with the galactopyranosyl fragment decreases in the following order (Fig. 2.8 [10]): 1Dc = 1Ca > 1Ac = 1Bc > 3cc = standard (di-*n*-propyl disulfide) > 1Ec. This means that the conjugates with *cis*-myrtanyl, myrtenyl, neomenthyl, and isobornyl moieties are the most active. The lowest activity was observed for the disulfides with *trans*-verbenyl fragment.

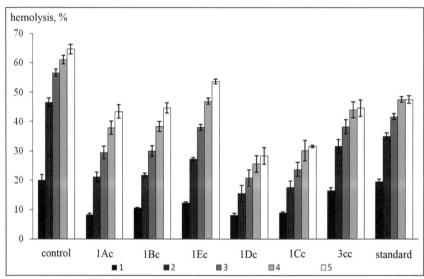

FIGURE 2.8   Influence of compounds 1Ac, 1Bc, 1Ec, 1Dc, 1Cc, 3cc and a standard (Pr$^n_2$S$_2$) in the concentration 10 μM on the degree of the H$_2$O$_2$-induced hemolysis of erythrocytes after 1–5 h of incubation.

The structurally similar compounds containing fructopyranose moiety exhibited a lower activity as compared to disulfides with galactopyranose fragment (Fig. 2.9). The calculations showed that the replacement of the galactopyranose fragment with fructopyranose one led to a 1.2–1.6 times decrease in the membrane-protective activity for most compounds, except a low-activity disulfide 1Ec containing *trans*-verbenyl fragment (see Figs. 2.8 and 2.9). Thus, in this case, erythrocytes were protected by the conjugates with galactopyranose fragment, the value of relative hemolysis was 44–69%, whereas when similar compounds with fructopyranose fragment were used, it was 69–91%. However, among the disulfides with fructopyranose moiety, the compounds containing myrtenyl, *cis*-myrtanyl, or neomenthyl fragments were found to be the most active. The regressive analysis indicates the presence of a statistically significant close posi-tive relationship between the activities of the corresponding galacto- and fructose-containing disulfides with five different terpene fragments in the composition ($R_S = 0.9$, $p = 0.037$).

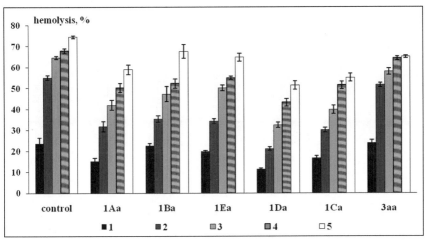

FIGURE 2.9 Influence of compounds 1Aa, 1Ba, 1Ea, 1Da, 1Ca, and 3aa in the concentration 10 μM on the degree of the H_2O_2-induced hemolysis of erythrocytes after 1–5 h of incubation.

For the symmetrical disulfides containing no sugar fragments (Figure 2.10 [10]), the highest membrane protective activity was found for compounds containing *cis*-myrtanyl 2DD, myrtenyl 2CC, and isobornyl 2BB moieties. The lowest activity was found in the disulfide with *trans*-verbenyl fragment 2EE. For the symmetrical neomenthane-derived disul-fide 2AA, a statistically significant membrane protective activity was not found (no data reported).

A comparative evaluation of the membrane-protective activity of unsymmetrical disulfides with galactose and fructose moieties, and symmetrical disulfides showed (Fig. 2.11 [10]) that for most disulfides, it depends on the structure of both the terpene and the carbohydrate frag-ments. Only for the conjugates with *trans*-verbenyl moiety, the presence and the character of the carbohydrate part of the molecule have no influence on the biological activity of the compounds. Generally, it can be stated that the highest ability to protect cells from the influence of hydrogen peroxide

is possessed by compounds containing *cis*-myrtanyl/myrtenyl and galactose fragments.

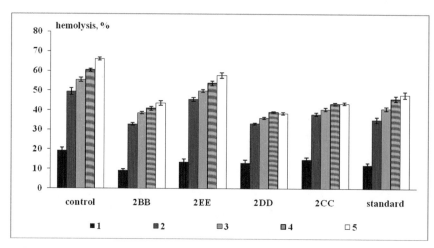

FIGURE 2.10 Influence of compounds 2(BB–EE) and a standard $(Pr^n_2S_2)$ in the
concentration 10 µM on the degree of the H_2O_2-induced hemolysis of erythrocytes after
1–5 h of incubation.

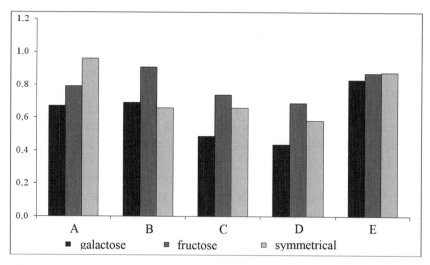

FIGURE 2.11 Relative membrane protective activity (the ratio of the hemolysis level in
the tested and control samples, 5 h after addition of H_2O_2) of the tested compounds with
galactose, fructose, and terpene fragments in the concentration 10 µM.

An ability of compounds to inhibit the process of LP in erythrocytes under conditions of the acute oxidation stress can be evaluated based on the data given in Figure 2.12 [10].

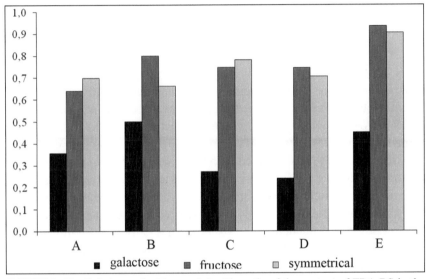

FIGURE 2.12   Relative antioxidant activity (the ratio of the content of TBA-RS in the tested and control samples, 5 h after addition of H_2O_2) of the tested compounds in the concentration 10 µM.

The disulfides with the *cis*-myrtanyl/myrtenyl fragments in the combination with the galactose moiety inhibited accumulation of LP secondary products most actively. A comparison reference and symmetrical disulfides with two carbohydrate fragments showed relatively low activity in this model system, inhibiting LP to 72–80% from the control data.

All 17 conjugates studied are characterized by the statistically significant positive dependence ($R_S = 0.56$, $p = 0.020$) between the ability to inhibit the accumulation of LP secondary products and to protect cells from damage.

Finally, most disulfides obtained in this work (except those containing *trans*-verbenyl fragment) exhibited the ability to inhibit transformation of erythrocyte oxyhemoglobin to methemoglobin upon treatment with hydrogen peroxide, which also indicates the presence of a pronounced antioxidant activity. The highest activity was found in the conjugates with the myrtenyl/*cis*-myrtanyl fragments (independent of the structure of the second substituent). The preincubation of erythrocytes with these disulfides led to a 2.0–2.4 times decrease in the level of accumulation of methemoglobin as compared with the control samples (containing no compounds tested), whereas a comparison agent inhibited the fotmation of methemoglobin only by 1.4 times.

2.4 CONCLUSIONS

Thus, the presence of antioxidant and membrane-protective activity of a number of the compounds synthesized have been investigated in the model of AAPH- and H_2O_2-induced erythrocyte hemolysis. High activity of the sulfoxides (S_s)-IVe, (R_s)-IVe, and (R_s)-IIIa under the action of both peroxide and AAPH as compared with the corresponding to them sulfides is probably due to the fact that, in this model system, sulfoxides form sulfenic acids, which are inhibitors of the hydroperoxide decomposition. The ability of these compounds to protect cells under acute oxidative stress conditions depended on the structure of a carbohydrate moiety. So, the sulfides and sulfoxides with galacto- (c,e) and fructopyranose (a,d) fragments appeared to be more active than the compounds with gluco-pyranose (b) fragment. The disulfides with *cis*-myrtanyl, myrtanyl, and neomenthyl fragments were the most active; nevertheless, the activity of galactopyranose-derived compounds were slightly higher than of similar structures having a fructopyranose fragment. For symmetrical disul-fides containing no sugar moieties, the highest activity was found for compounds containing *cis*-myrtanyl, myrtenyl, and isobornyl fragments. A low activity in this model system was found for disulfides of different structure containing *trans*-verbenyl fragment.

ACKNOWLEDGMENTS

The work was supported by the Russian Foundation for Basic Research, project no. 13-03-01312_a.

The studies were performed using equipment of the Center for Collec-tive Use (CCU) "Khimia "Molecular Biology" of the Institutes of Chem-istry and Biology, Komi Scientific Centre, Ural Division, RAS.

KEYWORDS

- **thioglycosides**
- **monosaccharides**
- **disulfides**
- **membrane-protective activity**
- **erythrocytes**
- **oxidative hemolysis**
- **antioxidants**

REFERENCES

1. Pouillart, P.; Ronco, G.; Cerutti, I.; Chany, C.; Villa P. Low Level Toxicity and Anti-tumor Activity of Butyric Mono- and Polyester Monosaccharide Derivates in Mice. *J. Biol. Regul. Homeost. Agents*, **1990**, *4*(4), 135–141.
2. Nicolaou, K. C.; Dai, W.-M. Chemistry and Biology of the Enediyne Anticancer Antibiotics. *Angew. Chem. Int. Ed. Engl.*, **1991**, *30*, 1387–1416.
3. Witczak, Z. J. Thio Sugars: Biological Relevance as Potential New Therapeutics. *Curr. Med. Chem.* **1999**, *6*(2), 165–178.
4. Aversa, M. C.; Barattucci, A.; Bonaccorsi, P. Glycosulfoxides in Carbohydrate Chemistry. *Tetrahedron*, **2008**, *64*(33), 7659–7683.
5. Andre, S.; Pei, Z.; Siebert, H.-C.; Ramstrom, O.; Gabius, H.-J. Glycosyldisulfides from Dynamic Combinatorial Libraries as *O*-Glycoside Mimetics for Plant and Endogenous Lectins: Their Reactivities in Solid-Phase and Cell Assays and Conformational Analysis by Molecular Dynamics Simulations. *Bioorg. Med. Chem.* **2006**, *14*(18), 6314–6326.
6. Duclos, Jr., R. I.; Lu, D.; Guo, J.; Makriyannis, A. Synthesis and Characterization of 2-Substituted Bornane Pharmacophores for Novel Cannabinergic Ligands. *Tetrahedr. Lett.* **2008**, *49*(39), 5587–5589.
7. Ovchinnikova, L. P.; Rotskaya, U. N.; Vasyunina, E. A.; Sinitsina, O. I.; Kandalint-seva, N. V.; Prosenko, A. E.; Nevinskii, G. A. Antioxidant Activity Thiophane [bis-[3-(3′,5′-di-*Tret*-Butyl-4-Hydroxyphenyl)Propyl]Sulfide]. *Russ. J. Bioorg. Chem.* **2009**, *35*(3), 379–384 (In Russian).
8. Izmest'ev, E. S.; Sudarikov, D. V.; Shevchenko, O. G.; Rubtsova, S. A.; Kutchin, A. V. The Synthesis and Membrane Protective Properties of Sulfanyl Imines Derived from Neomenthane and Isobornane Thiols. *Russ. J. Bioorg. Chem.* **2015**, *41*(1), 77–82 (In Russian).

9. Belykh, D. V.; Buravlev, E. V.; Chukicheva, I. Iu.; Tarabukina, I. S.; Shevchenko, O. G.; Pliusnina, S. N.; Kuchin, A. V. Synthesis of Novel Terpenophenol-chlorin Conjugates and Evaluation of Their Membranotropic and Membrane Protective Properties. *Russ. J. Bioorg. Chem.* 2012, *38*(5), 558–564 (In Russian).

10. Pestova, S. V.; Izmest'ev, E. S.; Shevchenko, O. G.; Rubtsova, S. A.; Kuchin, A. V. Synthesis and Membranoprotective Properties of New Disulfides with Monoterpene and Carbohydrate Fragments. *Russ. Chem. Bull.* **2015,** *64*(3), 723–731 (In Russian).

11. Henkelman, S.; Rakhorst, G.; Blanton, J.; Oeveren, W. Standardization of Incubation Conditions for Hemolysis Testing of Biomaterials. *Mater. Sci. Eng.: C.* **2009,** *29*(5), 1650–1654.

12. Wang, C.; Qin, X.; Huang, B.; He, F.; Zeng, C. Hemolysis of Human Erythrocytes Induced by Melamine–cyanurate Complex. *Biochem. Biophys. Res. Commun.*, **2010,** *402*(4), 773–777.

13. Banerjee, A.; Kunwar, A.; Mishra, B.; Priyadarsini, K. I. Concentration Dependent Antioxidant/pro-oxidant Activity of Curcumin: Studies from AAPH induced Hemolysis of RBCs. *Chem.-Biol. Interact.* **2008,** *174*(2), 134–139.

14. Asakawa, T.; Matsushita, S. Coloring Conditions of Thiobarbituric Acid Test for Detecting Lipid Hydroperoxides. *Lipids*, **1980,** *15*(3), 137–140.

15. Pestova, S. V.; Sudarikov, D. V.; Rubtsova, S. A.; Kutchin, A. V. Synthesis and Asymmetric Oxidation of Thioglycosides Derived from Neomenthanethiol and α-D-Galactose. *Russ. J. Org. Chem.*, **2013,** *49*(3), 366–373 (In Russian).

16. Pestova, S. V.; Izmest'ev, E. S.; Rubtsova, S. A.; Kuchin, A. V. Synthesis and Oxidation of Thioglicosides Underlain by Neomenthanethiol, D-Glucose, and D-Fructose. *Russ. J. Org. Chem.* **2014,** *50*(5), 670–677 (In Russian).

17. Ko, F. N.; Hsiao, G.; Kuo, Y. H. Protection of Oxidative Hemolysis by Demethyldiisoeugenol in Normal and β-thalassemic Red Blood Cells. *Free Rad. Biol. Med.*, **1997,** *22*(1–2), 215–222.

18. Lopez–Revuelta, A.; Sanchez–Gallego, J. I.; Hernandez-Hernandez, A.; Sanchez-Yague, J.; Llanillo, M. Membrane Cholesterol Contents Influence the Protective Effects of Quercetin and Rutin in Erythrocytes Damaged by Oxidative Stress. *Chem.-Biol. Interact.* **2006,** *161*(1), 79–91.

19. Costa, R. M.; Magalhaes, A. S.; Pereira, J. A.; Andrade, P. B.; Valentao, P.; Carvalho, M.; Silva, B. M. Evaluation of Free Radical-scavenging and Antihemolytic Activities of Quince (*Cydonia oblonga*) Leaf: A Comparative Study with Green Tea (*Camellia sinensis*), *Food Chem. Toxicol.* **2009,** *47*(4), 860–865.

20. Takebayashi, J.; Chen, J.; Tai, A. A Method for Evaluation of Antioxidant Activity Based on Inhibition of Free Radical-induced Erythrocyte Hemolysis. In: Advanced Protocols in Oxidative Stress II, Methods In Molecular Biology; Armstrong, D., Ed.; Heidelberg; Humana Press: New York, London, 2010; Vol. 594, pp. 287–296.

21. Koelewijn, P.; Berger, H. Mechanism of the Antioxidant Action of Dialkyl Sulfoxides. *Ser. Chem. Works Netherlands*, **1972,** *91*(11), 1275–1286.

22. Barnard, D.; Bateman, L.; Cain, M.E.; Colclough, T.; Cunneen, J. I. The Oxidation of Organic Sulphides. Part X. The Co-oxidation of Sulphides and Olefins. *J. Chem. Soc.* **1961,** 5339–5344.

23. Bateman, L.; Cain, M.; Colclough, T.; Cunneen, J. I. Oxidation of Organic Sulphides. Part XIII. The Antioxidant Action of Sulphoxides and Thiolsulphinates in Autoxidizing Squalene. *J. Chem. Soc. (Resumed)*, **1962,** 3570–3578.

24. Barnard, D.; Cole, E. R.; Cunneen, J. I. Sulfoxides, and Thiolsulfinates as Inhibitors of Autoxidation and Other Free Radical Reactions. *Chem. Ind.* **1958,** *29,* 918–919.

25. Prosenko, A. E.; Terakh, E. I.; Kandalintseva, N. V.; Pinko, P. I.; Gorokh, E. A.; Tolstikov, G. A. Synthesis and Antioxidative Properties of New Sulfur-Containing Derivatives of Sterically Hindered Phenols. *Russ. J. Appl. Chem.* **2001,** *74*(11), 1899–1902 (In Russian).

26. Zhao, F.; Liu, Z.-Q.; Wu, D. Antioxidative Effect of Melatonin on DNA and Erythrocytes against Free-radical-induced Oxidation. *Chem. Phys. Lipids* **2008,** *151*(2), 77–84.

CHAPTER 3

THE SYNTHESIS AND PROPERTIES OF NEW OXYGEN- AND NITROGEN-CONTAINING TERPENE ACID DERIVATIVES

MAKSIM P. BEI* and ANATOLIJ P. YUVCHENKO

The Institute of Chemistry of New Materials of the National Academy of Science of Belarus, 36, Skoriny St., Minsk 220141, Republic of Belarus

Corresponding author. E-mail: beymaksim@gmail.com; mixa@ichnm.basnet.by; aspirin55@yandex.ru

CONTENTS

ABSTRACT

This chapter is devoted to the development of efficient methods of synthesis of new derivatives of maleopimaric acid (MPA) (the most available diterpenoid substance, isolated from maleated rosin)—esters, amides, imides, imido amides, diimido acids, imido esters, fumaropimaric acid monoamides, and of oxygen- and nitrogen-containing derivatives of previously unknown citraconopimaric acid. The reaction of maleopimaric and citraconopimaric acids with some secondary amines was investigated, and formation of N-substituted cyclic imides of citraconopimaric and MPAs was found.

3.1 INTRODUCTION

In the last decades, the intensive studies were directed toward the development of new methods for valuable chemical products preparation from renewable raw materials, including wood chemistry product—rosin. The individual substances that can be isolated or synthesized from rosin are of special interest [1–7]. The isolation of individual resin acids from rosin is a complicated and laborious process. The most promising approach to the preparation of individual terpene derivatives is to modify rosin in particular by active dienophiles (maleic anhydride, fumaric acid, citraconic acid, 1,4-benzoquinone, etc.) followed by isolation from modified rosin of individual substances. The most available individual substance thus obtained is maleopimaric acid (MPA), isolated from maleated rosin. This compound is used for the production of printing inks, alkyd resins, lubricants, and paper, and its derivatives show pronounced biological activity—bactericidal, nematicidal, immunomodulatory and hepatoprotective properties, and low toxicity [8–10].

The development of efficient methods for the preparation of MPA derivatives and its structural analogues and determination of structure–properties relations of obtained terpene compounds for preparation of new high-technology materials refer to the intensively developing "green chemistry" and are of great scientific and technological importance. The new MPA and its structural analog derivatives may find applications as industrial polymer additives, monomers for the synthesis of polymers and copolymers, optically active dopants to LC-materials, new bioactive compounds for pharmacy (immunomodulators, hepatoprotektors, etc.).

3.2   THE SYNTHESIS OF NEW MALEOPIMARIC ACID DERIVATIVES

MPA, 1 contains reactive carboxylic and anhydride groups that make it convenient synthon for the preparation of a new terpenoid compounds with a wide range of properties and applications.

Among the unsaturated esters of MPA vinyl and diallyl esters [11, 12] and esters on the basis of MPA epoxy-derivatives and acrylic, methacrylic acids [13, 14] have been described. Diallyl maleopimarate has been synthesized by the reaction of diallyl maleate with gum rosin at 165–180°C followed by extraction of the adduct formed from the reaction mixture [11, 12].

We developed the synthesis and determined physicochemical properties of allyl and propargyl esters of MPA 2a, b, including film-forming and light-sensitivity properties [15]. The modification of the only carboxylic group by acid-catalyzed reaction of MPA with allyl alcohol is quite problematically; also, we were unable to prepare MPA allyl ester 2a by acylation of allyl alcohol with MPA chloride—only a complex mixture was obtained containing products of the anhydride ring cleavage. The target allyl and propargyl esters 2a, b were synthesized in a high yield (98–99%) by the reaction of MPA with allyl or propargyl bromide in N,N-dimethylformamide (DMF) in the presence of potassium carbonate. This reaction proceeds selectively at carboxylic group maintaining anhydride ring intact (Fig. 3.1) [15].

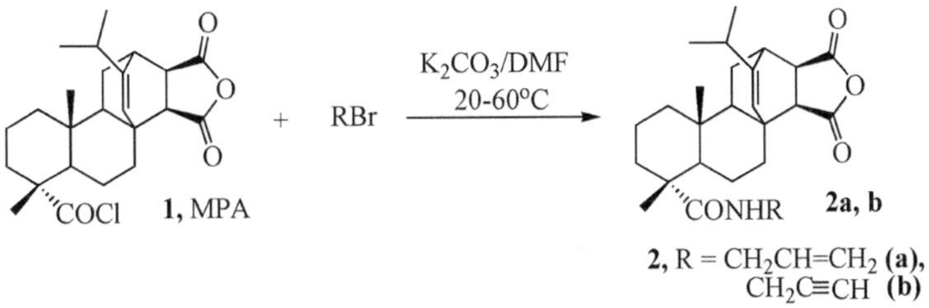

FIGURE 3.1   Synthesis of maleopimaric acid allyl and propargyl esters 2a, b

Thermal stability of esters 2a, b was determined by derivatography method. It was found that allyl ester 2a begin to decompose with a noticeable rate at 300°C, which is 15°C above the temperature of the MPA decomposition start. Thermal stability of propargyl ester 2b (decomposition point ~ 290°C) was the same as that of MPA.

Esters 2a, b and MPA, 1 were used for thermal vacuum deposition (TVD) to obtain colorless transparent glossy films of thickness 0.9–1.4 μm with good adhesion to the surface of various substrates (silicon monocrystal, quartz, polymer layers, etc.). The TVD-films from compounds 2a, b may be of practical interest for the manufacture of photomask for wet etching of topology in the transparent conductive layer of indium–tin oxide [15]. The allyl ester 2a is a viscosity stabilizer of low-density polyethylene modified with itaconic acid. The adhesive with increased adhesion to polar substrates and melt flow index was developed from ethylene–vinyl acetate copolymer and allyl ester 2a [16].

A number of MPA aliphatic imides and its derivatives have been described, including imides with antimicrobial and immunomodulatory properties [1, 17–19].

We developed the synthesis of previously unknown MPA N-alkylimides 3a–e, including long-chain dodecyl- and octadecylimides 3d, e by reaction of primary aliphatic amines with MPA in the melt or in solution [20]. The introduction of dodecyl- or octadecyl moieties in compound greatly increases its lipophilic properties.

It was established that the reaction of MPA and butylamine in refluxing benzene for 6–8 h at molar ratio of reagents 1:1 gives MPA N-butylimide 3a in 98% yield. In reaction of MPA with hexylamine, the use of refluxing toluene and excess amine (20%) is preferred. The reaction completes within 6–8 h to give MPA N-hexylimide 3b in 97% yield.

It should be noted that the rate of the imide formation decreases at the increased length of n-alkyl chain of the amine: Butylamine completely reacts with the acid 1 in boiling benzene over 6–8 h, whereas octadecylamine completely reacts under these conditions over 40–48 h.

The reaction of MPA with alkylamines in the melt was investigated. The reaction of MPA with dodecyl-, octadecylamine (in equimolar ratio) at 140–160 °C within 10–12 h leads to the formation of MPA imides 3d, e with up to 99% yields. The method of obtaining of long-chain MPA imides in the melt is preferred over the method using a solvent for the synthesis of imides 3d, e: It reduces the reaction time from 40–50 to 6–12 h, an excess of reagent is not required, and an additional step to remove unreacted reagents after the reaction is not needed. The reaction of MPA with hexyl- and octylamine (in equimolar ratio) as well proceeds smoothly without a solvent at 130–160°C within 4–6 h (hexylamine) and 10 h (octylamine) to give MPA imides 3b, c with up to 97% yields (Fig. 3.2) [20].

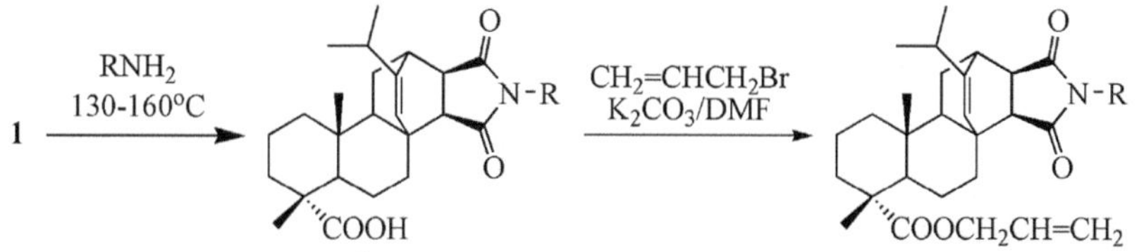

$$3, R = n\text{-}C_4H_9 \text{ (a)}, n\text{-}C_6H_{13} \text{ (b)}, n\text{-}C_8H_{17} \text{ (c)},$$
$$n\text{-}C_{12}H_{25} \text{ (d)}, n\text{-}C_{18}H_{37} \text{ (e)}$$

$$R = n\text{-}C_6H_{13} \text{ (4)}, n\text{-}C_8H_{17} \text{ (5)},$$
$$n\text{-}C_{18}H_{37} \text{ (6)}$$

FIGURE 3.2   The synthesis of maleopimaric acid N-alkylimides 3a–e and its allyl esters 4–6.

In the synthesis of imides 3b–e in the melt, not only individual MPA can be used, but also its solvate with acetic acid, formed in the process of the acid 1 isolation from maleated rosin by crystallization from acetic acid [15], thus avoiding a further stage of the solvate decomposition. Acetic acid is distilled off in the process of reaction, and the final product does not contain it.

The presence in the compounds 3a–e of a carboxyl group enables their further modification. By treatment of imides 3b, c and 3e with allyl bromide in DMF in the presence of potassium carbonate allyl esters of the MPA, imides 4–6 were synthesized in 96–98% yields (see Fig. 3.2) [20].

By the method of derivatography, we evaluated the thermal stability of imides 3a–e. Compounds 3a–e are thermally stable and begin to decompose with a noticeable rate at 300–320°C, which is 15–35°C above the temperature of the MPA decomposition start.

N-Alkylimides 3a–e can be used as additives to the polyethylene to increase its adhesion properties. The additives of 3–5% of imides 3b, c are the most effective: They don't decrease strength, elastic, and rheological properties of polymer composition and increase adhesion strength of the modified polyethylene to wood in 1.6–1.8 times [21].

We developed method of synthesis of MPA diimide 5, previously obtained from levopimaric acid and N,N'-hexamethylenedimaleimide [22], directly from MPA and 1,6-hexandiamine [15]. It was established that the heating of solution of MPA and 1,6-hexandiamine in molar ratio 1:0.45–0.48 in 1,2-dichlorobenzene at 150 °C for 6 h leads to the preparation of MPA diimide 5 with 51% yield. Subsequent treatment of the diimide 5 with allyl bromide in DMF in the presence of potassium carbonate at 18–20 C for18 h leads to the formation of diallyl ester 6 with 86% yield (Fig. 3.3) [15].

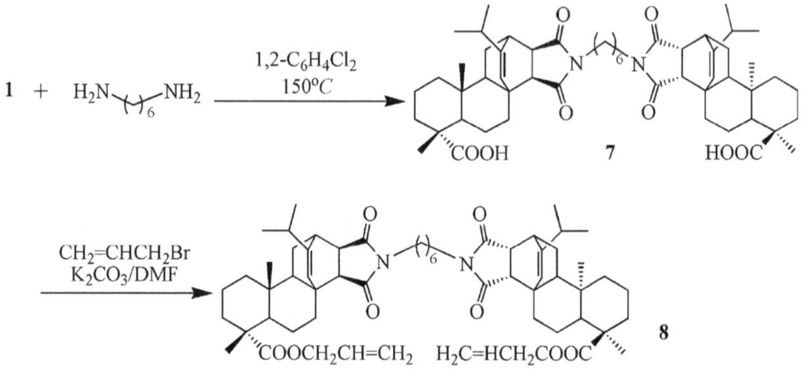

FIGURE 3.3   The synthesis of *N*, *N*-hexamethylenediimide MPA 7 and its diallyl ester 8.

Only a few representatives of MPA aromatic imides have been known: 4-carboxyphenyl imide, 4-aminophenyl imide converted into thermally stable poly(amido)imides [23], *N*-(2-methyl-α-naphthyl)imide displaying antineoplastic activity [24]. We developed the synthesis of some previously unknown arylimides from MPA and available primary aromatic amines: aniline, *p*-toluidine, *p*-hydroxyaniline, *p*-bromoaniline, *m*-aminobenzoic acid and investigated reaction of MPA with *p*-nitroaniline [25].

It was established that reaction of MPA with aniline (10 mol.% excess) in DMF at 130 °C within 6–8 h not only leads to the formation of *N*-phenyl-limide 9a but also some minor substances were detected; therefore, we used further more inert benzene and toluene as a reaction medium. Reaction was conducted by refluxing of solution of MPA and 10 mol.% excess of aniline. Toluene is more preferable solvent in comparison with benzene because of its higher boiling point. Reaction of MPA with amines in boiling toluene is completed within 4–16 h to give MPA *N*-arylimides 9a–d in a high yields (64–84%) (Fig. 3.4) [25].

The refluxing of MPA with *p*-nitroaniline or *m*-aminobenzoic acid in toluene doesn't lead to *N*-arylimide formation. Also the fusion of MPA with *p*-nitroaniline at 170–190 °C for 6 h doesn't result in the imide formation: The reaction mixture contains only initial substances. The refluxing of MPA and *m*-aminobenzoic acid in *p*-xylene or toluene doesn't lead to the formation of the imide 9e, but interaction of these reagents in boiling pyridine for 8 h gives MPA N-(*m*-carboxyphenyl)imide 9e in 83% yield [25].

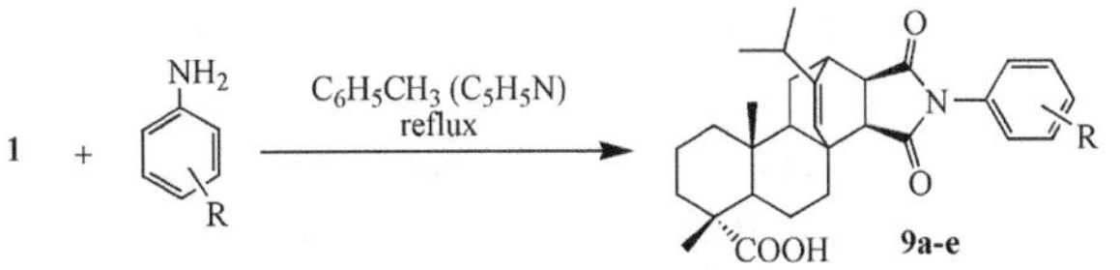

9, R = H (**a**), 4-Br (**b**), 4-OH (**c**), 4-CH_3 (**d**), 3-COOH (**e**)

FIGURE 3.4 The synthesis of maleopimaric acid *N*-arylimides 9a–e.

The study of prepared imides 9a–d properties (solubility, melting point) allows to suggest that the synthesis of these compounds as well as of other MPA *N*-arylimides is possible not only from individual MPA but also from commercially available rosin-maleic anhydride adduct (RMA).

The efficient method of synthesis of MPA *N*-arylimides from RMA without isolation of individual MPA was developed. The method is based on treatment of RMA (content of MPA is about 57%) with primary aromatic amines in toluene (refluxing) to form exclusively *N*-arylimides 9a–d, f–m as insoluble precipitates that could be easily separated off the unreacted resin acids, well soluble in toluene. The optimum reaction time for formation of MPA *N*-arylimides from RMA in refluxing toluene is 40 h (Fig. 3.5) [26].

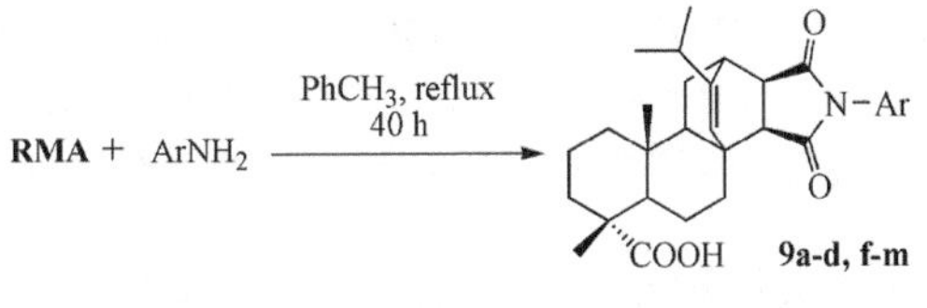

9f-m, Ar = *o*-$CH_3C_6H_4$ (**f**), *p*-$NH_2C_6H_4$ (**g**), *m*-$NH_2C_6H_4$ (**h**), *p*-$CH_3OC_6H_4$ (**i**), *p*-FC_6H_4 (**j**), *p*-ClC_6H_4 (**k**), *m*-ClC_6H_4 (**l**), *m*-IC_6H_4 (**m**)

FIGURE 3.5 The synthesis of maleopimaric acid *N*-arylimides 9a–d, f–m from adduct of rosin with maleic anhydride.

This preparation of *N*-arylimides avoids the stage of MPA isolation from the adduct of rosin and maleic anhydride, allowing to significantly reduce the process duration and amount of the consumed organic solvents; the target imides, including MPA phenylimide 9a, *p*-, *o*-methylphenylimides 9d, f, *p*-hydroxyphenylimide 9c, *p*-, *m*-aminophenylimides 9g, h, *p*-methoxyphenylimide 9i, 4-fluoro-(4-, 3-chloro-, 4-bromo-, 4-iodo) phenylimides 9b, j–m, could be isolated in 51–71% (based on rosin–maleic anhydride adduct) or 60–99% (based on MPA contained in the rosin–maleic anhydride adduct) yield (Table 3.1) [26].

TABLE 3.1   The Yields and Melting Points of Maleopimaric Acid N-Arylimides 9a–d, f–m.

Amine	Product	Yield, % based on RMA (based on MPA)	Melting point (°C)
$C_6H_5NH_2$	9a	50.9 (75.6)	305–307
p-$CH_3C_6H_4NH_2$	9d	62.7 (90.3)	302–304
p-$HOC_6H_4NH_2$	9c	60.1 (86.2)	302–304
p-$BrC_6H_4NH_2$	9b	67.4 (85.7)	275–278
o-$CH_3C_6H_4NH_2$	9f	52.8 (76.0)	279–281
p-$NH_2C_6H_4NH_2$	9g	69.2 (99.4)	349–351
m-$NH_2C_6H_4NH_2$	9h	61.2 (87.9)	288–290
p-$CH_3OC_6H_4NH_2$	9i	71.2 (99.3)	286–288
p-$FC_6H_4NH_2$	9j	55.2 (78.8)	306–308
p-$ClC_6H_4NH_2$	9k	57.4 (79.4)	296–298
m-$ClC_6H_4NH_2$	9l	58.9 (81.5)	273–275
p-$IC_6H_4NH_2$	9m	50.8 (59.6)	241–243

The synthesis of diimides from MPA and commercially available diamines is of special interest as the resulting products have two carboxyl groups and can be used for the preparation of poly (imido) amides and poly (imido) esters.

The reaction of amino imides 9g, h with MPA in solution or in the melt was studied. It was established that the reaction of MPA with *p*-aminophenylimide 9g (molar ratio 1.5:1) in 1,2-dichlorobenzene (reflux) within 3–8 h leads to the formation of diimido diacid 10 in 44–47% yield (Fig. 3.6) [26], which was previously prepared by reaction of levopimaric acid and *p*-phenylenedimaleimide [22]. The interaction of *p*-aminophenylimide 9g and MPA (molar ratio 1:5) in melt at 260–270 °C for 2 h does not increase the yield of diimido diacid 10 (37%) [25].

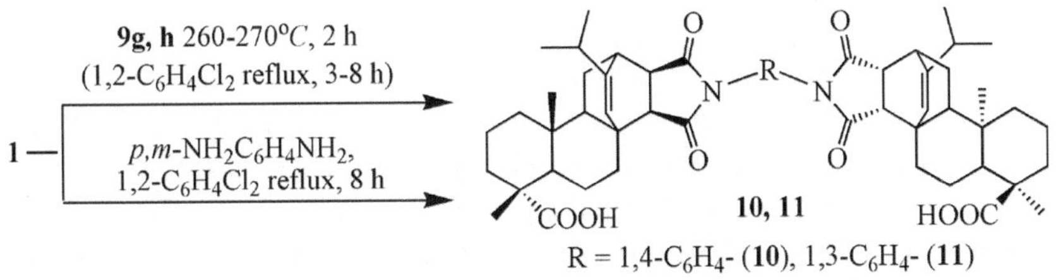

FIGURE 3.6 The synthesis of aromatic diimido diacids 10, 11 from maleopimaric acid.

The reaction of MPA with m-aminophenylimide 9h in the above conditions leads to the formation of inseparable mixture of diimido diacid 11 and starting compounds (from ^{1}H NMR data).

It was established that diimido diacids 10, 11 in higher yields can be obtained via the reaction of MPA and p- and m-phenylenediamines (molar ratio 3:1) in refluxing 1,2-dichlorobenzene: The yield of diimido diacid 10 reaches 68%, and diimido diacid 11 was separated in individual state by preparative thin-layer chromatography with 48% yield (Fig. 3.6) [26]. The higher yields of diimido diacid 10, 11 in reaction of MPA with p- and m-phenylenediamines in contrast to reactions of MPA with p- and m-aminophenylimides 9g, h are probably due to the oligomers formation at 180–260 °C from starting p- and m-aminophenylimides 9g, h [23].

Treatment of diimido diacid 10 with dimethyl sulfate or allyl bromide in DMF at 18–20°C in the presence of potassium carbonate afforded dimethyl ester 12a and diallyl ester 12b in 92 and 78% yield, respectively (Fig. 3.7) [26].

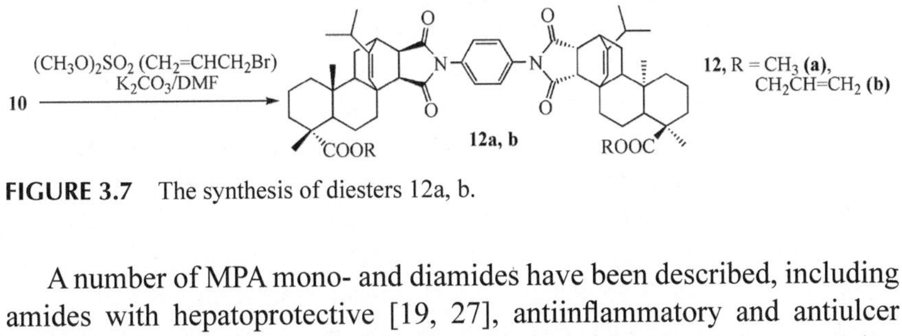

FIGURE 3.7 The synthesis of diesters 12a, b.

A number of MPA mono- and diamides have been described, including amides with hepatoprotective [19, 27], antiinflammatory and antiulcer activities [10]. New photoactive poly(amido)imides have been synthesized from MPA amides containing azo-dyes fragment [9].

We elaborate the synthesis of new aliphatic, aromatic, and heterocyclic MPA amides 14a–j (Fig. 3.8) [25]. It was established that optimum conditions for the reaction of maleopimaroyl chloride 13 with aliphatic (allyl-, hexyl-, octyl-, dodecylamine), aromatic (4-methylaniline, 4-bromoaniline, benzylamine), and heterocyclic (3-phenyl-5-aminoisoxazole, 3-(*p*-tolyl)-5-aminoisoxazol) amines are tetrahydrofuran (THF) as a solvent, triethylamine as hydrogen chloride acceptor and carrying out the reactions at 18–20 C for 8 h.

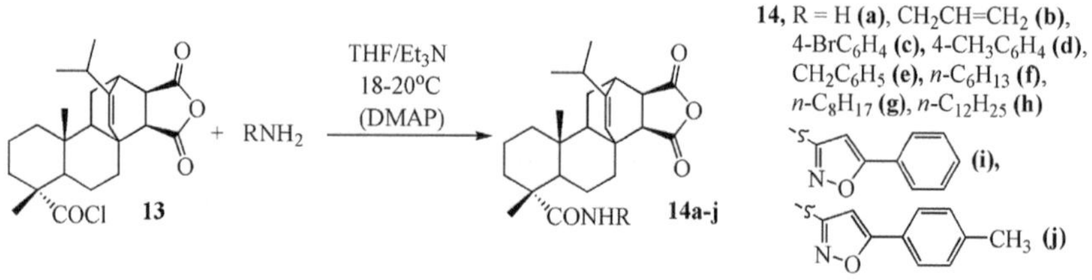

FIGURE 3.8   The synthesis of maleopimaric acid amides 14a–j.

MPA amide 14a was prepared by passing gaseous ammonia through the solution of the acid chloride 13 in THF. The target amide 14a was isolated in 32% yield. In the synthesis of aliphatic amides 14f–h, 4-(dimethylamino)pyridine (DMAP) was added to the reaction mixture as a catalyst. The amides 14b–j were synthesized in 52–95% yields, and the formation of anhydride ring opening products was not observed.

The compounds possessing *trans*-1,2-dicarboxylic fragment (*trans*-cyclohexane- and cyclohex-4-ene-1,2-dicarboxylic acids, tartaric acid acetonide, etc.) have a wide applications for the synthesis of ligands for the asymmetric catalysis [28], separation of chiral amines [29], for the synthesis of anticancer [30], and other pharmaceutical substances [31].

We elaborated the method of synthesis of aliphatic, aromatic, and heterocyclic fumaropimaric acid monoamides 15a–h containing *trans*-1,2-dicarboxylic fragment. The reaction was carried out by refluxing of MPA amides 14a–f, anilide 14k, and piperidine amide 14l for 2 h in water–methanol solution of potassium hydroxide (Fig. 3.9) [32]. Initially, potassium salts are formed by alkaline hydrolysis of anhydride cycle and then isomerize into more stable salts of *trans*-1,2-dicarboxylic acids. After the reaction product was acidified with 20% HCl solution, the mixture of *trans*-1,2-dicarboxylic acids 15a–h with impurities of *cis*-1,2-dicarboxylic acids precipitated. It was separated by filtration and boiling within 30–40 min in benzene solution, thus *cis*-1,2-dicarboxylic acids were dehydrated and started forming MPA monoamides 14a–f, k, l (^{1}H NMR data). For the separation of fumaropimaric acid monoamides 15a–h from mixtures, they were converted into water soluble salts by treatment with 2-amino-2-methylpropanol. Insoluble MPA monoamides 14a–f, k, l were filtered off, and the subsequent acidification of mother liquor gives the precipitates of target *trans*-1,2-dicarboxylic acids 15a–h. It should be mentioned that in elaborated reaction conditions the hydrolysis of amido group does not occur for all starting amides 14a–f, k, l. Amido diacids 15a–h were prepared in 82–98% yields as a colorless crystal substances moderately soluble in conventional organic solvents [32].

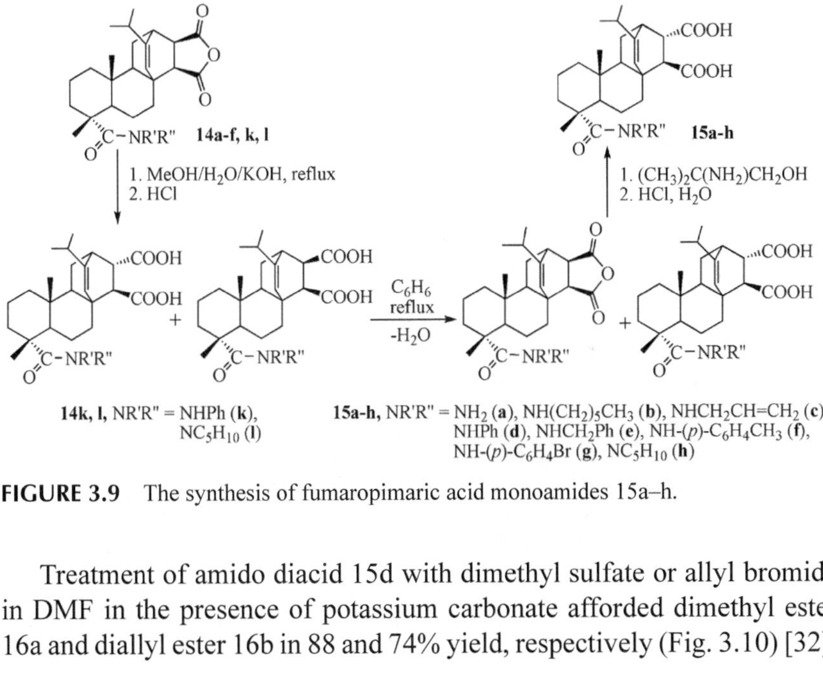

FIGURE 3.9   The synthesis of fumaropimaric acid monoamides 15a–h.

Treatment of amido diacid 15d with dimethyl sulfate or allyl bromide in DMF in the presence of potassium carbonate afforded dimethyl ester 16a and diallyl ester 16b in 88 and 74% yield, respectively (Fig. 3.10) [32].

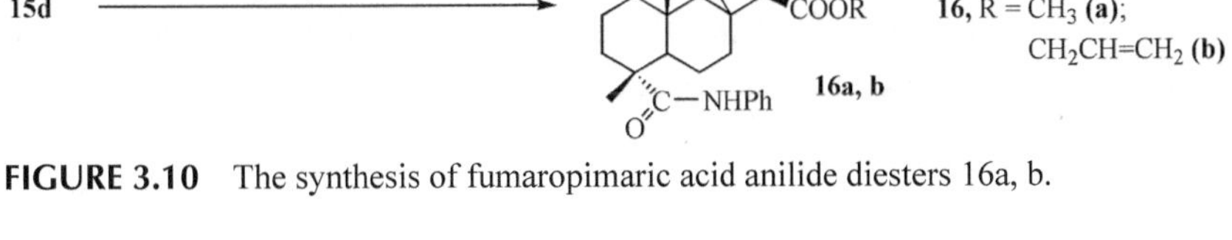

FIGURE 3.10   The synthesis of fumaropimaric acid anilide diesters 16a, b.

Starting from fumaropimaric acid monoamide 15h, we developed the synthesis of 1,4-diiodide and 1,4-diamine as a potential starting compounds for the asymmetric catalysis ligands synthesis. Selection of the starting amide 15h is determined by formation of tertiary amine after subsequent carboxamide group reduction.

Treatment of amido diacid 15h with diethyl sulfate in DMF in the presence of potassium carbonate afforded diethyl ester 17 in 99% yield, it was converted into diol 18 by reduction with LiAlH$_4$ in 58% yield. One of the most effective methods of alcohol transformation into halide is a preparation of sulfonates (in particular, tosylates) and subsequent nucleophilic substitution of p-toluenesulfonyl group by halide ion. For the synthesis of ditosylate 19 from diol 18, we used two well-known methods: reactions of alcohol with tosyl chloride in pyridine or in methylene chloride in the presence of organic base and catalytic amounts of DMAP [33]. The use of the first method did not give a satisfactory result: Isolated ditosylate 19 contains large amounts of impurities, and only the second method gave the target ditosylate 19 almost in quantitative yield (98%) (Fig. 3.11).

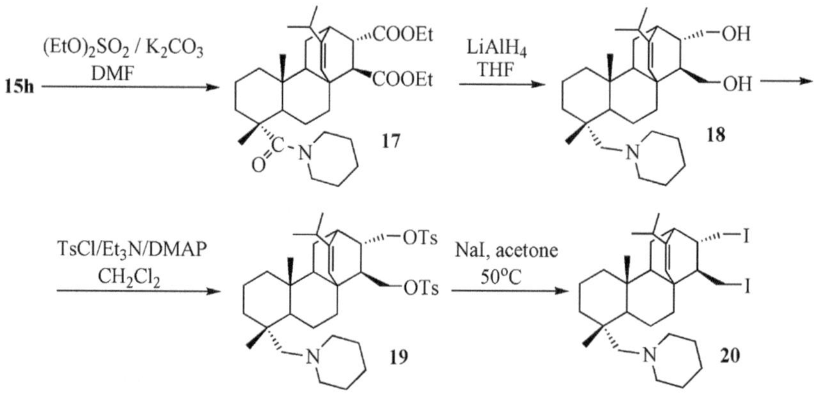

FIGURE 3.11 The synthesis of 1,4-diiodide 20 from fumaropimaric acid monoamide.

Treatment of ditosylate 19 with sodium iodide for 50 h at 50 °C gave 1,4-diiodide 20 with quantitative yield (see Fig. 3.11). This compound may be used as a useful synthon for a new terpene derivatives synthesis. Diethyl ester 17 and 1,4-diiodide 20 are viscous oils, diol 18 and ditosylate 19 are colorless crystals.

Reactions of ditosylate 19 with diethyl- and benzylamines were studied, and it was established that refluxing of ditosylate 19 in solution of these amines for 5 h doesn't lead to tosyl groups substitution. Initial ditosylate 19 was separated from solution in diethylamine quantitatively. The stability of ditosylate 19 toward amines is probably due to steric and dipole shielding of close p-tosyl groups [34].

To synthesize the 1,4-diamines from diol 18, we used another synthetic route. The treatment of diol 18 with methanesulfonyl chloride in pyridine gave dimesylate 21 89% yield, which was converted into diazide 22 by reaction with sodium azide (yield 88%). Reduction of diazide 22 with LiAlH$_4$ in THF gave diamine 23 in 89% yield (Fig. 3.12).

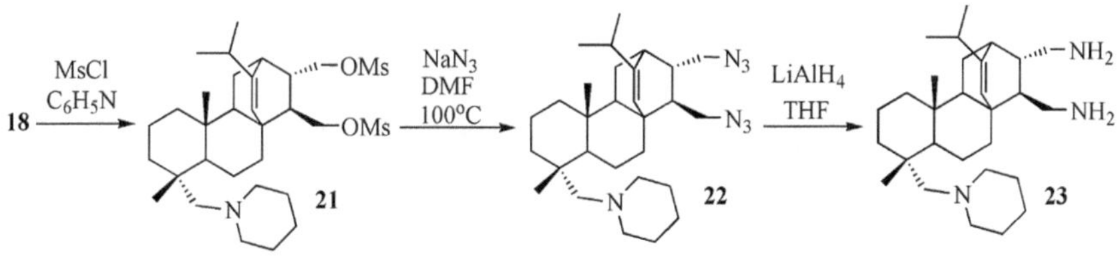

FIGURE 3.12   The synthesis of terpene 1,4-diamine 23.

Among the MPA, imido amides aliphatic and heterocyclic imido amides have been described including substances with hepatoprotective activity [19]. Aliphatic and aromatic poly(amido)imides with enhanced thermal stability were used for the preparation of light-sensitive, anticorrosive, and hydrophobic films [23, 35–37].

It can be expected that conversion of carboxylic and anhydride groups of MPA into N-aryl amide and cyclic N-aryl imide groups will enhance thermal stability of imido amides and provide the use of these substances in high-temperature processes, in particular in modification of polymer products. Aromatic imido amides can be useful also for the studying of its biological properties.

The method of N-(aryl)aralkyl MPA imido amides 24a–k synthesis from aromatic amides 14c–e, k and amines (aniline, p-toluidine, p-anisidine, p-fluoro-, p-chloro-, p-bromoaniline, benzylamine, and 2-pycolylamine) was developed (Fig. 3.13) [38]. Optimum reaction conditions were established as: molar ratio amide:amine 1:3, refluxing in p-xylene for 40 h. The yields of imido amides 24a–k reach 41–94%. The conducting of this reaction at lower temperatures (refluxing in toluene), and the use of only 20–100% excess of amine led to the substantial decrease of target product yields. For example, reaction of MPA anilide 14k with aniline (molar ratio 1:1.2, refluxing in toluene for 40 h) gave imido amide 24a with only 20% yield, and when the solvent was replaced by *p*-xylene, the imido amide 24a was prepared with 31% yield. The use of 100% excess of aniline (refluxing in *p*-xylene for 40 h) gave imido amide 24a with 50% yield, and only the use of three equivalents of aniline allowed synthesizing imido amide 24a with 70% yield. The increase of imido amide yield with the increase of amine excess is probably due to higher concentration of amine and polarity of reaction system [39].

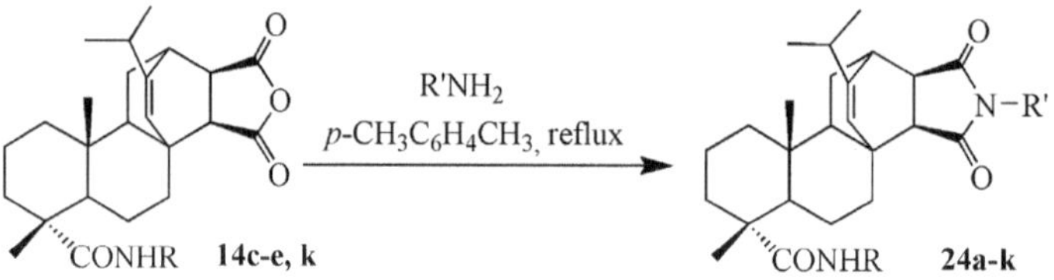

24a-h, R = C_6H_5, R' = C_6H_5 (**a**), *p*-$CH_3C_6H_4$ (**b**), *p*-$CH_3OC_6H_4$ (**c**), *p*-FC_6H_4 (**d**), *p*-ClC_6H_4 (**e**), *p*-BrC_6H_4 (**f**), $CH_2C_6H_5$ (**g**), CH_2-(2-C_5H_4N) (**h**); **24i-k**, R = *p*-$CH_3C_6H_4$, R' = *p*-$CH_3C_6H_4$ (**i**); R = *p*-BrC_6H_4, R' = *p*-BrC_6H_4 (**j**); R = $CH_2C_6H_5$, R' = $CH_2C_6H_5$ (**k**)

FIGURE 3.13 The synthesis of maleopimaric acid *N*-aryl(aralkyl)imido amides 24a–k.

The compounds 24a–k are colorless crystal substances soluble in DMF, dimethyl sulfoxide and insoluble in methanol, diethyl ester, and hexane.

By the method of derivatography, we evaluated the thermal stability of imides 24a, d–g, j, k. Compounds 24a, d, e, g are thermally stable and begin to decompose with a noticeable rate at 325–335°C, which is 30–40°C above the temperature of the starting MPA anilide 14k and 40–50 °C above the temperature of MPA decompositions start.

3.3 THE SYNTHESIS OF CITRACONOPIMARIC ACID AND ITS DERIVATIVES

Apart from maleic anhydride, in the preparation of rosin diene adducts fumaric acid, 1,4-benzoquinone, citraconic acid, and its anhydrides are used. The reaction of resin acids methyl ester with citraconic anhy-dride produces three isomeric citraconopimaric acid methyl esters [40]. Synthesis and properties of individual citraconopimaric acid (CPA) have not been described in the literature.

We studied the preparation of CPA from resin acids and citraconic anhydride formed from readily available itaconic acid on heating above the melting point (172 °C). It was established that the heating of mixture of rosin and itaconic acid at 180–200 C for 8–12 h leads to the reaction of citraconic anhydride with levopimaric acid 25 and to the formation of complex mixture containing isomeric citraconopimaric acids 26a, b and unreacted resin acids (IR, ^{1}H NMR data). IR spectrum of the reaction product contains bands at 1850, 1785 cm^{-1} characteristic for cyclic anhy-dride C=O bonds. In the NMR, ^{1}H spectrum peaks of abietic, neoabietic, palustric acids (5.78 ppm, 6.21 ppm, and 5.40 ppm, respectively) almost disappear, peaks of dehydroabietic, pimaric, and iso-pimaric acid (7.20 ppm, 5.15 ppm, and 5.33 ppm, respectively) maintain intact [41], and there are new peaks of vinyl protons and CH$_3$-groups.

The crystallization of the reaction product from CCl$_4$ gives precipitate of solvate CPA with CCl$_4$, and its thermal decomposition at 135°C gives citraconopimaric acid as a mixture of two isomers 26a, b (Fig. 3.14). It should be mentioned that instead of MPA, citraconopimaric acid does not form the solvate with acetic acid.

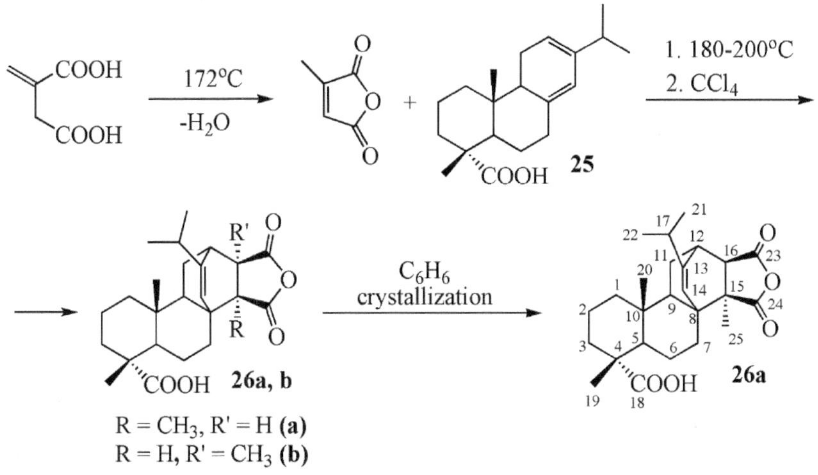

FIGURE 3.14 The synthesis of citraconopimaric acid 26a.

The formation of citraconic anhydride from itaconic acid is indicated by separation of water from the reaction mixture on heating above 170°C. Citraconic anhydride reacts with levopimaric acid formed by isomerization of abietic type resin acids (abietic, neoabietic, and palustric) at 180–200°C (see Fig. 3.14) [42].

Instead of citraconopimaric acid methyl ester [40], thin-layer chromatography of citraconopimaric acid isomers does not lead to its separation. By partial crystallization of mixture of isomers 26a and 26b from benzene, the isomer 26a was separated in 36% yield (based on mixture of isomers 26a, b) [42].

With an aim of unambiguous determination of isomer 26a structure, we studied it by two-dimensional NMR spectroscopy and performed full assignments of peaks in ^{1}H and ^{13}C NMR spectra [43]. Assignments of the hydrogen atoms were made by analysis of the ^{1}H NMR spectra and the COSY and NOESY correlation spectra. Assignments of the carbon atoms in the ^{13}C spectra were made based on analysis of the ^{13}C NMR spectra, DEPT, HSQC, and HMBC correlation spectra. It was established that the formula 26a corresponds the structure of isolated citraconopimaric acid isomer. The chemical shifts found for the ^{1}H and ^{13}C atoms in compound 26a are given in Table 3.2.

TABLE 3.2   Chemical Shifts (ppm) of Atoms in Citraconopimaric Acid Isomer 26a [43].

C-atom	13C shift	H-atom	1H shift
1	38.46	1α	0.94
		1β	1.41
2	17.09	2	1.48; 1.54
3	36.90	3α	1.71
		3β	1.62
4	46.96	–	–
5	49.12	5	1.68
6	21.60	6α	1.34
		6β	1.48
7	29.89	7α	1.74
		7β	2.27
8	42.28	–	–

TABLE 3.2 *(Continued)*

C-atom	13C shift	H-atom	1H shift
9	46.97	9	1.59
10	37.74	–	–
11	27.52	11α	1.61
		11β	1.28
12	36.60	12	3.00
13	147.83	–	
14	127.75	14	5.56
15	53.23	–	–
16	54.90	16	2.58
17	32.71	17	2.21
18	185.73	–	–
19	16.59	19	1.14
20	16.38	20	0.59
21/22	20.45	21/22	0.96
21/22	19.99	21/22	0.96
23	172.84	–	–
24	175.34	–	–
25	18.60	25	1.40

Note: β refers to the protons located above the plane of the ring, that is, on the side where the angular methyl groups are located; α refers to the protons located below the plane of the ring.

The stereochemistry of addition and the configuration of methyl group 25 follows from the NOESY spectra. The proton at C^{14} has a cross peak with protons of methyl group 20, which indicates location of the bridge above the plane of the molecule. The protons of methyl group 25 and $C^{16}H$ are coupled with 9αH and 11αH, respectively. This fact is evidence in favor of their α configuration. For the methyl group, we also observe coupling with the protons at C^7, which is additional support for the correctness of the assignment.

We have established that the compound obtained is the addition product of citraconic anhydride and levopimaric acid (citraconopimaric acid) and is a close structural analog of MPA, containing a CH_3 group in the α position of the anhydride group.

It can be assumed that the chemical properties of CPA will be similar to that of MPA, and reactions of CPA with alkyl halides, amines will proceed depending from conditions only at the carboxyl or anhydride group or at both groups simultaneously. We developed reactions of CPA with allyl and propargyl bromides, primary amines, thionyl chloride, etc.

The allyl and propargyl esters 27a, b were synthesized in a high yields (95–99%) by the reaction of CPA with allyl or propargyl bromide in DMF in the presence of potassium carbonate (Fig. 3.15) [15].

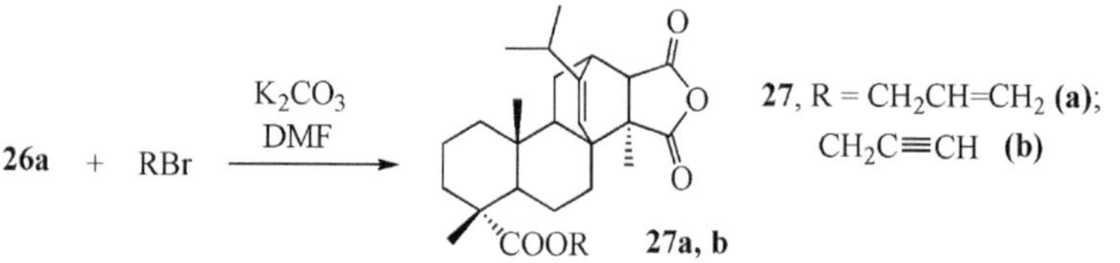

FIGURE 3.15   The synthesis of citraconopimaric acid allyl and propargyl esters 27a, b.

The treatment of the acid 26a with thionyl chloride gave citraconopimaric acid chloride 28 in quantitative yield (Fig. 3.16), which is well soluble in diethyl ether instead of MPA chloride. The treatment of acid chloride 28 with aniline or N-(2-methyl-5-aminophenyl)-4-(pyrid-3-yl) pyrimidine-2-amine at 60–65°C for 2 h afforded anilide 29a (yield 74%) and heterocyclic amide 29b (yield 48%) [44].

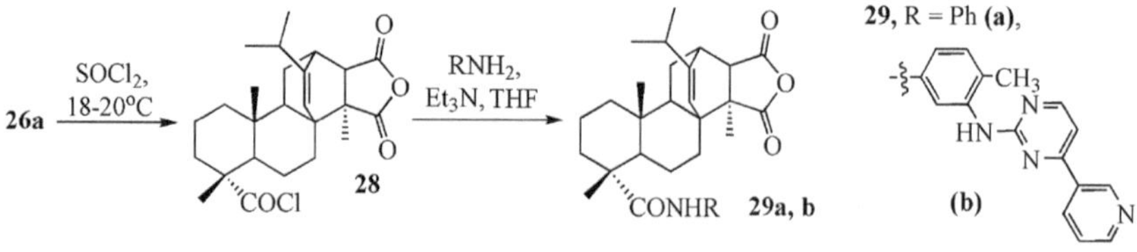

FIGURE 3.16   The synthesis of citraconopimaric acid amides 29a, b.

We developed a method for the synthesis of citraconopimaric acid imides. It was established that instead of MPA [17], CPA 26a does not react with ammonia (33% solution in water) at 50°C during 22 h. However, the heating of citraconopimaric acid 26a with n-butylamine in autoclave, *n*-octylamine and ethanolamine at 125°C for 16 h led to the formation of corresponding *N*-substituted imides 30a–c in high yields (87–100%) (Fig. 3.17) [45].

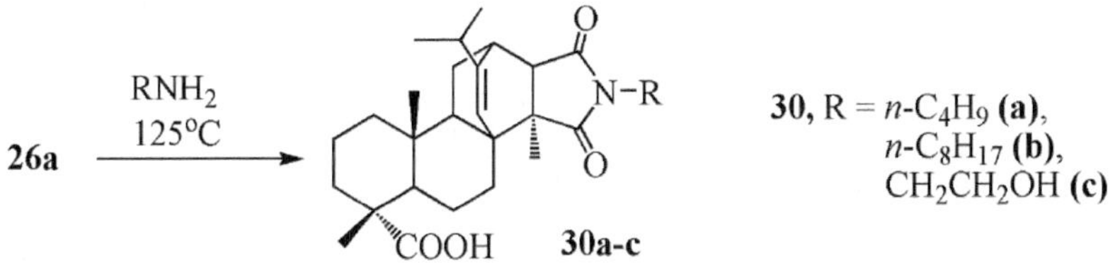

FIGURE 3.17 The synthesis of citraconopimaric acid imides 30a–c.

The treatment of *N*-(2-hydroxyethyl)imide 30c with dimethyl sulfate in DMF in the presence of potassium carbonate afforded methyl ester 31 in 73% yield (Fig. 3.18).

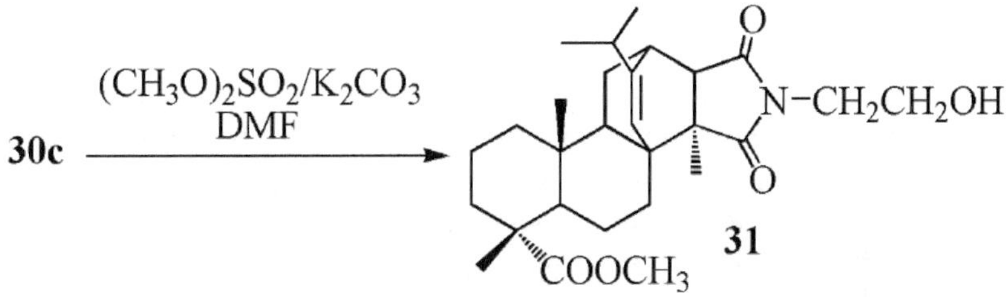

FIGURE 3.18 The synthesis of citraconopimaric acid *N*-(2-hydroxyethyl)imide 31.

The treatment of citraconopimaric acid *N*-octylimide 30b with thionyl chloride gave acid chloride 32 in quantitative yield, and its reaction with aromatic amines (aniline, *p*-chloro-, *p*-bromoaniline) afforded citraconopimaric acid imido amides 33a–c in 68–89% yields (Fig. 3.19).

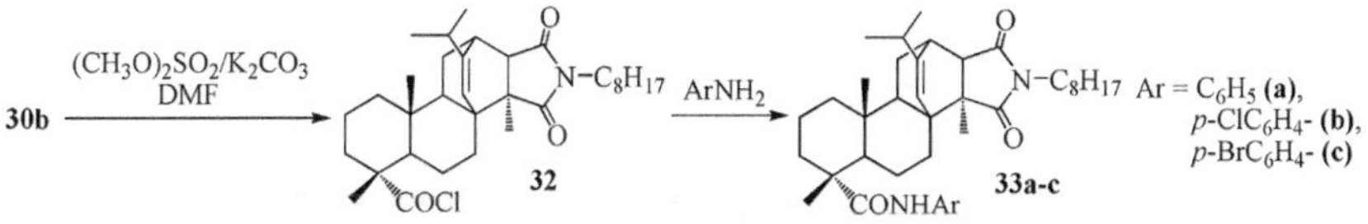

FIGURE 3.19 The synthesis of citraconopimaric acid imido amides 33a–c.

3.4 THE INVESTIGATION OF THE REACTION OF MALEOPIMARIC AND CITRACONOPIMARIC ACIDS WITH SECONDARY ALIPHATIC AMINES

Secondary amines generally react with compounds containing cyclic anhydride group giving anhydride ring opening products—amido acids [46]. Also, there is the only example of conversion of phthalic anhydride by fusion at 200 C with secondary amine (di(2-cyanoethyl)amine) to cyclic imide—phthalic anhydride 2-cyanoethylimide [47].

The reaction of citraconopimaric acid 26a with some secondary aliphatic amines (diethyl-, di-*n*-propyl-, di-*n*-butyl-, methyl-(2-hydroxyethyl)-, ethyl-(2-hydroxyethyl)amines) was investigated to study the steric influence of the CH_3 group at C^{15} of the acid 26a on regioselectivity of anhydride ring opening by nucleophilic agents and the possibility of formation of corresponding amido acids [48]. It was established after several experiments that the heating of citraconopimaric acid 26a solution in diethylamine in autoclave at 135°C for 40 h leads to the formation of the only product (81% in mixture according to ^{1}H NMR). The ^{1}H NMR spectrum of the product revealed only five additional proton signals that could be attributed to diethylamido group. Similar results were obtained when the acid 26a was treated in the same way with di-*n*-propylamine and di-*n*-butylamine: In both cases, the only product was formed, and the number of H-atoms in ^{1}H NMR spectra associated with dialkylamido group was only half of expected. Based on spectral data (IR, ^{1}H and ^{13}C NMR, and mass-spectroscopy) and elemental analysis of the obtained compounds, it was established that the reaction of the acid 26a with secondary amines instead of amido acids leads to the formation of cyclic citraconopimaric acid *N*-ethyl-, *N*-propyl-, *N*-butylimides 30a, 34a, b (Fig. 3.20, Table 3.3).

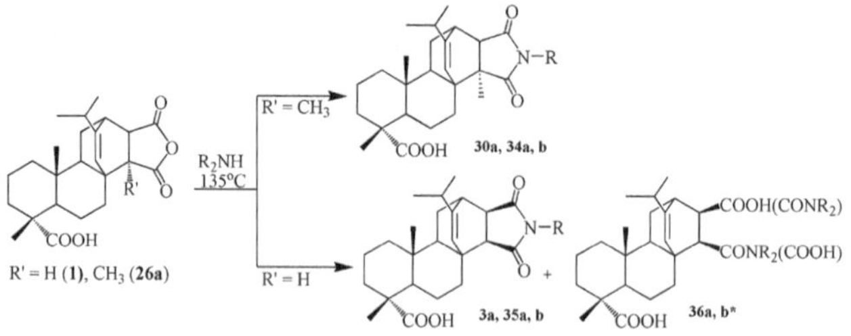

$R = C_2H_5 \text{ (34a, 35a, 36a)}, n\text{-}C_3H_7 \text{ (34b, 35b, 36b)}, n\text{-}C_4H_9 \text{ (3a, 30a)}$

FIGURE 3.20 Reactions of maleo- and citraconopimaric acids with secondary amines.
Note: *Denotes exact location of the amide group in compounds 36a, b is not defined.

TABLE 3.3 Reaction Time, Temperature, Conversion, and Yields in Reaction of Itraconopimaric Acid 26a with Secondary Amines [48].

Amine	Reaction time, h (temperature, °C)	Conversion of acid 26a, %[1]	Product, (yield, %)[1]
$(CH_3CH_2)_2NH$	40 (56)	0	–
$(CH_3CH_2)_2NH$	40 (135)	81	34a (81)
$(CH_3CH_2CH_2)_2NH$	40 (135)	73	34b (73)
$(CH_3CH_2CH_2CH_2)_2NH$	60 (135)	100	30a (100)
$CH_3NHCH_2CH_2OH$	24 (135)	100	37 (54); 30c (46)
$CH_3CH_2NHCH_2CH_2OH$	24 (135)	100	34a (38); 30c (62)

[1]Determined by ^{1}H NMR.

In contrast to reactions of citraconopimaric acid 26a, reactions of MPA 1 with diethyl-, dipropyl-, dibutylamines gave one product (MPA N-butyl-imide 3a) only for dibutylamine. In reactions of MPA with diethyl-, dipropylamines together with corresponded MPA N-ethyl-, N-propylimides 35a, b (yields 33–84%) (Table 3.4) amido acids 36a, b were formed and isolated as individual substances with 10–15% yields (Table 3.4, see also Fig. 3.20).

TABLE 3.4 Reaction Time, Temperature, Conversion, and Yields in Reaction of Maleopimaric Acid 1 with Secondary Amines [48].

Amine	Reaction time, h (temperature, °C)	Conversion of acid 1, %[1]	Product, (yield, %)[1]
$(CH_3CH_2)_2NH$	30	43	35a (33)
$(CH_3CH_2)_2NH$	60	81	35a (58)
$(CH_3CH_2CH_2)_2NH$	30	70	35b (57)
$(CH_3CH_2CH_2)_2NH$	60	95	35b (84)
$(CH_3CH_2CH_2CH_2)_2NH$	30	82	3a (72)
$(CH_3CH_2CH_2CH_2)_2NH$	60	100	3a (100)
$CH_3NHCH_2CH_2OH$	24	100	38 (56); 39 (44)
$CH_3CH_2NHCH_2CH_2OH$	24	100	35a (49); 39 (51)

[1]Determined by ^{1}H NMR.

When citraconopimaric 26a or maleopimaric 1 acid was treated with asymmetric methyl-(2-hydroxyethyl)-, ethyl-(2-hydroxyethyl)amines at 135°C for 24 h, the formation of a mixture of two CPA or MPA imides almost in equal proportions was observed [48]. The reaction of the acid 26a with methyl-(2-hydroxyethyl)amine provided a mixture of N-methylimide 37 and N-(2-hydroxyethyl)imide 30c of citraconopimaric acid in molar ratio 1:0.85 (according to ¹H NMR), and the treatment with ethyl-(2-hy-droxyethyl)amine gave a mixture of N-ethylimide 34a and N-(2-hydroxy-ethyl)imide 30c of citraconopimaric acid in molar ratio 1:1.63 (Fig. 3.21, see also Table 3.3).

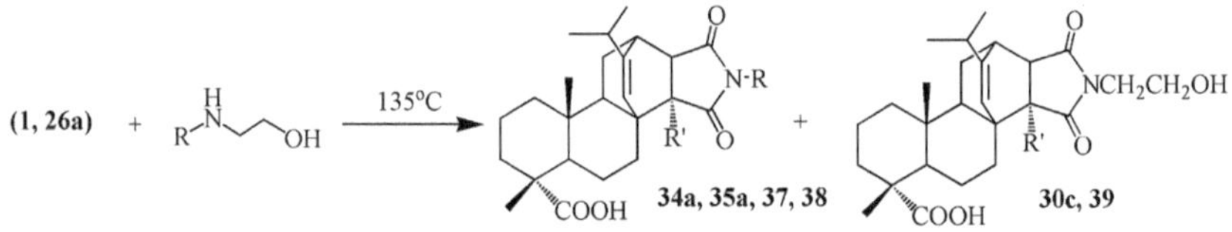

R' = CH₃ (30c, 34a, 37)	R = Me (37)	1	:	0.85*⁾
	R = Et (34a)	1	:	1.63*⁾
R' = H (35a, 38, 39)	R = Me (38)	1	:	0.78*⁾
	R = Et (35a)	1	:	1.04*⁾

FIGURE 3.21   Reaction of citraconopimaric 26a and maleopimaric 1 acids with asymmetric aliphatic amines.

Note: *Denotes ratio of products was determined by analysis of integration curve in ¹H NMR spectra.

The acid 1 in the same conditions reacts with methyl(ethyl)-(2-hy-droxyethyl)amines giving mixtures of N-methyl(ethyl)imide 38 (35a) and N-(2-hydroxyethyl)imide 39 (molar ratio 1:0.78 and 1:1.04, respectively) (see also Fig. 3.21, Table 3.4). These mixtures were separated with isola-tion of individual compounds by preparative thin-layer chromatography.

Due to the reaction conditions (a large excess of secondary amines which were used as solvents, ca. 2% solution of the acids 1, 26a in amine), there was possibility of imides formation by reaction of the acids 1, 26a with primary amines impurities (ethyl-, n-propyl-, n-butylamine) in corre-sponding secondary amines. We performed the reaction of the acid 1, 26a with di-n-butylamine (purity 99.9%) in a higher concentration of the acids 1, 26a (ca. 15% solution of the acids 1, 26a in amine) to check such option. After the mixture was being heated during 70 h at 135°C, the acids 1, 26a, as expected, completely converted into the imides 3a and 30a respectively.

In addition, the treatment of MPA with equimolar amounts of di-*n*-dibutylamine in diglyme at 135°C for 60 h gave the mixture of 55% starting MPA and 45% MPA *N*-butylimide 3a. Therefore, the obtained result clearly witnessed that impurities of primary amines were not involved in the cyclic imide formation.

The proposed mechanism for the formation of the *N*-alkylimides of citraconopimaric and MPAs in the reactions of amines with the acids 1, 26a is shown in Figure 3.22. The first step involves the nucleophilic attack by the amine on the carboxyl C-atom of the anhydride group resulting in formation of the amido acid (A). Intramolecular cyclization of the amido acid (A) leads to the cyclic intermediate (B) that is stabilized by alkyl group migration to form the intermediate (C). Elimination of the corresponding alcohol (C) generates the cyclic *N*-alkylimide (D). The mechanism of this reaction is almost similar to the formation of *N*-substituted imides in reaction of cyclic anhydrides with primary amines [49, 50]. In the case of unsymmetrical secondary amines reactions with the acids 1 or 26a, the formation of two different products results from migration of one from two different *N*-alkyl groups of the intermediate (B).

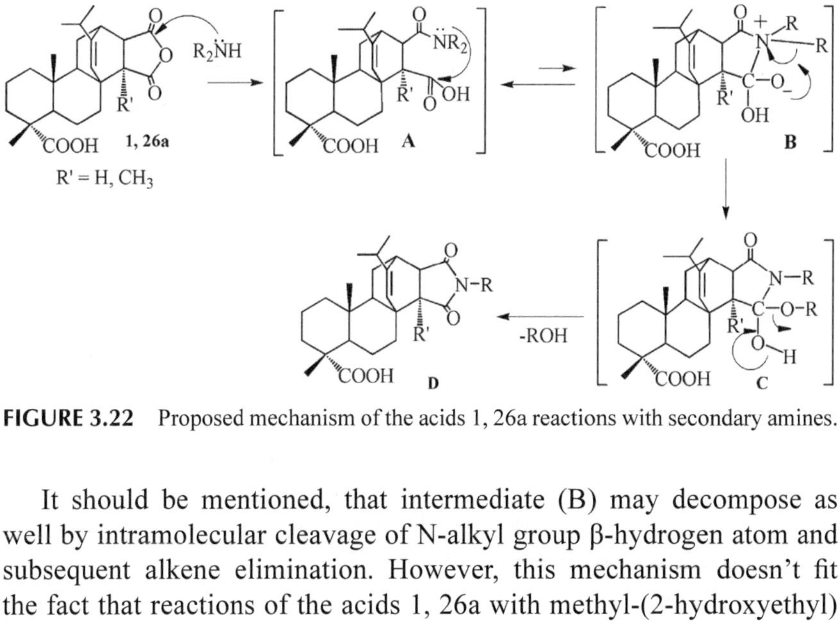

FIGURE 3.22 Proposed mechanism of the acids 1, 26a reactions with secondary amines.

It should be mentioned, that intermediate (B) may decompose as well by intramolecular cleavage of N-alkyl group β-hydrogen atom and subsequent alkene elimination. However, this mechanism doesn't fit the fact that reactions of the acids 1, 26a with methyl-(2-hydroxyethyl) amine provided mixtures of two imides (Tables 3.3 and 3.4) because of the absence of β-hydrogen atoms in methyl group. In the case of MPA reaction with diethyl-, dipropylamines, we assume initial formation of two possible amido acids, and one of them is more stable to the intramolecular cyclization. This suggestion is confirmed by the fact that isolated imido acid 36b did not form cyclic imide by heating in starting secondary amine (dipropylamine) or diglyme at 135 C for 60 h. The heating of imido acid 36b at 220°C for 2 h under argon converts it to N-propylimide 35b [48].

3.5 CONCLUSION

The needs of environmental protection and sustainable development require a wide use of renewable resources instead of petroleum-based raw materials for the manufacture of synthetic products. Fine chemicals based on bio-based materials are often more biodegradable and less toxic. As natural-based fine chemicals, rosin-based substances are important because of availability of raw material. We developed a series of effective methods for chemical transformations of MPA that is one of the most available synthons prepared from rosin and developed the methods for citraconopimaric acid and its derivatives synthesis. A number of new unsaturated esters, imides, amides, imidoesters of MPAs, and terpene diimido acids were synthesized. We developed an improved method of synthesis of long-chain (C_{12} and C_{18}) MPA imides in melt; the effective method of synthesis of MPA N-arylimides from maleated rosin; the synthesis of new terpenoid substances, fumaropimaric acid monoamides, with trans-1,2-dicarboxylic fragment; the method of preparation of MPA structural analogue—citraconopimaric acid as a mixture of two isomers and isolation of the isomer with methyl group at C-15, synthesis of esters, amides, imides, imidoester of citraconopimaric acid. The formation of cyclic N-alkylimides in reaction of maleo- and citraconopimaric acids with secondary amines was determined for the first time, and the effect of structure of terpenoid acid and amine on reaction selectivity was established: For the case of citraconopimaric acid, N-alkylimides are the only products, for the case of MPA together with cyclic imides, up to 10–15% of amidoacids are obtained.

KEYWORDS

- **rosin**
- **terpenoids**
- **maleopimaric acid**
- **citraconopimaric acid**
- **esters**
- **amides**
- **imides**

REFERENCES

1. Wang, J.; Chen, Y. P.; Yao, K.; Wilbon, P. A.; Zhang, W.; Ren, L.; Zhou, J.; Nagar-katti, M.; Wang, C.; Chu, F.; He, X.; Decho, A. W.; Tang, C. Robust Antimicrobial Compounds and Polymers Derived From Natural Resin Acids. *Chem. Commun.* **2012,** *6*(48), 916–918.

2. Galin, F. Z.; Flehter, O. B.; Tret'yakova, E. V. Synthesis and Reactions of Diene Adducts of Resin Acids. *Chem. Comput. Simul.* **2004,** *2*(5), 1–21 (In Russian). Butlerov Communications.

3. Wiyono, B.; Tachibana, S.; Tinambunan, D. Characteristics and Chemical Composition of Maleopimaric and Fumaropimaric Rosins Made of Indonesian *Pinus merkusii* Rosin. *Pak. J. Biol. Sci.* **2007,** *18*(10), 3057–3064.

4. Xu, T.; Liu, H.; Song, J.; Shang, S. B.; Song, Z. Q.; Chen, X. J.; Yang, C. Synthesis and Characterization of Imide Modified Poly(dimethylsiloxane) with Maleopimaric Acid as Raw Material. *Chin. Chem. Lett.* **2015,** *5*(26), 572–574.

5. Hengshan, W.; Tian, X.; Yang, D.; Pan, Y.; Wu, Q.; He, C. Synthesis and Enantiomeric Recognition Ability of 22-crown-6 ethers Derived From Rosin Acid and BINOL. Tetrahedron: *Asymmetry*, **2011,** *4*(22), 381–386.

6. Tolstikov, A. H.; Tolstikova, O. V.; Khlebnikova, T. B.; Karpyshev, N. N. Higher Terpenoids in the Synthesis of Chiral Phosphor- and Nitrogen-Containing Ligands for Metal-Complex Catalysts of Asymmetric Reactions. *Chem. Comput. Simul.* **2002,** *7*(2), 1–8. Butlerov Communications.

7. Kazakova, O. B.; Tret'yakova, E. V.; Kukovinets, O. S.; Abdrakhmanova, A. R.; Kabalnova, N. N.; Kazakov, D. V.; Tolstikov, G. A.; Gubaidullin, A. T. Synthesis of Nontrivial Quinopimaric Acid Derivatives by Oxidation with Dimethyldioxirane. *Tetrahedron Lett.* **2010,** *14*(51), 1832–1835.

8. Nazyrov, T. I.; Tret'yakova, E. V.; Kazakova, O. B. Regioselective Oxidation of the Methyl Ester of Maleopimaric Acid by Dimethyldioxirane. *Chem. Nat. Compd.* **2013,** *6*(48), 1002–1003.

9. Kim, S. J.; Kim, B. J.; Jang, D. W.; Kim, S. H.; Park, S. Y.; Lee, J. H.; Lee, S. D.; Choi, D. H. Photoactive Polyamideimides Synthesized by the Polycondensation of Azo-dye Diamines and Rosin Derivative. *J. Appl. Polym. Sci.* **2001,** *4*(79), 687–695.

10. Kazakova, O. B.; Tret'yakova, E. V.; Kukovinets, O. S.; Tolstikov, G. A.; Nazyrov, T. I.; Chudov, I. V.; Ismagilova, A. F. Synthesis and Pharmacological Activity of Amides and the Ozonolysis Product of Maleopimaric Acid, *Russ. J. Bioorg. Chem.* **2010,** *6*(36), 832–840.

11. Bicu, I.; Mustata, F. Allylic Polymers from Resin Acids. *Die Angew. Macromol. Chem.* **1997,** *4273*(246), 11–22.

12. Bicu, I.; Mustata, F. Allylic Polymers from Resin Acids. Monomer Synthesis at High *Die Angew* Temperature, *Macromol. Chem.* 1998, *4446*(255), 45–51.

13. Atta, A. M.; El-Saeed, S. M.; Farag, R. K. New Vinyl Ester Resins Based on Rosin for Coating Application. *React. Funct. Polym.* **2006,** *12*(66), 1596–1608.

14. Lee, J. S.; Hong, S.I. Synthesis of Acrylic Rosin Derivatives and Application as Negative Photoresist. *Eur. Polym. J.* **2002,** *2*(38), 387–392.

15. Bei, M. P.; Azarko, V. A.; Yuvchenko, A. P. Synthesis, Film-forming, and Light-Sensitivity Properties of Allyl and Propargyl Maleopimarates and Cytraconopimarates. *Russ. J. Gen. Chem.* **2010,** *5*(80), 940–943.

16. Pesetskii, S. S.; Krivoguz, Yu. M.; Bei, M. P.; Yuvchenko, A. P. (2011). The Method for Preparation of Adhesive, Patent of the Republic of Belarus 14660 (In Russian).

17. Schuller, W. H.; Lawrence, R. V. Some New Derivatives of Maleopimaric Acid. *J. Chem. Eng. Data,* **1967,** *2*(12), 267–269.

18. Svikle, D. Ya.; Prikule, A. Ya.; Shuster, Ya. Ya.; Veselov, I. A. Biological Activity and Toxicity of Maleopimaric Acid Derivatives. *Pharm. Chem. J.* **1978,** *5*(12), 617–620.

19. Delevalee, F.; Deraedt, R.; Benzoni, J. Method of Inducing Immunostimulating Activity, Patent USA 4880803, 1989.

20. Bei, M. P.; Yuvchenko, A. P. Synthesis and Properties of Maleopimaric *N*-(*n*-Alkyl) Imides, *Russ. J. Gen. Chem.* **2010,** *2*(80), 253–257.

21. Petrushenia, A. F.; Reviaka, M. M.; Yatsenka, V. V.; Bei, M. P.; Yuvchenko, A. P. Analysis of the Influence of Secondary Terpenoid Products on the Adhesion of the PE. *Proc. Natl. Acad. Sci. Belarus, Phys.-Tech. Ser.* **2012,** *3*, 21–24 (In Russian).

22. Buynova, E. F.; Solncev, A. P.; Volozhin, A. I. The Synthesis of Diterpenic Imidocarboxylic Acids. *Proc. Natl. Acad. Sci. Belarus, Chem. Ser.* **1996,** *4*, 118–119 (In Russian).

23. Ray, S. S. Polymers from Renewable Resources. XII. Structure Property Relation in Polyamideimides from Rosin. *J. Appl. Polym. Sci.* **1988,** *6*(36), 1283–1293.

24. Yao, G.; Ye, M.; Huang, R.; Li, Y.; Zhu, Y.; Pan, Y.; Liao, Z.; Wang, H. Synthesis and Antitumor Activity Evaluation of Maleopimaric Acid *N*-Aryl Imide Atropisomers. *Bioorg. Med. Chem. Lett.* **2013,** *24*(23), 6755–6758.

25. Bei, M. P.; Yuvchenko, A. P. The Synthesis of New Maleopimaric Acid Amides and Imides. *Proc. Natl. Acad. Sci. Belarus, Chem. Ser.* **2010,** *1*, 78–82 (In Russian).

26. Bei, M. P.; Yuvchenko, A. P.; Puchkova, N. V.. Efficient Synthesis of Maleopimaric Acid *N*-Arylimides. *Russ. J. Gen. Chem.* **2015,** *5*(85), 1034–1039.

27. Taylor, J.B., Ramm, P.J. & Fried, F. Amide Derivatives of Maleopimaric Acid, Patent of England 1400481, 1975.

28. Lin, G. -Q.; Li, Y. -M.; Chen, A. S. *Principles and Applications of Asymmetric Synthesis*; John Wiley & Sons Inc. New York., 2001; 507p.

29. Mangeney, P.; Grojean, F.; Alexakis, A.; Normant, J.F. Improved Optical Resolution of *R,R-N,N'*-Dimethyl-1,2-Diphenyl Ethylene Diamine, *Tetrahedr. Letters.* **1988**, *22*(29), 2675–2676.

30. Timerbaev, A. R.; Hartinger, C. G.; Aleksenko, S. S.; Keppler, B. K. Interactions of Antitumor Metallodrugs with Serum Proteins: Advances in Characterization Using Modern Analytical Methodology. *Chem. Rev.*, **2006**, *6*(106), 2224–2248.

31. Huigens, R. W.; Richards, J. J.; Parise, G.; Ballard, T. E.; Zeng, W.; Deora, R.; Melander, C. Inhibition of *Pseudomonas aeruginosa* Biofilm Formation with Bromoageliferin Analogues. *J. Am. Chem. Soc..* **2007**, *22*(129), 6966–6967.

32. Bei, M. P.; Yuvchenko, A. P. The Synthesis of *trans*-1,2-Dicarboxylic Acids from Maleopimaric Acid Monoamides. *Proc. Natl. Acad. Sci. Belarus, Chem. Ser.* **2010**, *3*, 84–87 (In Russian).

33. Smith, A. B.; Bosanac, T.; Basu, K. Evolution of the Total Synthesis of (−)-Okilactomycin Exploiting a Tandem Oxy-Cope Rearrangement/Oxidation, a Petasis–Ferrier Union/Rearrangement and Ring-Closing Metathesis. *J. Am. Chem. Soc.* **2009**, *6*(131), 2348–2548.

34. Sinott, M. *Carbohydrate Chemistry and Biochemistry*; Royal Society Chemistry, 2007; p 748.

35. Bicu, I. Crosslinked Polymers from Resin Acids. *Die Angew. Macromol. Chem.* **1996**, *4085*(234), 91–102.

36. Bicu, I.; Mustata, F. Water Soluble Polymers from Diels-Alder Adducts of Abietic Acid as Paper Additives. *Macromol. Mater. Eng.,* **2000**, *1*(280–281), 47–53.

37. Atta, A. M.; Mansour, R.; Abdou, M. I.; Sayed, A. M. Epoxy Resins from Rosin Acids: Synthesis and Characterization. *Polym. Adv. Technol.* **2004**, *9*(15), 514–522.

38. Bei, M. P.; Yuvchenko, A. P.; Sokol, O. V.; Puchkova, N. V. Synthesis and Properties of N-aryl(aralkyl) Imidoamides of Maleopimaric Acid. *J. Gen. Chem..* **2016**, *4*(86), 634–638 (In Russian).

39. Kim, Y. J.; Glass, T. E.; Lyle, G. D.; McGrath, J. E. Kinetic and Mechanistic Investigations of the Formation of Polyimides under Homogeneous Conditions, *Macromolecules*, **1993**, *6*(26), 1344–1358.

40. Gastambide, B.; Langlois, L. Stereochemical etudes VII. Diene Synthesis in Resin Acids, Action of Peroxoacids and Hydrogenation of Double Bonds. *Helvetica Chim. Acta.* **1967**, *8*(51), 2048–2057 (In French).

41. Skakovskii, E. D.; Tychinskaya, L. Yu.; Gaidukevich, O. A.; Kozlov, N. G.; Klyuev, A. Yu.; Lamotkin, S. A.; Shpak, S. I.; Rykov, S. V. NMR Determination of the Composition of Balsams from Scotch Pine Resin, *J. Appl. Spectr.* **2008**, *3*(75), 439–443.

42. Bei, M. P.; Yuvchenko, A. P. The Method of Synthesis of Citraconopimaric Acid. Patent of the Republic of Belarus, 2009; 13646 (In Russian).

43. Bei, M. P.; Baranovskii, A. V.; Yuvchenko, A. P. Structure of a Rosin–Itaconic Acid Adduct From 2D NMR Spectroscopy Data. *J. Appl. Spectr.* **2009**, *4*(76), 603–606.

44. Koroleva, E. V.; Gusak, K. N.; Ignatovich, J. V.; Ermolinskaya, A. L.; Bei, M. P.; Yuvchenko, A. P. Synthesis of Maleopimaric and Citraconopimaric Acids N-[3-(pyrimidin-2-yl)aryl]amides. *Russ. J. Org. Chem.* **2012**, *8*(48), 1121–1125.

45. Bei, M. P.; Yuvchenko, A. P. The Synthesis, Properties and Use of New Secondary Terpenoid Products from Rosin. *Proc. Natl. Acad. Sci. Belarus, Chem. Ser.* **2013,** *4,* 68–74 (In Russian).
46. Atodiresei, I.; Schiffers, I.; Bolm, K. Stereoselective Anhydride Openings, *Chem. Rev.* **2007,** *12*(107), 5683–5712.
47. Chodroff, S.; Kapp, R.; Beckmann, C. The Reaction of Certain Secondary Amines with Phthalic Anhydride. A New Synthesis of β-Alanine, *J. Am. Chem. Soc.* **1947,** *2*(69), 256–258.
48. Bei, M. P.; Yuvchenko, A. P.; Baranovskii, A. V. Formation of *N*-Alkylimides in Reaction of Maleopimaric and Citraconopimaric Acids with Secondary Amines. *Proc. Natl. Acad. Sci. Belarus, Chem. Ser.* **2013,** 4, 104–108 (In Russian).
49. Vidal, T.; Petit, A.; Loupy, A.; Gedye, R. N. Re-examination of Microwave-Induced Synthesis of Phtalimide. Tetrahedron, **2000,** *30*(56), 5473–5482.
50. Yanquing, P.; Song, G.; Qian, X. Imidation of Cyclic Carboxylic Anhydrides Under Microwave Irradiation, *Synth. Commun.* **2001,** *12*(31), 1927–1933.

CHAPTER 4

STRUCTURAL AND CHEMICAL MODIFICATION OF CELLULOSE IN PHOSPHOTUNGSTIC ACID–FORMIC ACID SYSTEM AND SULFATION PREPARED DERIVATIVES

ELENA V. UDORATINA* and MICHAEL A. TORLOPOV

Institute of Chemistry of Komi Scientific Center of Ural Branch of the Russian Academy of Sciences, 48, Pervomayskya St., Syktyvkar 167982, Komi Republic, Russia

*Corresponding author. E-mail: udoratina-ev@chemi.komisc.ru

CONTENTS

ABSTRACT

In this work, a controlled destruction of cellulose in formic acid catalyzed by phosphotungstic acid (concentration 1.0–0.1 mol.%, reaction time 1–15 min) was demonstrated. Resulting powder products were obtained with a degree of polymerization of 140–270. The degradation of the polysaccharide is followed by total or partial esterification of formic acid and obtained cellulose formate (CF) having a degree of substitution of 0.8–2.2. The synthesized CFs are soluble in organic solvents and are used for sulfation by aminosulfonic acid in a homogeneous medium. Water-soluble cellulose sulfates with a degree of substitution above 2.0 were prepared. The transformation products of cellulose have been studied using chemical analysis, FTIR, NMR spectroscopy, X-ray diffraction, and SEM.

4.1   INTRODUCTION

Modern approaches in the field of deep processing of polysaccharides are as subsequent optimization step the molecular weight of the polymers and their functionalization. The molecular weight of polysaccharide affected the ease and manufacturability of process modifications, mechanical and rheological properties of the products and determines the toxicity of polymers. Among the known chemical methods to reduce the molecular weight, polysaccharides, including cellulose, presently preferred methods of acid catalyst and catalytic thermal destruction using a mineral acid in aqueous and aqueous-organic media, acidic catalysts in organic solvents, for example, Lewis acids [4, 5] and heteropolyacids (HPA) [6].

Keggin-type HPA possess unique physical and chemical properties, high Bronsted acidity, heat stability, and nontoxicity. HPA are solid acids with good solubility and stability in water and oxygen-containing solvents, capable of technologically acceptable regeneration; they are estimated as environment-friendly and economically feasible catalysts [7, 8]. It confirms that use of HPA for receiving valuable products of the directed destruction of cellulose causes interest.

Earlier [9, 10], we investigated the cellulose destruction using HPA catalyst in water and acetic acid. The results show that destruction of cellulose in acetic acid is more effective in comparison with traditional approaches, such as the hydrolysis catalyzed by hydrochloric, nitric, or sulfuric acids because low molar concentration of HPA is used and time decreases from 120 to 15 min [1–3]. Besides, it shows that destruction of cellulose in acetic acid is followed by an esterification of macromolecule with formation of cellulose acetates [9]. Research of destruction of cellulose in wider homological number of carbonic acids is of interest. It is supposed that acid and catalytic impact on cellulose in these environments will lead to the similar combined result—destruction with a simultaneous esterification of polysaccharide.

Products of acid—catalytic destruction of cellulose are the powder cellulose (PC), including microcrystalline cellulose (MCC), they are widely applied in various industries. PC and its modifications are used as sorbents, fillers, carriers, thickeners in the food and pharmacological industry [11, 12]. For preparation of various derivatives, lowered molecular weight cellulose is preferred, which is easily soluble that allows carrying out synthesis in homogeneous conditions. Homogeneous acid—catalyzed reactions of cellulose—allows to increase uniformity of polymers modification and to use softer conditions for cellulose modification. Moreover, in this case, softer conditions of modification and specific reagents are used. For dissolution of cellulose at modification complex solvents [13], ionic liquids [14] can be used. Sometimes, derivative cellulose such as esters can be used for transformations in solution.

The limited number of the works devoted to receiving and application of cellulose formates (CFs) is known [15, 16]. CFs can be rather easily synthesized, but it is unstable which limits its practical application. At the same time, CFs are dissolved in organic solvents and therefore are interesting as an intermediate product to receiving valuable and demand derivatives of cellulose in homogeneous conditions [17]. One of such derivatives is cellulose sulfate (CS). CS is the water-soluble polysaccharide in many repeating structure of natural polysaccharides sulfates possessing biological activity [18].

In this work, results of the combined transformation of the cellulose including catalytic destruction and esterification by phosphotungstic acid (PTA)–formic acid system were described, and use obtaining derivatives (CFs) with various molecular mass and degree of substitution for synthesis of CSs was also demonstrated.

4.2 MATERIALS AND METHODOLOGY

4.2.1 MATERIALS

The bleached softwood pulp (Mondi Corporate, Syktyvkar) was used as starting material. The average degree of polymerization (DP) is 910, crystallinity index (C.I.) is 0.79, and lignin content is 0.6% (sulfuric acid method [19]). Before using, cellulose was washed in 0.2-M HCl and distilled water until neutral reaction on Cl⁻-ions, and then of acetone, and dried at 100°C. Chemical for reaction: formic acid (85%, Vecton), PTA ($H_3PW_{12}O_{40}$, chemically pure, Vecton), aminosulfonic acid (NH_2SO_3H, Sigma-Aldrich), sodium hydroxide (chemically pure, Vecton), acetone, and ethanol were purified by distilled dimethylformamide (DMF) dehydration molecular sieves. Working solutions were prepared just before the experiment.

4.2.2 METHODS OF ANALYSIS

The DP was calculated from the intrinsic viscosity of cellulosic solutions in cadoxen [19]. The amount of formil groups (HCO–) in cellulose was determined by chemical method based on saponification of ester groups with a 0.5M NaOH, then back titration of excess alkali with a mineral acid. Degree of substitution of the CFs was calculated using the following equation: $DS_{HCO} = 162\omega_{FC}/(4600-28\omega_{FC})$, where ω_{FC} is the amount of bound formic acid, mass%.

The sulfur amount in the sulfated cellulose (CS) was determined by elemental analysis. Elemental analysis was carried out in an equipped CHNS–Elemental Analyzer (Vario Micro Cube "Elementar"). Before the analysis, samples were burned in a stream of oxygen with a chromatographic registration of product. Degree of substitution (DS) of the CSs was calculated by amount of sulfur in cellulose derivatives using the following equation: $DS_S = 162\omega_S/(3200-102\omega_S)$, where ω_S is the amount of sulfur, mass%.

FTIR spectra were recorded on a Prestige 21 "Shimadzu" FTIR spectrometer in KBr tablets at the range of 400–4000 cm⁻¹. NMR spectra were recorded on a spectrometer "Bruker Avance II 300" (300 MHz) in D_2O. X-ray diffraction (XRD) analysis was carried out on a XRD-600 "Shimadzu" equipment. Diffraction intensity was measured at 2θ

diffraction angles from 5° to 40° with 0.1° increment. The C.I. was calculated by the ratio of the intensities of reflexes at 2θ diffraction angles 22 and 19° (Segal's method) [20]. Samples microstructure was studied on a TESCAN VEGA 3 SBU scanning electron microscope (Czech Republic) in BSE regime at 20-kV accelerating voltage. Samples were placed on carbon adhesive tape.

4.2.3 METHODS

4.2.3.1 PREPARATION OF THE CELLULOSE FORMATES (CF)

The starting wood pulp (1.0 g, 6.2 mmol) was placed in a flask with reflux condenser. Then, the sample was saturated with a 40-mL PTA solution in formic acid (concentration PTA 0.1–1.0 mol.% from cellulose). The reaction mixture was heated at 100.8°C for 5–15 min. After completion, the reaction was stopped by adding cold water (100 mL). After the CF was washed by deionized water and hot ethanol (50 mL), it was dried in vacuum at 60°C. Yield CF 63–85%, DS_{HCO} = 0.8–2.2. FTIR spectra CF (KBr, cm^{-1}) (1.0 mol.% PTA, DS_{HCO} 2.2): 3350 v(OH), 2900 v(CH$_2$), 1728 v(CO), 1430–1200 δ(C–H); 1170–1100 v(C$_6$–O–H),1070 v(C$_3$–O and C–C), 900–930 v(C$_1$–H).

4.2.3.2 PREPARATION OF THE MICROCRYSTALLINE CELLULOSE

CF preparing in the destruction of cellulose with formic acid in the presence PTA was treated with 0.5M NaOH solution (50 cm^3) at 20°C for 1 h, then separated, washed, and dried as in the previous procedure. Yield MCC 63–85%, DP = 140–270, C.I. = 0.80–0.88. FTIR spectra MCC (1.0 mol.% PTA): (KBr, cm^{-1}): 3354 v(OH), 2899 v(CH$_2$), 1430–1200 δ(C–H); 1160–1100 v(C$_6$–O–H), 1060 v(C$_3$–O and C–C), 898 v(C$_1$–H).

4.2.3.3 PREPARATION OF THE CELLULOSE SULFATE (CS) FROM CF

The prepared and dried CF was added to DMF (30 mL), and the mixture was stirred for 10 min. Then, 2.43 g NH$_2$SO$_3$H (25 mmol) was added, and mixture was heated at 50–100°C for 5 h. Next, the reaction mixture was cooled and dissolved in 1M NaOH. The resulting mixture was purified by dialysis against distilled water (membrane CelluSep), then was freeze-dried. Yield CS 63–85%, $DS_s = 1.1$–2.5. FTIR spectra (KBr, cm^{-1}): 3480 ν(OH), 2950, 2900 ν(CH$_2$), 1640 (oxidized acid salt COO$^-$), 1380 δ(C–H); 1240 cm^{-1} ν(S=O); 1070, 1030 ν(C$_3$–O and C–C), 810 cm^{-1} ν(C–OS). ЯМР ^{13}C (D$_2$O), δ, p.m. (I, Hz): 102.8 (C$_1$), 81.2 (C$_2$–OSO$_3^-$), 80.1 (C$_2$), 77.2–75.4 (C$_3$–C$_5$), 68.8 (C$_6$–OSO$_3^-$), 61.2 (C$_6$).

4.3   RESULTS AND DISCUSSION

4.3.1   STRUCTURAL AND CHEMICAL TRANSFORMATION OF CELLULOSE

Transformation of cellulose was carried out in formic acid in the presence of H$_3$PW$_{12}$O$_{40}$, showing the greatest catalytic activity from a number of the HPA (H$_3$PW$_{12}$O$_{40}$, H$_3$PMo$_{12}$O$_{40}$, and H$_4$SiMo$_{12}$O$_{42}$) studied earlier [9].

Figure 4.1 shows the general scheme of cellulose macromolecule transformation in the presence of the catalyst PTA (0.1 to 1.0 mol.%) at a solvent boiling temperature for 5–15 min. Destruction and esterification of cellulose macromolecule are observed. Results of the combined influence of reagents are given in Table 4.1.

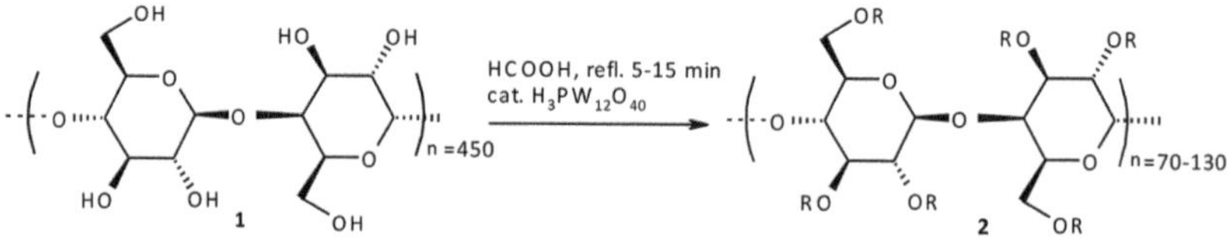

FIGURE 4.1   Destruction and esterification of cellulose, obtaining CFs (2), R = H, CHO.

In the given conditions, the smaller values of degree polymerization (at 3.4–6.6 times less, in comparison with control) confirm destruction of cellulose. The products with a DP 270–140 depending on concentration of the catalyst in system of 0.1–1.0 mol.%, respectively, are received.

Also, cellulose esterification is confirmed by FTIR spectroscopy and chemical analysis. The amount of the connected formic acid in the derivative cellulose is defined, it makes 19–45 mass% samples prepared in the presence of 0.1–1.0 mol.% PTA and DS_{HCO} are 0.8–2.2, respectively. The esterification of cellulose takes place in the formic acid without catalyst; the percentage of the formil group is 40.1 mass%. However, product retains the fibrous structure even when the reaction time increases to 55 min, wherein DP = 450 (see Table 4.1, No. 2).

TABLE 4.1 Destruction and Esterification of Cellulose in HCOOH in the Presence of $H_3PW_{12}O_{40}$.

No.	$H_3PW_{12}O_{40}$, mol.%	Degree of polymerization	Amount of HCO-group, mass%	Degree of substitution
1	Initial cellulose	910	–	–
2*	0	450	40.1	1.87
3	0.1	270	19.2	0.77
4	0.5	210	26.5	1.11
5	1.0	140	45.2	2.20

Note: Reaction conditions: $t = 15$ min, $T = 100.8°C$. *$t = 55$ min. Degree of polymerization is determined after saponification of ester groups with a 0.5M NaOH, HCO–formil groups.

Probable cleavage of glucosidic bond in the reaction conditions is followed by formylation, resulting degradation of hemiacetal hydroxyl. The presence of the formil group in CF was further verified by using FTIR spectroscopy. The characteristic band C–O at 1728 cm^{-1} in the spectra of CFs (Fig. 4.2, spectrum 3) clearly indicates the formation of formil group [17]. Other absorption bands are characteristic of initial cellulose (see Fig. 4.2, spectrum 2).

At transformation of cellulose in the presence of HPA, there is a question of a possibility of introduction $H_3PW_{12}O_{40}$ molecules in a cellulose macromolecule. There are published data on the sorption HPA of cellulosic materials from acidic aqueous solutions (pH = 1) [21]. Analysis of the FTIR data $H_3PW_{12}O_{40}$ spectrum (see Fig. 4.2) showed that PTA contains four characteristic bands within a range of 800–1100 cm^{-1}, namely 1080, 987, 889, and 823 cm^{-1}, assigned to the modes of P–O, W=O, and W–O–W, respectively. Among these, peaks of Keggin anion PTA in the spectrum of CF confirms that acid adsorption on the cellulose surface is not observed. It is established that working solution $H_3PW_{12}O_{40}$–HCOOH exhibits activity when reusing, in particular both degraded and esterifying abilities. So, as a result of exposure to filter 1.0 mol.% $H_3PW_{12}O_{40}$–HCOOH solution on the cellulose to obtain the product with DP = 260 and DS_{HCO} = 2.2, the reaction time increases to 30 min. This may be due to partial inactivation of HPA associated with acid restoration by reaction products (formation of phosphotungstic blue complex).

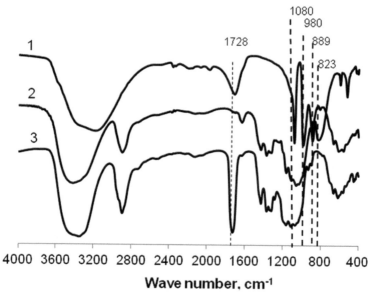

FIGURE 4.2   FTIR spectra of $H_3PW_{12}O_{40}$ (1), initial cellulose (2), and the cellulose formate (3).

4.3.2   DYNAMICS OF CELLULOSIC ESTERIFICATION AND DESTRUCTION IN THE FORMIC ACID–PTA SYSTEM

We investigated the effect of reaction time on characteristics of CFs. For that purpose, it regulates the esterification of cellulose and obtaining of CFs with defined degree of substitution. Figure 4.3 shows the number of HCO-groups dependent on the duration of the esterification reaction. Longer esterification reaction in the presence of 0.5-mol.% PTA from 1 to 10 min smoothly raises the content of formyl groups in the cellulose from 16 to 27 mass% (DS_{HCO} = 0.7–1.3) (Table 4.2).

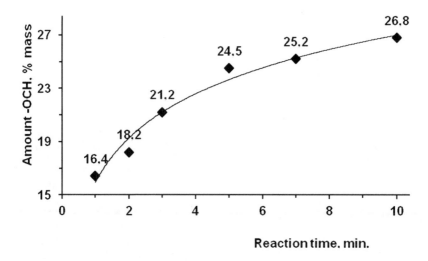

FIGURE 4.3   Changing the contents of formil groups depending on the reaction time of the cellulosic esterification in OH–0.5 mol.% $H_3PW_{12}O_{40}$.

TABLE 4.2   Dynamics of Cellulosic Esterification and Destruction of of a system HCOOH–0.5mol.% $H_3PW_{12}O_{40}$.

Reaction time, min	Yield, %	Amount of HCO groups, mass %	Degree of substitution	DP	C.I.
1	94	16.4	0.64	350	–
2	85	18.2	0.72	–	–
3	89	21.2	0.86	220	0.78
5	80	24.5	1.01	–	0.76
7	75	25.2	1.05	210	–
10	72	26.8	1.13	180	0.72

Note: Initial cellulose—Degree of polymerization—DP = 910, Crystallinity index—C.I. = 0.79.

The degree of destruction of CFs is also dependent on the duration of the reaction. In the first minute of reaction, there is a rapid decrease of the DP from 910 of the initial value to 350 units (see Table 4.2). Further, DP value sequentially decreases by 37% (3 min), 40% (7 min), and 49% (10 min) from the previous value. The reaction time has an effect on yield of CF. Reaction time affects the yield as well as the output is reduced by 6–28%

of the initial value. Reduced C.I. with increasing reaction time from 0.79 (the initial value) to 0.72 (10 min) indicates a gradual decrease of crystalline phase in the CFs and hence amorphization of its macromolecule.

Figure 4.4 shows photomicrographs demonstrating the dynamics destruction of cellulose fiber structure in formic acids–PTA system within 1–7 min.

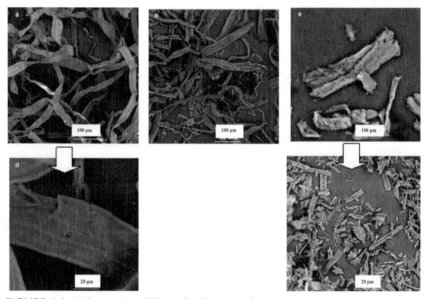

FIGURE 4.4 Micrographs (SEM) of cellulose surface at destruction in a HCOOH–0.5 mol.% $H_3PW_{12}O_{40}$ system at varying time, min: 1 (a, d); 3 (b); 7 (c, e).

Individual fibrils tape width of 44 μm with a smooth surface is presented in micrograph 4a (1 min of destruction). These fibrils gradually increase the reaction time to disintegrate first to fragments with the size of 30–230 μm in length and an average of 30 mm in width (4b), followed by obtaining fragments with the size of 16–76 μm in length and 18–20 μm in width (4c) with a rough, damaged surface (4e).

4.3.3 PREPARATION OF MCC

Treatment of CF by aqueous solutions of NaOH (0.5 M, 1 h, 20°C) leads to hydrolysis of ester linkages as evidenced by the absence of the absorption band at the FTIR spectrum of the cellulose at the 1728 cm^{-1} (C–O and HCO group) (Fig. 4.5). The content of the ordered parts in cellulose macromolecule increases. The type of the respective diffraction patterns and values of C.I. confirm this (Fig. 4.6).

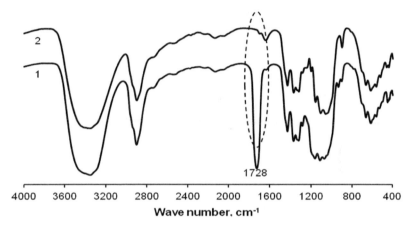

FIGURE 4.5 FTIR spectra of cellulose formates in a system HCOOH–1.0 mol.% $H_3PW_{12}O_{40}$ to (1) and after (2) treating the 0.5-M NaOH.

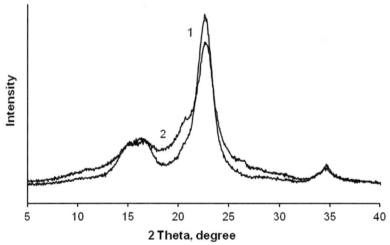

FIGURE 4.6 The X-ray diffraction spectra of esterifies cellulose in the HCOOH–1.0 mol.% $H_3PW_{12}O_{40}$ to (1) and after (2) treatment of 0.5M NaOH. $I_{kr.}$ (1) = 0.67, $I_{kr.}$ (2) = 0.81.

Thus, the MCC is obtained by catalytic thermal destruction of cellulose of 0.1–1.0 PTA in formic acid for 15 min followed by treatment with alkaline solution. MCC is described by the value of DP 140–270, high C.I. 0.8–0.9, also does not contain additional functional groups (Table 4.3). It should be noted that the known methods for obtaining MCC, such as hydrolytic degradation in aqueous solutions of mineral acids, involve the use of an excess of reagent (3–5 M to 1 M cellulose) and long reaction time (to 2 h) [1, 2].

TABLE 4.3   Characteristic Microcrystalline Cellulose.

No.	$H_3PW_{12}O_{40}$, mol.%	Degree of polymerization	Crystallinity index	Yield, %
1	Initial cellulose	910	0.79	100
2	0	450	–	97
3	0.1	270	0.80	85
4	0.5	210	0.83	72
5	1.0	140	0.88	63

Reaction conditions: At first—100.8°C, 15 min, then—0.5M NaOH, 20°C, 1 h.

4.3.4   SULFATING OF CELLULOSE FORMATES

The obtained CF with set values of DP and DS_{HCO} were exposed to to sulfation. Sulfation of CFs was performed with aminosulfonic acid in DMF at five-fold molar excess sulfating agent at a temperature of 50, 70, and 100°C for 5 h (Fig. 4.7).

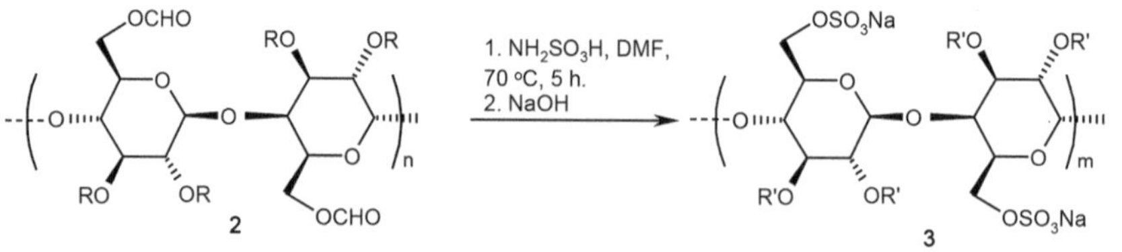

R=SO₃Na, CHO, H; R'= SO₃Na, H;

FIGURE 4.7   Scheme of cellulose formate sulfation (2) and preparation of cellulose sulfates (3).

Sulfation takes place under homogeneous conditions as the CF with a degree of substitution above 1.0 and sulfation agent—NH_2SO_3H is dissolved in DMF. It is found that the reaction ability of cellulose increases with raise sulfation temperature, resulting in a cellulose derivative (CS) with higher degrees of substitution (Table 4.4).

TABLE 4.4 The Conditions and Results of Sulfating of Cellulose Formate in NH_2SO_3H–DMF.

No.	Degree of substitution of cellulose formate	Degree of polymerization	Temperature, °C	Degree of substitution of cellulose sulfate
1	2.8	280	100	2.5
2	3.0	400	70	1.1
3	2.2	160	70	2.2
4	2.2	160	50	2.3
5	1.4	140	70	2.1
6	1.4	140	50	1.8

In most cases, the CF sulfation succeeded to introduce a cellulosic unit average that is more than two sulfate groups ($DS_s > 2$). An exception is example 2 (see Table 4.4) which has a relatively low degree of esterification ($DS_s = 1.1$) and is connected with its lesser solubility in the reaction mixture, most likely due to the relatively high (400 units) of the sample molecular weight. CS with low DS_s ($DS_s = 1.8$) is also obtained at a relatively low temperature (50 C) in the case of sulfation FC with the lowest degree of formylation (example 6), affecting the homogeneity of the reaction medium. Thus, the main influence in the degree of substitution of the CS synthesized from FC is an ability of the FC to form a homogeneous solution, which in turn depends on the degree of substitution and molecular weight of the sample.

The FTIR spectrum of the synthesized CS (Fig. 4.8) contains the characteristic absorption bands at 1240 cm^{-1} correlated with stretching vibrations S=O and 810 cm^{-1} (C–OS).

In the ^{13}C NMR spectrum (Fig. 4.9), the signals are observed in the field of 102.8 ppm (C1), signals 81.2 and 80.1 ppm are correlated with esterified and unsubstituted C2-atoms, respectively, group signals in the 77.2–75.4 ppm (C3–C5) and 68.8 ppm are for atoms C6 in sulfate group, and 61.2 ppm is for the nonsubstituted C6.

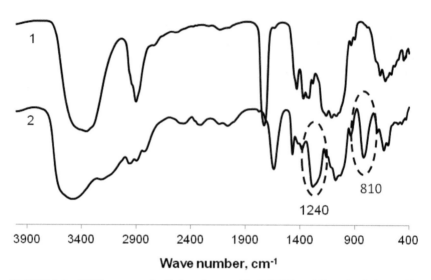

FIGURE 4.8 FTIR spectra of cellulose formate (1) (DSHCO = 2.2) and cellulose sulfate (2) (DS$_s$ = 2.3).

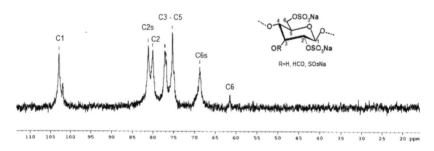

FIGURE 4.9 ^{13}C NMR spectra of cellulose sulfate (DS$_s$ = 2.3).

Preparation of sulfated derivatives based on CF is previously described in the paper [17], which the authors have used for the treatment of highly substituted formate SO$_3$ complexes with DMF or pyridine, and ClSO$_3$H in DMF. CS with DS, no higher than 0.88, was obtained at 25°C and three-fold molar excess sulfating agent.

As follows from the data presented, sulfation of CF in DMF has led to transesterification of part formate groups regardless of the nature sulfating agent compared with other esters, for example, cellulose acetates. Sulfating of CF by NH$_2$SO$_3$H can be obtained CS with varying degrees of substitution as well as highly substituted ($DS_s = 3$). However, it requires a considerable amount of sulfating agent, and the reaction temperature gets increased.

4.4 CONCLUSIONS

The study found that adjusting the conditions of a complex transformation of cellulose in formic acid medium in the presence of a TPA may be produced by either high-quality MCC (DP = 140, C.I. = 0.88) or reactive esters cellulose in a powder form (CF, $DS_{HCO} = 2.2$ and DS = 140). In turn, CFs are promising product for further esterification, for example, sulfation. Water-soluble CS with a maximum degree of substitution of sulfate groups at 2.5 were obtained by reacting CFs with aminosulfonic acid. Thus, the combination of polymer reactions, namely the esterification and destructions, is a new perspective method of complex transformation of polysaccharides.

ACKNOWLEDGMENT

The researches were carried out using the equipment of the Center for Collective Use "Chemistry" Institute of Chemistry, Komi Science Centre, Ural Branch of the Russian Academy of Sciences.

KEYWORDS

- acid-catalytic destruction
- heteropolyacids
- microcrystalline cellulose
- formolis
- sulfation
- phosphotungstic acid
- cellulose derivatives

REFERENCES

1. Shcherbakova, T. P.; Kotelnikova, N. E.; Byhovtsova, Yu. V. Comparative Studying of Powder and Microcrystalline Cellulose of Various Natural Origin. Physical and Chemical Characteristics. *Chem. Plant Raw Mater.* **2011**, *3*, 31–40 (In Russian).
2. Kargarzadeh, H.; Ahmad, I.; Abdullah, I.; Dufresne, A.; Zainudin, S. Y.; Sheltami R. M. Effects of Hydrolysis Conditions on the Morphology, Crystallinity, and Thermal Stability of Cellulose Nanocrystals Extracted from Kenaf Bast Fibers. *Cellulose*, **2012**, *19*(3), 855–866.
3. Udoratina, E. V.; Demin, V. A. Degradation of Hardwood Sulfate Pulp in Aqueous Dioxane. *Russ. J. Appl. Chem.* **2005**, *78*(8), 1333–1336 (In Russian).
4. Kuvshinova, L. A.; Frolova, S. V. Acid and Catalytic Transformation of Biopolymers. In: *Structure and Physical and Chemical Properties of Cellulose and Nanocomposites on Their Basis*; Aleshina, L. A., Gurtova, V. A., Melesch, N. V., Eds.; Petrozavodsk, chapter 3, 2014, pp. 30–97 (In Russian).
5. Frolova, S. V.; Kuvshinova, L. A.; Kutchin, A. V. Way of Obtaining Powder Cellulose. Patent RU No. 2478664, 2013 (In Russian).
6. Tian, J.; Wang, J.; Zhao, S.; Jiang, C.; Zhang, X.; Wang, X. Hydrolysis of Cellulose by the Heteropoly acid $H_3PW_{12}O_{40}$. *Cellulose*, **2010**, *17*, 587–594.
7. Cheng, M.; Shi, T.; Wang, S.; Guan, H.; Fan, C.; Wang, X. Fabrication Of Micellar Heteropolyacid Catalysts for Clean Production of Monosaccharides from Polysaccharides. *Catal. Commun.* **2011**, *12*, 1483–1487.
8. Timofeeva, M. N. Acidity and Catalytic Properties of Homogeneous and Heterogeneous Systems Based on Geteropoliconnections. Thesis for the degree of DSc of chemistry (02.00.15), kinetic and catalysis, Institute of Catalysis. GK Boreskov, Siberian Branch of Russian Academy of Sciences, Novosibirsk, 2010, p. 301 (In Russian).
9. Udoratina, E. V.; Torlopov, M. A. Partial Destruction of Cellulose in Water and in Acetic Acid in the Presence of Heteropolyacids, News of Higher Education Institutions. *Chem. Chem. Technol.* 2013, *56*(6), 69–74 (In Russian).
10. Torlopov, M. A.; Kutchin, A. V.; Udoratina, E. V. Way of Obtaining Microcrystalline Cellulose. Patent RU No. 2528261, 2014 (In Russian).
11. Siro, I.; Plackett, D. Microfibrillated Cellulose and New Nanocomposite Materials: A Review. *Cellulose*, **2010**, *17*, 459–494.
12. Xiong, R.; Zhang, X.; Tian, D.; Zhou, Z.; Lu, C. Comparing Microcrystalline with Spherical Nanocrystalline Cellulose from Waste Cotton Fabrics. *Cellulose*, **2012**, *19*(4), 1189–1198.
13. Kotelnikova, N. E.; Bykhovtsova, Yu. V.; Mikhailidi, A. M.; Mokeev, M. V.; Saprykina, N. N.; Lavrent'ev, V. K.; Vlasova, E. N. Solubility of Lignocellulose in *N,N*-Dimethylacetamide/Lithium Chloride. waxs, 13cp/MAS NMR, FTIR and SEM Studies of Regenerated from the Solutions Samples. *Cell. Chem. Technol.* **2014**, *48*, 643–651.
14. Zhang, Z.; Zhao, Z. K. Solid Acid and Microwave-Assisted Hydrolysis of Cellulose in Ionic Liquid. *Carbohydr. Res.* **2009**, *344*, 2069–2072.
15. Sun, Y.; Lin, L.; Deng, H.; Peng, H.; Li, J.; Sun, R.; Liu, S. Hydrolysis of Bamboo Fiber Cellulose in Formic Acid. *Front. For. China*, **2008**, *3*(4), 480–486.

16. Fujimoto, T.; Takahashi, S.; Tsuji, M.; Miyamoto, T.; Inagaki, H. Reaction of Cellulose with Formic Acid and Stability of Cellulose Formate. *J. Polym. Sci. Part C: Polym. Lett.* **1986**, *24*(10), 495–501.

17. Philipp, B.; Wagenknecht, W.; Nehls, I.; Ludwig, Y.; Schnabelrauch, M.; Klemm, D. Untersuchungen zur sulfatirung von celluloseformiat im vergleich zu celluloseacetat unter homogenen reaktionsbedingugen. *Cell. Chem. Technol..* **1990**, *24*(6), 667–678.

18. Udoratina, E. V.; Torlopov, M. A.; Drozd, N. N.; Makarov, V. A. Oxyethylated Cellulose Sulfates. *Polym. Sci., Ser. B.*, 2012, *54*(3–4), 175–182 (In Russian).

19. Obolenskaya, A. V.; Yel'nitskaya, Z. P.; Leonovich, A. A. Laboratory Work on the Chemistry of Wood and Cellulose Textbook for High Schools. *Ecology*, 1991, *320* (In Russian).

20. Park, S.; Baker, J. O.; Himmel, M. E.; Parilla, P. A.; Johnson, D. K. Research Cellulose Crystallinity Index: Measurement Techniques and Their Impact on Interpreting Cellulase Performance. *Biotechnol. Biofuels*, **2010**, 3–10.

21. Zuy, O. V.; Zaitsev, V. N.; Alekseev, S. A.; Trachevsky, V. V. Sorption of Heteropolyacids on Cellulose Sorbents. *Chem. Phys. Surf. Technol.* **2012**, *3*(1), 66–73 (In Russian).

CHAPTER 5

EFFECT OF COMPLEXATION WITH PHOSPHOLIPIDS AND POLARITY OF MEDIUM ON THE REACTIVITY OF PHENOLIC ANTIOXIDANTS

LYUDMILA N. SHISHKINA*, LIDIYA I. MAZALETSKAYA, KSENIYA M. MARAKULINA, YULIYA K. LUKANINA, IRINA G. PLASHCHINA, and NATALIYA I. SHELUDCHENKO

Emanuel Institute of Biochemical Physics of the Russian Academy of Sciences, Kosygin st., 4, Moscow 119334, Russia

Corresponding author. E-mail: shishkina@sky.chph.ras.ru

CONTENTS

ABSTRACT

The influence of the structure of isobornylphenols on their kinetic characteristics in the initiated oxidation of methyl oleate or ethylbenzene and autooxidation of methyl oleate in thin layer in mixtures with lecithin is studied, for example, 2-isobornyloxyphenol (I) and 2-isobornylphenol (II). The destruction of the intramolecular H-bond in I molecule while carrying out reactions in polar medium results in its increased ability to interact with peroxy radicals and the stable free radical diphenylpicryl-hidrazyl (DPPH). Effect of lecithin on the inhibitory efficiency of investigated isobornylphenols is found to depend on the initiation rate in the oxidation process and the ability of substances to form complexes with lecithin. In presence of isobornylphenols aggregates of lecithin both the size distribution of particles and the diameter of the main fraction depending on a polarity of medium change. Thus, isobornylphenols act upon the different parameters of the oxidation processes in the complex model systems. It allows us to suppose that the biological activity of isobornylphenols is due to their ability, both to influence on the different stages of the physicochemical regulatory system of the lipid peroxidation in tissues and to affect the structural state of the cell membranes. The data obtained help in checking the properties of various compounds on different model systems to carry out the preliminary choice of the most promising biologically active substances for practice use the medium polarity.

5.1 INTRODUCTION

Efficiency of the phenolic antioxidants (InH) essentially depends on many factors, including the properties of the medium [1]. This results in the difference of their activity as inhibitors of the free radical processes including in the model systems. Among factors causing differences in the inhibitory efficiency of InH in systems with the different oxidation extent the ability of InH to form complexes with phospholipids (PLs) [2–5] and the medium polarity [1, 6–8] play an important role. Besides, inhibitory efficiency of antioxidant and PL mixtures depends on the concentrations of both antioxidants and PL, the reaction rate of the radical initiation in the oxidation process and the nature of antioxidants [2, 5, 9, 10]. At present, terpenephenols synthesized on the basis of natural raw materials and containing as alkyl substituents isobornyl (1,7,7-trimethylbicyclo[2,2,1] hept-*exo*-2yl) group in the *o*- and *o,o'*-positions are regarded as promising biologically active substances which are characterized by high antiradical and antioxidant activities in chemical experiments [11] and in animal model systems [12–14]. As shown, the biological activity of compounds under an injection in organism is determined by their ability to affect the parameters of the physicochemical regulatory system of the lipid peroxidation in tissues including the structural state of the biological membranes [15]. It is due to the necessity to investigate the properties of isobornylphenols at the different complex model systems.

The aim of this work was to study the influence of the structure of isobornylphenols, for example, 2-isobornyloxyphenol and 2-isobornylphenol on their kinetic characteristics in the oxidation reactions under the different oxidation conditions in the complex systems and their ability both to form complexes with lecithin and to affect its aggregation, depending on the polarity of medium.

5.2   MATERIALS AND METHODOLOGY

The synthesis of 2-isobornyloxyphenol (**I**) and 2-isobornylphenol (**II**) was carried out by scientific workers in Institute of Chemistry of Komi Scientific Center of the Ural Branch of the Russian Academy of Sciences (Head is Corresponding member of RAS A.V. Kutchin), according to the procedures described in works [16, 17]. Substances were used without the additional treatment. The structures of the InH **I** and **II** are given in Figure 5.1. As source of natural PL, we used a 10% alcohol solution of soy-bean lecithin (Antigen, Ukraine). After the alcohol was removed, solutions of lecithin in different solvents are prepared.

The kinetic characteristics of InH and their mixture with lecithin (L) were determined by following models. The inhibitory efficiency was determined by model of the autooxidation of methyl oleate by the atmospheric oxygen in a thin layer at 323 °K. The course of oxidation was followed by the accumulation of hydroperoxides (ROOH), the concentration of which was determined by iodometric titration. Details of the analysis were published in Ref. [18]. The antiradical activity of compounds, that is, values of the rate constant of its reaction with peroxy radicals (k_7), was determined by using initiated azobisisobutironitrile oxidation both of methyl oleate in a mixture (1:1) with an inert solvent (chlorobenzene) and ethylbenzene at 333 °K. The initial rate of oxidation as well as the duration of the induction period (τ) were obtained from the kinetic curves of oxygen uptake. Details of analysis were presented in works [10, 11]. The interaction of InH with the stable free radical diphenylpicrylhidrazyl (DPPH) was studied in toluene and ethanol at 298 °K. The course of reaction was followed by the kinetics of the DPPH consumption. The concentration of DPPH was measured by the spectrophotometrical method at $v = 19.3 \times 10^3$ cm^{-1}. The initial rate of the DPPH consumption (W_c) was determined from the initial sections of kinetic curves.

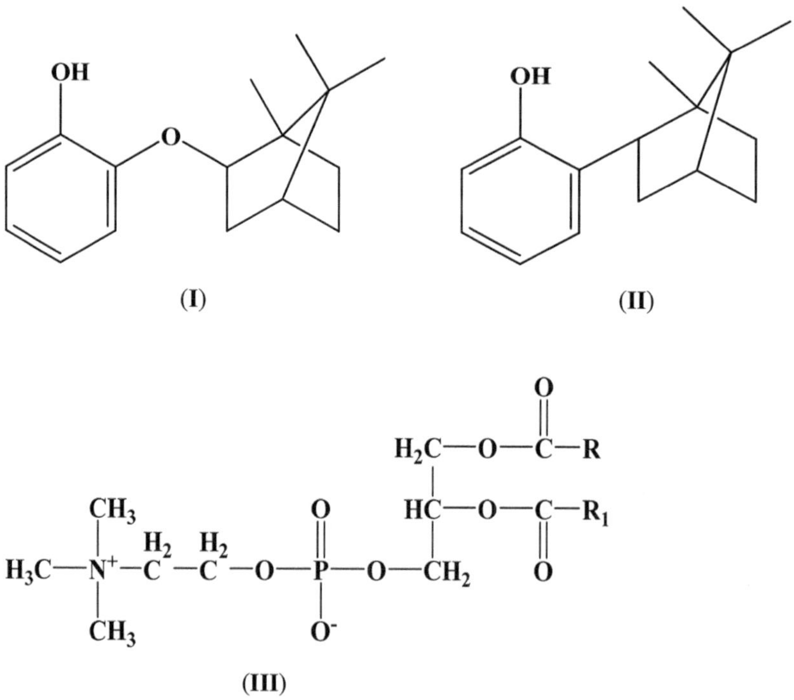

FIGURE 5.1   Structural formula of 2-isobornyoxylphenol (I), 2-iosobornylphenol (II), and phosphatidylcholine (III).

The qualitative and quantitative composition of PL in lecithin was determined by thin-layer chromatography (TLC) [19]. It was used type G silica gel (Sigma, USA), glass plates 9 × 12 cm and mixture of solvents chloroform:methanol:glacial acetic acid:distilled water in a volume ratio of 12.5:7.5:2:1 as a mobile phase. The separated PL fractions were developed in iodine vapor. The amount of inorganic phosphorus was determined spectrophotometrically (KFK-3, Russia) at $\lambda = 800$ nm according to the formation of a colored-phosphorus–molybdenum complex in the presence of ascorbic acid. Details of the procedure were described in the work [20].

Complexation of lecithin with InH was studied by UV- and IR-spectroscopy. The UV-spectra were recorded at wavelengths from 190 to 400 nm on a UNICO 2800 spectrophotometer (UNICO, United States). The composition of the complexes was determined by isomolar series method adapted by us for colorless solutions [21]. The deviation of the optical density of the mixture from the corresponding characteristics of the overall spectrum at λ in the range of 275 ± 5 nm was evaluated depending on the L:InH molar ratio. The overall spectrum is understood as the algebraic sum of the spectra of the individual components with the same concentrations.

The IR-spectra of the solutions of the individual components in the nonpolar solvent and their mixtures were recorded on a Spectrum 100 FT-IR spectrometer (Perkin-Elmer, Germany) equipped zinc selenide as an optic material in the frequency range from 4000 to 850 cm^{-1}, n-hexane as solvent, [InH] = [L] = 1×10^{-2} mol/L.

Effect of isobornylphenols **I** and **II** on the aggregation of lecithin in the nonpolar (hexane) and polar (0.8% aqueous solution of ethanol) media was studied by dynamic light scattering method using a ZetaSizer Nano ZS analyzer (Malvern Instruments, Great Britain). It was equipped with a He–Ne laser (4 mW) and an automated program for data processing at 25°C and a fixed scattering angle of 173°. The hydrodynamic radius of particles formed in the polar and nonpolar media of both L and its equimolar mixtures with InH was measured. The L concentration was 30 μg/mL and 35 μg/mL, and the InH concentration was 4.3×10^{-5} mol/L and 5×10^{-5} mol/L in hexane and aqueous-ethanol solution, respectively. The size distribution of particles in each sample was determined at least five times. The relative error of measurements was less than 2%.

The experimental data were processed by the commonly used variation statistics method using the Excel program package. The results are presented as arithmetic mean values indicating root-mean-square errors of the arithmetic mean ($M \pm m$).

5.3   RESULTS AND DISCUSSION

As known, one of the most important properties of InH is their antiradical activity, and the quantitative parameters of this activity allowing comparison of InH of the different nature and structure are the rate constant of its interaction with peroxy radicals (k_7) and the stoichiometric coefficient of inhibition (f). Isobornylphenols, due to low volatility and rather low toxicity, are considered as promising InH both for stabilization of various organic compounds and biological active substances with a wide range of applications. As already mentioned, they are characterized by high antiradical and antioxidant activities in chemical model systems. Moreover, earlier it is shown that the reactivity of o-alkyl substituted isobornylphenols is substantial higher compared with o-alkoxy substituted analogs [11, 22, 23]. As found, InH **I** possesses the lowest reactivity in respect to ethylbenzene peroxy radicals among isobornylphenols. So, in nonpolar medium, the inhibitory efficiency of InH **I**, that is, the fk_7 value, is 6.5 times less active as compared to **II** in the reaction with the ethylbenzene peroxy radicals due to the formation of an intramolecular H-bond in the **I** molecule between the H atom of OH group and oxygen atom of the alkoxy substituent [11]. Using IR-spectroscopy, the formation of the intramolecular H-bond is confirmed by a shift of OH group stretching vibrations both of o-alkoxy isobornyl and o-alkoxy isocamphyl substituted phenols more than 50–60 cm^{-1} to lower frequencies compared with their o-alkyl substituted analogs [11, 22]. However, differences in the reactivity of InH with peroxy radicals decrease with increasing the polarity of medium. Besides, the inhibition parameter fk_7 both for o-alkyl and o-alkoxy substituted phenols has for certain more low value under conditions of the initiated oxidation of methyl oleate [24]. Disrupting the intramolecular H-bond in InH **I** molecule under carrying out the reaction with the stable free radical DPPH in the polar medium (ethanol) results in the inversion of effect compared with data obtained in toluene the physicochemical properties of which are similar to ethylbenzene as solvent. It follows from the comparison of data presented in Figures 5.2 and 5.3 that **I** characterized a higher reactivity than **II** under carrying out the reaction in ethanol. The increased reactivity of **I** in polar medium is satisfactory agreement with the electron donor capacity of alkoxy and alkyl substituents. The absence of the intramolecular H-bond in InH **I** molecule under using ethanol as solvent and the formation of the intermolecular

H-bonds between molecules **I** or **II** and ethanol were confirmed by means of IR-spectroscopy in work [24]. Besides, the ratio of $fk_7(\mathbf{I})/fk_7(\mathbf{II})$ is equal 1.65 under the initiated oxidation of methyl oleate, as shown in Ref. [24]. As methyl oleate is less polar solvent in comparison with ethanol, perhaps, in this case, the influence of the intermolecular H-bond on the reactivity **I** is less substantially compared with **II**. This assumption is supported by results of IR-spectroscopy for solutions of InH **I** and **II** in methyl oleate presented in work [25]. However, the destruction of the intramolecular H-bond in molecule **I** in polar solvent results in the practically equal reactivity of both isobornyloxy- and isobornylphenols during the methyl oleate autooxidation in a thin layer under the diffusion condition. This conclusion is confirmed by data presented in Figure 5.4. Earlier, the ability **I** to inhibit significantly both the accumulation of secondary products of the lipid peroxidation in erythrocytes and the oxidation of oxyhaemoglobin to methaemoglobin was described in Ref. [26]. It also indicated the increased ability of **I** to trap free radicals in a biological system which is characterized by its polarity.

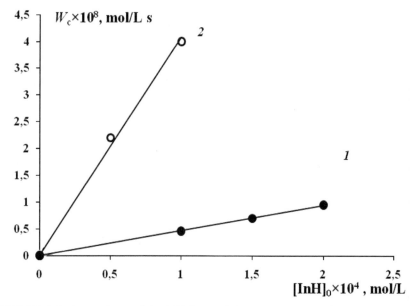

FIGURE 5.2 Dependence of the initial rate of the DPPH consumption on the initial **I** concentration in different solvents: 1—toluene, 2—ethanol; 298 K; [DPPH] = 1 × 10^{-4} mol/L.

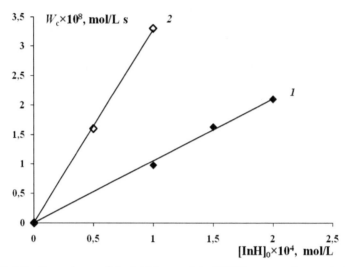

FIGURE 5.3 Dependence of the initial rate of the DPPH consumption on the initial **II** concentration in different solvents: 1—toluene, 2—ethanol; 298 K; [DPPH] = 1×10^{-4} mol/L.

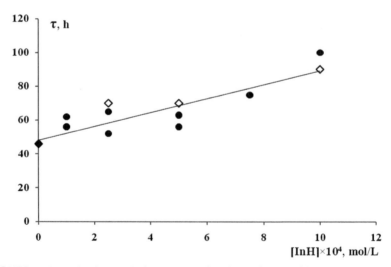

FIGURE 5.4 Induction period (τ) as a function of the initial concentration of isobornylphenols **I** (points) and **II** (rhombs).

Widely used food additive lecithin is a mixture of natural lipids, and the main fraction among PL in lecithin is phosphatidylcholine (PC), the structural formula of which is also presented in Figure 5.1. According to

TLC data, in used samples of lecithin, the share of PL among total lipids was 56–81%, and the relative content of PC in the PL composition varied from 80 to 92%. This agrees with the literature data about evident differences in the PL share and fraction composition of PL from lot to lot in commercial natural preparations [27, 28].

As shown in works [5, 9, 10], the influence of the lecithin additives on the inhibitory efficiency of the natural InH is due to both a scale of the interaction of components in the mixture and the rate of the chain initiation in the oxidation process. The data obtained agree with those available in literature. So, in the initiated oxidation, the inhibitory efficiency of InH **II** in mixture with lecithin is revealed to depend on the substrate nature. Although the lecithin additives to **II** had no influence on the reaction rate of initiated oxidation of methyl oleate, the inhibitory efficiency **II** decreased in initiated ethylbenzene oxidation in the lecithin presence. This follows from the data in Figure 5.5. However, the lecithin additives to **II** had also no influence on the accumulation of hydroperoxides in the methyl oleate autooxidation. This conclusion is followed from data presented in Figure 5.6. Obviously, a substantial decrease of the inhibitory efficiency of InH **II** under conditions of the oxidation in polar medium is due to the formation of the intermolecular H-bonds between molecules of **II** and methyl oleate, an existence of which was shown in work [25].

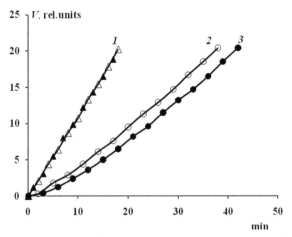

FIGURE 5.5 Kinetic curves of oxygen uptake in the inhibitory oxidation of methyl oleate (333 K, $W_i = 1 \times 10^{-7}$ mol/L/s) in the presence of 1×10^{-3} mol/L **II** or its mixture with 1.25 mg/mL lecithin (1) and ethylbenzene (333 K, $W_i = 5 \times 10^{-8}$ mol/L/s) in the presence of the 1×10^{-4} mol/L **II** and 1.25 mg/mL lecithin mixture (2) and 1×10^{-4} mol/L **II** (3).

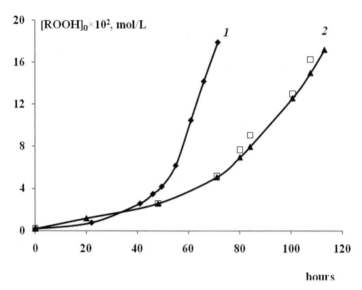

FIGURE 5.6   Kinetic curves of the accumulation of hydroperoxides under methyl oleate autooxidation without additives (1) and in the presence of **II** or its mixture with lecithin (2); 323 K; $[\textbf{II}] = 1 \times 10^{-3}$ mol/L; $[\text{L}]_0 = 5$ mg/mL.

It is also needed to note that under using hexane as solvent, the marked changes in UV- and IR spectra for studied InH are detected in the lecithin presence. As one of the complexation parameters, we considered the molar proportion of PL at which the deviation of the optical density of the spectrum of the mixture from the corresponding value of the overall spectrum was maximal. In Figure 5.7, these deviations are given in percent of the optical density of the overall spectrum. As seen, the most deviation is revealed by 60% and 70% molar share of lecithin in case of **I** and **II**, respectively. However, in addition to a main maximum, there are the significant deviations of the optical density for mixtures of the InH **I** and **II** with lecithin under 20% and 40% molar share of lecithin for **I** and **II**, respectively. This suggests the formation of molecular complexes of variable composition, which was also earlier shown for mixtures of luteolin with lecithin [4]. As can be seen from data in Figure 5.7, the relatively weak differences are revealed in UV-spectra for both phenols with lecithin, although these changes for **II** are a little more than of **I**. Perhaps, the existence of an intramolecular H-bond in the molecule **I** in nonpolar medium, as already mentioned, decreases a possibility of **I** to take part in the formation of complexes with lecithin.

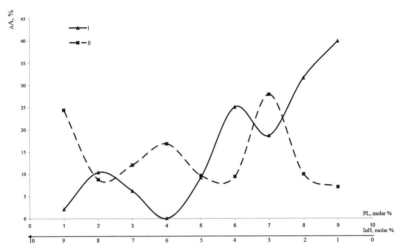

FIGURE 5.7 UV-spectra: mixtures of antioxidants **I** (triangles) and **II** (squares) with lecithin in dependence on the proportion of components.

The complexation was confirmed not only by changes in the UV spectral characteristics, but also by means of IR spectroscopy under the IR spectra analysis of the equimolar mixtures InH and L relative to the corresponding characteristics of the individual components. These data are presented in Figure 5.8.

The lecithin influence on the intensity of the stretching vibration of the phenol OH-group is revealed to be significantly higher for **II** as compared with **I**. This conclusion is supported by results in Figure 5.8. Thus, the formation of the intramolecular H-bond between the hydrogen atom of the hydroxyl group and the oxygen atom of the *o*-substituent accounts for both the low antiradical activity of **I** and its decreased ability to form complexes with lecithin.

The medium polarity influence is detected under the analysis of the InH surfactant properties in the model of the lecithin aggregation depending on the medium polarity. In used samples of L for these experiments, the PL share in the total lipid content had 71% ± 4% and 79.9 % ± 0.9% and the PC share in the PL composition was 80% ± 1% and 84.7% ± 0.6% in the nonpolar and polar solutions, correspondingly. As seen from data in Figure 5.9, while the diameter of the main fraction of lecithin is practically independent on the medium polarity (102.2 ± 6.1 nm and 114.8 ± 8.6 nm in hexane and aqueous–ethanol solution, respectively), its share in polar medium is 1.5 times more as compared with than in hexane (see Fig. 5.10).

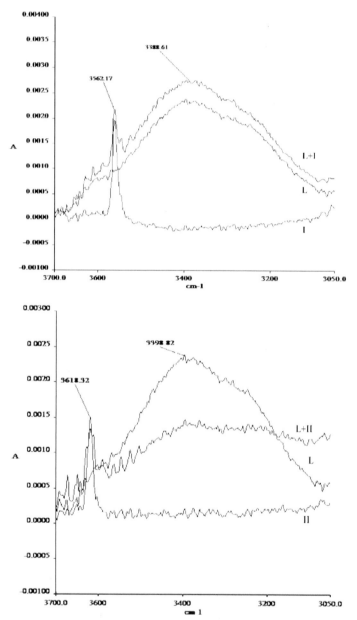

FIGURE 5.8 IR-spectra of lecithin, antioxidants **I** and **II**, and their mixtures in the region of stretching vibrations of the free OH-groups for phenols in solution of hexane; [L] = [InH] = 10^2 mol/L; the PC share in the PL composition for this sample of lecithin is 85% ± 1%.

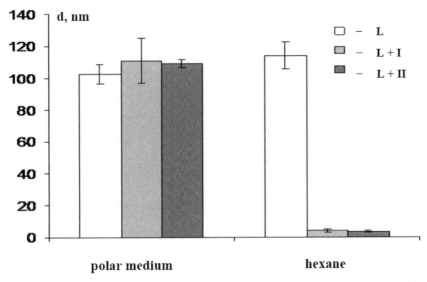

FIGURE 5.9 Influence of the medium polarity on the size of micelles of lecithin and its mixtures with phenolic antioxidants **I** and **II**.

Besides, in hexane solution, the second fraction of L is presented by small particles with $d = 3.2 \pm 0.1$ nm, but in aqueous–ethanol medium, it consists from large micelles having $d = 1750 \pm 200$ nm (Fig. 5.10). In the presence of InH, there are changes both percentage distribution and sizes of micelles of L depending on the medium polarity. So, the size of the main fraction of lecithin has not for certain changed in the polar solution (111 ± 14 nm for mixture of L with **I** and 108.9 ± 2.8 nm for **II**, correspondingly) and is almost 30 times reduced in hexane (4.0 ± 1.1 nm and 3.5 ± 0.7 nm for mixture of L with **I** and **II**, respectively), as seen from the data in Figure 5.9. Besides, the second fraction of L in mixtures with **I** and **II** consists also of the large particles in the aqueous–ethanol solution, but in hexane, there are three fractions of L of various diameters in the presence of investigated compounds (Fig. 5.10). The identity of the particle diameters for mixtures of lecithin with these phenols under its relative low concentration agrees with their slight differences to form the molecular complexes with lecithin in hexane (see Fig. 5.7). However, there are substantial differences in surfactant properties of InH **I** and **II** when the L concentration in hexane is 55 µg/mL, as shown in work [29].

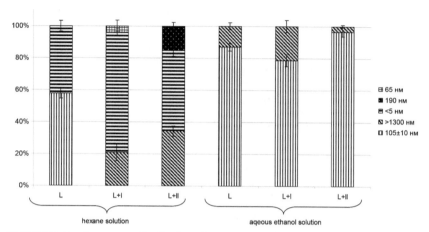

FIGURE 5.10   Percentage distribution of aggregates of the various diameters in solutions of lecithin and its mixtures with antioxidants depending on the medium polarity. Size of aggregates is denoted by marking.

First, the main fraction of L under its increased concentration in non-polar medium has the substantial less diameter of particles which was equal 15.3 ± 0.5 nm. Second, while InH **I** don't practically cause changes in the percentage distribution of micelles to their diameters as compared with that for L under its concentration of 55 µg/mL, isobornylphenol **II** is characterized by membrane-active properties, under these conditions result in decreasing both the share of the main fraction of L 1.6 times and diameter of micelles 2.5 times as compared with **I**. Hence, the ability of isobornyl-phenols to influence the aggregation of L in nonpolar solvent is due to not only a structure of its molecule, but also the L concentration in medium.

5.4   CONCLUSIONS

Data obtained and those available in literature about the influence of isobor-nylphenol structure on their kinetic characteristics in the model oxidation processes depending on the medium polarity allow us to conclude the following conclusions.

The increasing ability of *o*-isobornyloxyphenol (**I**) to interact with peroxy radicals and DPPH in the polar medium because of the destruction of the intramolecular H-bound in its molecule and the formation of the intermolecular H-bond between molecules of the polar solvent and **I** are revealed. There is no difference in the antioxidant activity of alkoxy- and alkylphenol under conditions of the autooxidation of methyl oleate in thin layer.

Effect of lecithin on the inhibitory efficiency of investigated isobornyl-phenols is found to depend on the initiation rate in the oxidation process and the nature of the oxidation substrate. So, in the initiated oxidation, lecithin does not affect on the inhibitory efficiency of InH **II** when the methyl oleate is an oxidation substrate, but substantially decreases this parameter of **II** in the ethylbenzene oxidation.

Studied phenols are shown to form the molecular complexes with lecithin which have a variable composition. The ability of substances to form complexes with lecithin depending on the presence of intermolecular OH-bond in the molecule structure is interrelated with their reactivity.

In presence of isobornylphenols, the lecithin aggregates change both the size distribution of particles and the diameter of the main fraction depending on the medium polarity and the lecithin concentration.

Thus, isobornylphenols act upon the different parameters of the oxidation processes in the complex model systems. It allows us to suppose that the biological activity of isobornylphenols is due to their ability both to influence on the different stages of the physicochemical regulatory system of the lipid peroxidation in tissues and to affect on the structural state of the cell membranes. The data obtained result in the necessity of checking the properties of various compounds on different model systems for the preliminary choice of the most promising for practice use.

ACKNOWLEDGMENTS

The authors sincerely thank scientific workers of Institute of Chemistry of Komi Scientific Center of the Ural Branch of the Russian Academy of Sciences Corresponding-member of RAS A.V. Kutchin, Doctor of Science in Chemistry, Professor of RAS I.Yu. Chukicheva, & PhD I.V. Fedorova for kindly placing us isobornylphenols for investigations.

This work was financially supported by the Fundamental Research Program of the Presidium of the Russian Academy of Sciences "Fundamental Sciences to Medicine" (2012–2014).

The work was carried out using scientific equipment of the Center of shared usage "New Materials Technology" of Emanuel Institute of Biochemical Physics of the Russian Academy of Sciences.

KEYWORDS

- **isobornylphenols**
- **structure**
- **reactivity**
- **autooxidation**
- **initiated oxidation**
- **DPPH**
- **lecithin**
- **complexation**
- **medium polarity**
- **surface-active properties**

REFERENCES

1. Emanuel, N. M.; Zaikov, G. E.; Maizus, Z. K. *Oxidation of Organic Compounds. Effect of Medium*. Pergamon Press: New York, 1984; p 612.
2. Burlakova, E. B.; Mazaletskaya, L. I.; Sheludchenko, N. I.; Shishkina, L. N. Inhibitory Effect of the Mixtures of Phenol Antioxidants Phosphatidylcholine. *Russ. Chem. Bull.* **1995,** *44*(6), 1014–1020.
3. Mazaletskaya, L. I.; Sheludchenko, N. I.; Shishkina, L. N. Role of the Nitrogen-Containing Moiety of Phosphatidylcholines in the Mechanism of Inhibiting Action of Their Mixtures with Natural Synthetic Antioxidants. *Petrol. Chem.* **2008,** *48*(2), 105–111.
4. Xu, K.; Liu, B.; Ma, Yu.; Du, J.; Li, G.; Gao, H.; Zhang, Yu.; Ning, Z. Physicochemical Properties Antioxidant Activities of Luteolin-Phospholipid Complex. *Molecules,* **2009,** *14*, 3486–3493.
5. Mazaletskaya, L. I.; Sheludchenko, N. I.; Shishkina, L. N. Lecithin Influence on the Effectiveness of the Antioxidant Effect of Flavonoids α-Tocopherol. *Appl. Biochem. Microbiol.* **2010,** *46*(2), 135–139.
6. Burlakova, E. B.; Krashakov, S. A.; Khrapova, N. G. Kinetic Peculiarities of Tocopherols as Antioxidants. *Chem. Phys. Rep.* **1995,** *14*,(10), 1657–1690.
7. Cuvelier, M. E.; Bondet, V.; Berset, C. Behavior of Phenolic Antioxidants in Pertitioned Medium: Structure–Activity Relationship. *J. Am. Oil Chem. Soc.* **2000,** *77*, 819–823.
8. Amorati, R.; Cavalli, A.; Corbini, L.; Zanbonin, L.; Land, L. Solvent and pH Effect on the Antioxidant Activity of Caffeic and Other Phenolic Acids. *J. Agric. Food Chem.* **2006,** *54*, 2932–2937.

9. Mazaletskaya, L. I.; Sheludchenko, N. I.; Shishkina, L. N. Effect of Lecithin on the Inhibitory Efficiency of α-Tocopherol during Methyl Oleate Oxidation. *Biophysics*, **2010**, (1), 18–23.

10. Mazaletskaya, L.; Sheludchenko, N.; Shishkina, L. Inhibitory Efficiency of Antioxidant and Phospholipid Mixtures under the Different Oxidation Extent of Methyl Oleate. *Chem. Chem. Technol.* **2012**, *6*(1), 35–41.

11. Mazaletskaya, L. I.; Sheludchenko, N. I.; Shishkina, L. N.; Kutchin, A. V.; Fedorova, I. V.; Chukicheva, I. Yu. Kinetic Parameters of the Isobornylphenols with Peroxy Radicals. *Petrol. Chem.* **2011**, *51*(5), 348–353.

12. Plotnikov, M. B.; Smol'yakova, V. I.; Ivanov, I. S.; Kutchin, A. V.; Chukicheva, I. Yu.; Krasnov, E. A. Antithrombogenic and Antiplatelet Activity of Ortho-Isobornyl-pjenol. *Bull. Exp. Biol. Med.* **2008**, *145*(3), 296–299.

13. Plotnikov, M. B.; Chernysheva, G. A.; Smol'yakova, V. I.; Ivanov, I. S.; Kutchin, A. V.; Chukicheva, I. Yu.; Krasnov, E. A. Neuroprotective Effects of Dibornol and Mechanism of Action Under Cerebral Ischemia. *Vestnik RAMN.* **2009**, (11), 12–17 (in Russian).

14. Buravlev, E. V.; Chukicheva, I. Yu.; Fedorova, I. V.; Grishko, V. V.; Anikina, L. V.; Lumpov, A. E.; Vikharev, Yu. B.; Kutchin, A. V. Anti-Inflammatory Activity of Isobornylphenol Derivates. *Chem. Nat. Compd.* **2010**, *46*(4), 478–480.

15. Burlakova, E. B. Bioantioxidants: Yesterday, Today, Tomorrow. In: *Chemical and Biological Kinetics. New Horizons*. Vol 2: *Biological Kinetics*; Burlakova, E. B., Varfolomeev S. D., Eds.; VSP: Leden, Boston: VSP, 2005; pp. 1–33.

16. Chukicheva, I. Yu.; Kutchin, A. V. Natural and Synthetic Terpenophenols. *Ross. Khim. Zhurn.* **2004**, *48*(3), 21–38 (in Russian).

17. Chukicheva, I. Yu.; Timusheva, I. V.; Spirikhin, L. V.; Kutchin, A. V. Alkylation of Pyrocatechol and Resorcinol by Camphene. *Chem. Nat. Compd.* **2007**, *43*(3), 245–249.

18. Mazaletskaya, L. I.; Sheludchenko, N. I.; Shishkina, L. N. Influence of the Initiatuin rate of Radicals on the Kinetic Characteristics of Quercetin and Dihydroquercetin in the Methyl Oleate Oxidation. In *Chemical Reactions in Gas, Liquid and Solid Phases: Synthesis Properties and Application*; Zaikov, G.E., Kozilowski, R.M., Eds.; NSP: New York, 2012; pp. 11–20.

19. In Biological Membranes: A Practical Approach. Findlay, J.B.C., Evans W.H., Eds.; Moscow: Mir. 1990; 424 p. (Russian version).

20. Shishkina, L. N., Kushnireva, Ye. V.; Smotryaeva, M. A. Combined Effect of Surfactant and Acute Irraqdiation at Low Dose on Lipid Peroxidation Process in Tissues and DNA Content in Blood Plasma of Mice. *Oxidation Commun.* **2001**, *24*(2), 276–286.

21. Marakulina, K. M.; Kramor, R. V.; Lukanina, Yu. K.; Kozlov, M. V.; Shishkina, L. N. Application of UV- and IR-Spectroscopy to Analyze the Formation of Complexes between Sphingomyelin and Phenolic Antioxidants. *Moscow Univ. Chem. Bull.* 2012, *67*(4), 185–191.

22. Mazaletskaya, L. I.; Sheludchenko, N. I.; Shishkina, L. N.; Kutchin, A. V.; Fedorova, I. V.; Chukicheva, I. Yu. Inhibiting Activity of Isocamphyl Substituted Phenols and Their Mixture with 2,6-Di-*tert*-Butylphenol in the Initiated Oxidation of Ethylbenzene. *Russ. J. Phys. Chem. A.* **2012**, *86*(6), 929–934.

23. Shishkina, L. N.; Mazaletskaya, L. I.; Marakulina, K. M.; Lukanina, Yu. K.; Plash-china, I. G.; Sheludchenko, N. I.; Buravlev, E. V.; Fedorova, I. V.; Chukicheva, I. Yu.; Kutchin, A. V. Kinetic Characteristics and Physicochemical Properties of Isobornyl Phenols with Different Alkyl Substituents in the *Ortho*-Position. *Russ. Chem. Bull. Intern. Ed.* **2014,** *63*(9), 2007–2012.

24. Mazaletskaya, L. I.; Sheludchenko, N. I.; Lukanina, Yu. K.; Shishkina, L. N. Effect of the Medium Polarity and the Hydrogen Bond on Reactivity of the *o*-Alkyl and *o*-Alkoxy Phenols in Different Model System. *Chem. Phys. Rep.* **2013,** *32*(3), 31–34 (in Russian).

25. Mazaletskaya, L. I.; Sheludchenko, N. I.; Shishkina, L. N.; Kutchin, A. V.; Fedorova, I. V.; Chukicheva, I. Yu. Influence of Lecithin on the Inhibitory Efficiency of Isobor-nylphenols. *Russ. J. Phys. Chem. A.* **2013,** *86*(9), 1532–1538 (in Rissian).

26. Shevchenko, O. C.; Plyusnina, S. N.; Shishkina, L. N.; Chukicheva, I. Yu.; Fedorova, I. V.; Kuchin, A. V. Membrane-Protective Properties of Isobornylphenols – A New Class of Antioxidants. *Biochem. (Moscow). Suppl. Ser. A: Membr. Cell Biol..* **2013,** *7*(4), 302–312.

27. Sotirhos, N.; Herslof, B.; Kenne, L. Quantitative Analysis of Phospholipids by ^{31}P-NMR. *J. Lipid Res.* **1986,** *27,* 386–392.

28. Hielscher, R.; Hellwig, P. Specific Far Infrared Spectroscopic Properties of Phospho-lipids. *Spectr.: Internet. J.* **2012,** 27, 525–532.

29. Shishkina, L. N.; Kozlov, M. V.; Marakulina, K. M.; Plashchina, I. G.; Plusnina, S. N.; Shevchenko, O. G.; Fedorova, I. V.; Chukicheva, I. Yu.; Kutchin, A. V. Surface Active Properties of Isobornylphemols in Systems with Different Degree of Complexity. *Biophysics.* **2012,** *57*(6), 786–791.

CHAPTER 6

BIOCATALYTIC CONVERSION OF LIGNOCELLULOSE MATERIALS TO FATTY ACIDS AND ETHANOL WITH SUBSEQUENT ESTERIFICATION

SERGEY D. VARFOLOMEEV[1,2], MARINA A. GLADCHENKO[1], SERGEY N. GAYDAMAKA[1], VALENTINA P. MURYGINA[1], VIOLETTA B. VOLIEVA[2], NONA L. KOMISSAROVA[2*], FARID M. GUMEROV[3], RUSTEM A. USMANOV[3], and ELENA V. KOVERSANOVA[4]

[1]*Department of Chemistry, M.V. Lomonosov Moscow State University, 1, Leninsky gory, Moscow 119991, Russia*

[2]*Emanuel Institute of Biochemical Physics of Russian Academy of Science, 4, Kosygin St., Moscow 119334, Russia*

[3]*Kazan State Technology University, 68, Karl Marks St., Kazan 420015, Russia*

[4]*Semenov Institute of Chemical Physics of Russian Academy of Science, 4, Kosygin St., Moscow 119334, Russia*

Corresponding author. E-mail: komissarova@polymer.chph.ras.ru

CONTENTS

ABSTRACT

A complex study of nonpolluting energy- and resource-saving processing of renewable feedstock with production of methane, ethanol, monosaccharides, biodiesel, and its light analogues is a base for a new technology with multipurpose use of biomass for producing valuable materials for fuel production. Biocatalytic processes with specialized microbe associations play an important role in processing carbohydrate raw materials from agricultural and woodworking wastes, such as straw, sawdust, etc. Biocatalytic methanogenesis is the most widely known process already implemented in practice. This process is complicated by generating additional organic compounds: ethanol, volatile fatty acids (VFAs)—acetic, propionic, butyric acids—toxic for microorganisms generating methane. This undesirable trend has triggered an alternative bioprocess aimed at obtaining the methanogenesis byproducts. A selection of microorganisms producing ethanol and VFA was carried out, pH and temperature conditions optimal for their functioning were determined, and the influence of these parameters on the yield and the ratio of biosynthesis products were studied.

Ethanol and VFA are in a low-concentration water solution state during the bioprocess, thus, their extraction method is needed. The best results are obtained by common extraction of ethanol and VFA by halogenated hydrocarbons. This is explained by the generation of ethanol–VFA complexes, more lipophilic than the individual alcohol, the latter being practically impossible to extract from water solutions.

Traditional methods of acid esterification by alcohols include using homogeneous or heterogeneous acid catalysts. Both variants have a number of limitations related to a ratio of reagents, process duration, catalyst type and cost, its removal from reaction sphere, and product purification. Alternative approaches minimizing these problems are being developed in the world. Recently, supercritical fluid (SCF) techniques have gained much attention. Application of SCF considerably enables, up to minutes, reduction in process time, dispenses with the catalyst, and, finally, reduces the total energy consumption.

6.1 INTRODUCTION

A fundamental task of producing fuels from renewable agricultural or technical feedstock is a conversion of structures of mostly carbohydrate nature to hydrocarbons or molecules containing hydrocarbonic fragments: carbohydrates → alcohols, carbohydrates → fatty acids → esters, and carbohydrates → biomass (bio-oil). The biomass anaerobic conversion into biogas can be considered as a perfect process in this regard. In this process, the carbohydrate component is disproportioned with the main energy content passing to methane:

$$(CH_2O)_2 \rightarrow CH_4 + CO_2 \qquad (6.1)$$

A widespread technique of biogas generation is based on the methanogenesis occurring actively in natural conditions. It has a number of shortcomings:

(1) Relatively low rate of functioning of complex methanogenerating microorganism consortium.
(2) Limitation in most cases due to biodestruction of biopolymers (lignocellulosic polymer complexes). Lignin doesn't undergo methanogenesis in this case.
(3) Process instability related to transition of the system into the state of hyperproduction of organic acids, pH decrease, and methanogenesis blocking.

A detailed study of methanogenetic microbe association functioning [1, 2] showed a possibility of using the first stage of methanogenesis (generation of liquid acids) to obtain components for the synthesis of high-energy motor fuel.

In a stable stationary state with lowered pH, the biomass conversion can occur with butyric acid generation according to stoichiometric equation:

$$(CH_2O)_6 \rightarrow 0.5 \; C_3H_7COOH + C_2H_5OH + 2 \; CO_2 + H_2 \qquad (6.2)$$

In real conditions, the process is more complex and is accompanied by the accumulation of small quantities of acetic and propionic acids, but it is rather a stable and intense process, if optimal conditions are created.

Conversion of fatty acids to a fuel form can be attained by several ways, but the VFA and ethanol generated together in the biocatalyst process are an ideal feedstock for the synthesis of ethyl esters. The esterification of organic acids is very efficient and fast in supercritical alcohols [3]. Thus, the VFA esterification was tested under conditions of supercritical ethanol fluids generation.

The main goals of this work were a detailed kinetic study of fatty acids and ethanol generation from different samples of carbon-containing raw materials including biomass, the optimization of process rate and target product yields, obtaining the esters of fatty acids under condition of supercritical ethanol, and testing the products in fuel compositions. Oxidative catalytical depolymerization and hydrolysis were used as a common means of preprocessing of solid raw materials. It is shown experimentally that a combined preprocessing enables a 10 times higher methanogenesis rate, and biogazification of lignin, in particular.

6.2 MATERIALS AND METHODOLOGY

The objectives of the study were modeling two medium compositions (Table 6.1) prepared on a mineral medium.

TABLE 6.1   The Chemical Composition of Model Streams.

The number of the model stream	The composition of the model stream	
	Glucose, g/L	Serum*, g/L
No. 1	From 1 to 4	–
No. 2	–	From 3 to 6

Note: *Serum contains 83% lactose, 13% protein, and 1% fat; g/L—grams/liter.

Mineral medium prepared from four solutions which had the following composition:

Solution no. 1: (g/L): NH_4Cl—170; $CaCl_2 \cdot 2H_2O$—8; $MgSO_4 \cdot 4H_2O$—9;

Solution no. 2: (mg/L): $FeCl_3 \cdot 4H_2O$—2000; $CoCl \cdot 6H_2O$—2000; $MnCl_2 \cdot 4H_2O$—500; $CuCl_2 \cdot 2H_2O$—30; $ZnCl_2$—50; H_3BO_3—50; $(NH_4)6Mo_7O_2 \cdot 4H_2O$—90; $Na_2SeO_3 \cdot 5H_2O$—100; $NiCl_2 \cdot 6H_2O$—50.

Solution no. 3: (g/L): KH_2PO_4—86.40.

Solution no. 4: (g/L): Na_2HPO_4—60.

To prepare 1 L of mineral medium, 10 mL solution no. 1, 1.4 mL solution no. 2, 100 mL solution no. 3 solution, and 10 cm^3 solution no. 4 were used; after finishing, water up to 1 L pH 5.6 was added.

Pretreatment of biomass (straw, sawdust, and lignin) for subsequent conversion into VFA and ethyl alcohol (EtOH) was performed according to the procedures described earlier via acid hydrolysis [4], as well as via catalytic oxidation using copper salts as a catalyst (alkali medium, a temperature range from room temperature to 90°C) [5, 6]. The wheat straw, which was preliminarily, dispersed on a ball mill, contained organic substances (OS) at a concentration of 17.6 g COD/L after acidic hydrolysis [4] including soluble OS 6.9 g COD/L; the sum of the reducing sugars was 2.4 g/L. The sawdust from softwood and lignin after oxidative depolymerization [5, 6] contained OS with concentrations of 44 and 94 g COD/L, respectively.

6.2.1   BIOCATALYST

In this work, acid-producing microorganisms that were isolated as the result of selection work from anaerobic methanogenic sludge taken from an acting anaerobic methanogenic reactor from waste treatment facilities of a plant for the production of crisps (corporation "Frito Lay", Kashira town, Russia) were used. The initial characteristics of anaerobic sludge, as determined according to the procedure described below, are presented in Table 6.2.

TABLE 6.2   Characteristics of the Biocatalyst.

Biocatalyst	Methanogenic activity, g COD*/g AFB**/day	Acid-producing activity, g COD*/g AFB**/day	Dry weight, g/L	Ash content, %	AFB, g/L
Initial	252.3	500.0	57.2	38.0	35.5

Note: *COD chemical oxygen demanded—rate of content of OSs; ** ash-free biomass (AFB) of the substances.

6.2.2   DESCRIPTION OF BIOREACTORS

The selection of acid-producing microorganisms for subsequent conversion of depolymerized biomass was performed in a 1.25-L anaerobic reactor (with an upflow anaerobic sludge blanket (UASB) design), which was 0.5 m in height at two temperature modes: mesophilic (35°C) and submesophilic (22–28°C). A total of 750 mL of anaerobic methanogenic sludge were used as an inoculum for the reactor (two-thirds of the total volume of the reactor). The selection of the biocatalyst was performed under flow cultivation at initial pH values below 5.6, which converted OSs of two model streams containing glucose or serum as a carbohydrate source (Table 6.1).

6.2.3 DETERMINATION OF SPECIFIC ACTIVITIES OF BIOCATALYSTS (INITIAL AND SELECTION)

The measurement of the acid-producing activity of the biocatalysts (initial and selection), calculation of its specific value, and the study of the efficiency and rate of formation of VFA on various substrates were performed in accordance procedures described below at four temperature modes (35, 30, 20, and 15°C).

For this purpose, the 50-mL model stream diluted with mineral medium up to a concentration of OS of 6 g COD/L was introduced to the flasks (120 mL), and 5 mL of an acid-producing biocatalyst was added from the working reactor (9% out of the total volume of the liquid phase); after this, the gas space of the reactor was filled with argon (anaerobic conditions).

Through certain periods measured in each flask:

(1) In a gas phase: the general pressure and concentration of hydrogen, carbon dioxide for confirmation of process of conversion of OS in to VFA.

(2) In a liquid phase: concentration of glucose, ethanol, and VFA.

The specific activity of the biocatalyst (A) was calculated from the tangent of the angle of the linear portion of the curve on the graphs (concentration of substrate (glucose) or products (ethanol, acetic, propionic, and butyric acid)–time) by the following formula:

$$Y = \frac{C \times V_{\text{liquid phase}}}{AFB}$$

where Y is the concentration of substrate or product, g COD/g AFB, C is the concentration of substrate or product, g COD/L, AFB is the ash-free biomass (AFB) of the substances (g), and $V_{\text{liquid phase}}$ is the volume of the liquid phase of the flasks.

6.2.4 KINETIC RESEARCHES

All studies on the bioconversion of the pretreated biomass into VFA and EtOH were performed under batch conditions (periodic cultivation in flasks under stationary conditions) at two temperature modes (35 and 20°C). For this purpose, 50 mL sample under study diluted with mineral medium up to a concentration of OSs of 2.5–8.5 g COD/L was introduced to the flasks (120 mL), and 5 mL of an acid-producing biocatalyst was added from the working reactor (9% out of the total volume of the liquid phase); after this, the gas space of the reactor was filled with argon (anaerobic conditions). The efficiency of the conversion of OSs into EtOH (g/L) and VFAs was calculated according to the equations given in [7].

6.2.5 ANALYTICAL METHODS

The products in the gas phase (hydrogen (H_2), methane (CH_4), and carbon dioxide (CO_2) were analyzed by gas chromatography using an LKhM 8 MD model 3 chromatograph equipped with catarometer (the carrier gas was argon, the flow rate of the carrier gas was 20 cm³/min, the column was 2 m in length and was filled with porapak Q and the temperature of the thermostat of the columns was 50°C). Hydrogen comes after 26 s, methane—42 s, and carbon dioxide—65 s. As the observed chromatographic peaks are quite narrow, so the amount of gas was calculated from the peak height for each assay sample was taken 0.2 cm³ gas.

Analysis of the volatile products in liquid phase ethanol, acetic, propionic, and butyric acids determined chromatograph "GC-15A Shimadzu" with a flame ionization detector on a 1-m long-packed Chromosorb 101. Carrier gas—argon, carrier gas velocity—30 cm³/min. The temperatures of the thermostat of the columns, detector, and evaporator were 190, 210, and 220°C, respectively. The exit time (min) of described above products:

ethanol—1.43, acetic acid—3.45, propionic acid—6.55, and butyric acid—12.35. Observed chromatographic peaks were not always sufficiently narrow, so the amount of volatiles is calculated by an integrator. For each test sample in 1 mK dm^3 liquid phase, chromatograph calibration was performed as follows. The flasks were prepared solutions of ethanol, acetic acid, propionic acid, and butyric acid in a concentration range from 0.01 to 0.1% by volume. From each bottle was taken samples three times in volumes 1 mcL. Volatile component was fixed the amount of analyte on chromatograph on indications integrator. The values obtained were averaged and a calibration graph of the readings from the integrator component concentration was plotted.

The concentration of glucose in the model stream no. 1 was measured using a standard set of reagents (corporation "Impakt", Russia).

The concentrations of OSs in model stream no. 2 and subsequent samples of pretreated biomass (COD, g/L) were determined according to the method [8]. The content of the sum of reducing sugars was determined according to the procedure described [9].

Gas chromatography–mass spectrometry analysis of samples was performed on a set of instruments, including gas chromatograph Trace-1310 and ISQ (innovative single quad—mass spectrometer detector, Thermo Scientific). The samples were analyzed using silica capillary column TR-5MS with a stationary phase crosslinked 5% phenylmethyl-polysiloxane (15 m × 0.32 mm × 0.25 μm). The column temperature was raised from 40°C to 290°C, the temperature ramping rate was 15°C/min, and was held for 5 min at the final temperature. Interface and injector temperature was 250°C. The carrier gas was helium. Volume sample injection—1 μL, split ratio was 1:40. Mass spectra were recorded under electron impact ionization energy of 70 eV, scanning range 30–450 a.m.u. Identification and interpretation of the components of the mixture was carried out by comparing the mass spectra of the chromatographic peaks obtained from mass spectra electronic database NIST-2011 and mass spectra of standard substances.

The experiments with supercritical ethanol were carried out at Kazan State Technological University [3].

Octane number of fuel composition based of commercial gasolines and esters was carried out in I.M. Gubkin Russian State University of fuel and gas [10].

6.3   RESULTS AND DISCUSSION

6.3.1   CREATING OPTIMAL MICROBIAL ASSOCIATION FOR THE CONVERSION OF PRETREATED BIOMASS INTO ETHANOL AND VOLATILE FATTY ACIDS (VFAS)

One effective method for the selection and adaptation of microorganisms is a continuous flow process. Therefore, in order to create the optimal acid-producing microbial community, a UASB anaerobic reactor, which has a relatively simple design and does not require the use of loading materials, was used. Anaerobic methanogenic sludge from acting waste treatment facilities was used as an inoculum for the selection and growth of acid-producing microorganisms in the UASB reactor, as described in Section 6.1.2. The selection of the biocatalyst was performed under flow cultivation at initial pH values below 5.6 in mesophilic (35°C) and submesophilic (26–28°C) modes, which converted OS of two model streams containing glucose or serum as a carbohydrate source. Each mode was maintained up to the achievement of a quasistationary state, that is, up to the achievement of constant values of the efficiency and the rate of conversion of fatty acids with the replacement of more than 3–5 reactor volumes.

The optimal conditions for the preparation of the maximum amount of VFAs on glucose are a pH of 4.7 and a temperature range from 35 to 27°C. The mean efficiency of glucose conversion into VFA (acetic, propionic, and butyric) at both temperature modes (35 and 27°C) was 64%. In this case, the formation of acetic acid was 28 and 25%; propionic acid, 11 and 10%; and butyric acid, 25 and 29%, respectively. Thus, the average yield of butyric acid on glucose at 27°C was higher (1.22 g COD/L; this corresponds to 0.36 mol/mol glucose) than that at 35°C (1.05 g COD/L; this corresponds to 0.31 mol/mol glucose).

The selection of an acid-producing biocatalyst was followed on serum in the same pH ranges, although at reduced (from 27 to 23°C) temperatures of acid production. The results showed the increase in the efficiency of conversion of OS (with the concentration of 7.0 g COD/L) in VFA up to 67%. In this case, the decrease in the degree of conversion (up to 20% OS to acetic acid; up to 10% OS to propionic acid) and increase in the degree of conversion into butyric acid up to 37% OS (this corresponds to a yield of butyric acid of 2.51 g COD/L) was observed.

After selection (77 days) of the initial anaerobic biocatalyst, the specific acid-production activity values were obtained at four temperature modes. The results (Table 6.3) showed the increase of the acid-producing activity of the biocatalyst by 1.31 ± 0.07 times in all cases; this is represented more clearly in Figure 6.1.

TABLE 6.3 Specific Acid-producing Activity (A, g COD/AFB/day) of the Biocatalyst Prior to and After Selection.

Temperature,°C	Specific acid-producing activity of the biocatalyst, g COD/g AFB/day	
	Prior to selection	**After selection**
35	0.50 ± 0.01	0.63 ± 0.01
30	0.42 ± 0.01	0.58 ± 0.01
20	0.29 ± 0.01	0.39 ± 0.01
15	0.20 ± 0.01	0.25 ± 0.01

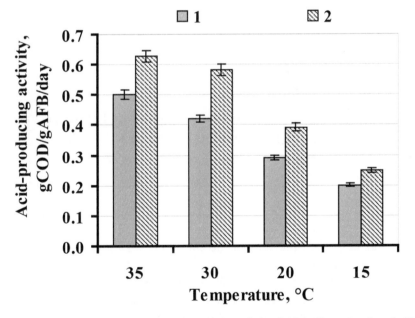

FIGURE 6.1 Specific acid-producing activity of the initial (1) and selected (2) biocatalysts obtained at several temperatures of cultivation.

For the acid-producing biocatalyst, after selection for 77 days, the characteristics were obtained via the same parameters as those for the initial case: AFB of the substances, 41 g/L (increased by 14%); dry weight, 63 g/L (increased by 9%), and ash content, 35% (decreased by 5%).

The kinetics of the formation of VFA under the action of an acid-producing biocatalyst on model streams with glucose and serum at four temperature values is given in Table 6.4. It follows that the maximum efficiency of conversion of OS into butyric acid (51–54%) was observed on both model streams at 15°C; however, longer contact of a biocatalyst with substrate was required (6 days).

TABLE 6.4 Kinetics of Formation of VFA and Ethanol From Glucose and Serum Depending on the Temperature and Time of Contact of Substrates With the Acid Producing Biocatalyst.

Temperature, °C	Time of contact with substrate, days	Effectiveness of conversion into ethanol, %	Effectiveness of conversion into acetic acid, %	Effectiveness of conversion into propionic acid, %	Effectiveness of conversion into butyric acid, %
Substrate is glucose with a concentration of 5.6 g/L (6.0 g COD/L)					
35	4	4.1 ± 0.2	26.6 ± 0.4	7.8 ± 0.2	43.0 ± 0.4
30	4	3.9 ± 0.1	22.3 ± 0.5	6.9 ± 0.2	45.0 ± 0.4
20	10	4.0 ± 0.2	19.1 ± 0.5	6.6 ± 0.1	47.2 ± 0.6
15	16	3.4 ± 0.1	14.4 ± 0.3	6.3 ± 0.1	50.6 ± 0.5
Substrate is serum with a concentration of 4.4 g/L (6.1 g COD/L)					
35	4	5.6 ± 0.3	22.8 ± 0.5	8.5 ± 0.4	44.5 ± 0.5
30	4	4.2 ± 0.2	18.9 ± 0.6	8.1 ± 0.2	48.5 ± 0.4
20	10	3.7 ± 0.2	14.2 ± 0.4	7.8 ± 0.3	51.7 ± 0.4
15	16	2.8 ± 0.1	9.7 ± 0.3	8.3 ± 0.4	53.8 ± 0.6

In this case, the efficiency of the conversion of substrates into acetic and propionic acids was low and corresponded to 14 and 6% (glucose) and 10 and 8%, respectively (serum). With an increase in the temperature of acid production, more intense formation of acetic acid was observed. The formation of propionic acid varied insubstantially, and the efficiency of the formation of butyric acid decreased (Table 6.3), while remaining sufficiently high at 20°C (47–52%). At 30 and 35°C, somewhat of a drop in the efficiency of conversion of substrates into butyric acid and an increase in the efficiency of the conversion of substrate into acetic acid was observed; however, the time of contact of biocatalyst with substrates was only 4 days (Table 6.3). This is confirmed by the values of the calculated specific activity of the formation of VFA under the action of a biocatalyst on glucose and serum, which are given in Figure 6.2.

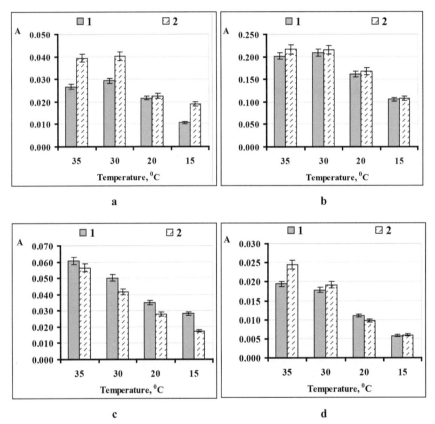

FIGURE 6.2 Calculated specific activity (A, g COD/g AFB/day) of the formation of (a) butyric, (b) propionic, (c) acetic acids, and (d) ethanol from glucose (1) and serum (2) under the action of the biocatalyst.

On glucose and serum, the calculated specific activity values for biocatalyst on the formation of butyric acid were at a maximum, from 0.2016 to 0.2153 g COD/g AFB/day at a temperature of 30–35°C. These results

showed that the conversion of the studied substrates can be conducted
with high efficiency both at 30–35°C and 20°C; however, the period of
conversion and high yield of undesirable side product (acetic acid) should
be considered. The temperature of 20°C with a time of contact of the
substrate with a biocatalyst of 10 days is considered most efficient in order
to obtain the maximum amount of butyric acid from the substrates under
study.

On batch experiments with glucose in flasks yields of ethanol and
VFA at various pH (4.2–6.0) at two temperature regimes were studied
(Fig. 6.3).

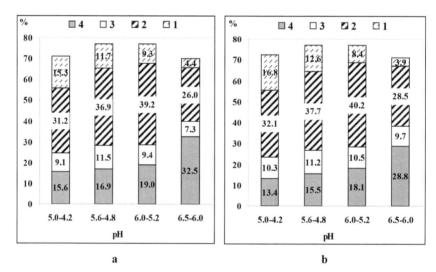

FIGURE 6.3   Effect of pH on the efficiency of ethanol (1), butyric (2), propionic (3), and
acetic (4) acids in two temperature regimes 35°C (a) and 20°C (b) under the action of the
acid-producing biocatalyst.

Maximum production of butyric and propionic acids were observed
at pH 4.8 to 5.2. This pH range is optimal for acid-producing biocatalyst which is capable to carry out the conversion of pretreated biomass
(sawdust, straw, and lignin) in propionic and butyric acid. At the same
time, the highest yield of ethanol is observed at the lowest of tested pH.

6.3.2   BIOCONVERSION PRETREATED LIGNOCELLULOSIC MATERIALS (BIOMASS)

After the end of the selection of biocatalyst, bioconversion was studied at two temperature modes (35 and 20°C) for three samples of pretreated biomass, namely, straw, sawdust, and lignin (see Section 6.1.2), with the use of a high activity acid-producing biocatalyst. Pretreated materials were kindly provided by A.M. Saharov and S.D. Razumovskiy (N.M. Emanuel Institute of Biochemical Physics). Substrates were characterized by different amounts of soluble organic compounds and reducing sugars.

These results are given in Tables 6.5–6.7. The efficiency of the conversion of soluble OS of hydrolyzed straw into butyric acid at 35°C at the sixth day of the experiment, as follows from the results given in Table 6.4, was from 40 to 43%; in this case, no dependence of the formation of VFA (butyric, propionic, and acetic) vs. initial concentration of soluble OS was detected. At 20 C, the effect of the initial concentration of OS on the yield of butyric acid was observed (the initial concentration of soluble OS did not affect the yield of other VFA). At the 10th day of experiment, the efficiency of the formation of butyric acid amounted to 40, 38, and 29% at three concentrations of soluble OS, respectively (2.5, 3.8, and 6.3 g COD/L) (Table 6.5). It should be noted that the concentration of EtOH was nearly 14% in all variants of the experiments (Table 6.5).

TABLE 6.5   Conversion of Hydrolyzed Straw into Ethyl Alcohol and VFA Depending on the Concentration of Substrate, Temperature, and Time of Contact with the Acid Producing Biocatalyst.

Initial concen-tration of soluble OS, g COD/L	Time of contact with the substrate, days	Effectiveness of conversion into ethanol, %	Effectiveness of conversion into acetic acid, %	Effectiveness of conversion into propionic acid, %	Effectiveness of conversion into butyric acid, %
		Temperature of conversion 35°C			
6.3	6	14.8 ± 0.3	22.4 ± 0.3	2.5 ± 0.1	40.6 ± 0.4
3.8	6	14.0 ± 0.3	21.2 ± 0.4	3.5 ± 0.1	41.6 ± 0.5
2.5	6	13.1 ± 0.2	21.4 ± 0.2	4.0 ± 0.2	42.5 ± 0.6
		Temperature of conversion 20°C			
6.3	10	13.5 ± 0.3	22.1 ± 0.2	4.9 ± 0.3	29.3 ± 0.3
3.8	10	13.9 ± 0.2	21.1±0.1	4.5 ± 0.2	38.3 ± 0.4
2.5	10	14.0 ± 0.4	21.5 ± 0.3	4.7 ± 0.3	39.6 ± 0.4

In Table 6.6, the results of the study of conversion of depolymerized sawdust into EtOH and VFA under the action of selected biocatalyst are given.

TABLE 6.6 Conversion of Depolymerized Sawdust into Ethyl Alcohol and VFA Depending on the Concentration of Substrate, Temperature, and Time of Contact with an Acid Producing Biocatalyst.

Initial concen- tration of soluble OS, g COD/L	Time of contact with the substrate, days	Effectiveness of conversion into ethanol, %	Effectiveness of conversion into acetic acid, %	Effectiveness of conversion into propi- onic acid, %	Effectiveness of conversion into butyric acid, %
		Temperature of conversion 35°C			
8.0	14	1.5 ± 0.1	18.8 ± 0.3	4.2 ± 0.2	18.7 ± 0.4
6.4	14	2.4 ± 0.1	21.3 ± 0.4	5.8 ± 0.1	20.0 ± 0.4
4.0	14	3.5 ± 0.2	23.9 ± 0.2	6.6 ± 0.1	22.4 ± 0.3
		Temperature of conversion 20°C			
8.0	18	1.4 ± 0.1	13.7 ± 0.2	3.3 ± 0.1	11.3 ± 0.3
6.4	18	1.8 ± 0.1	15.5 ± 0.3	4.2 ± 0.2	13.3 ± 0.3
4.0	18	3.0 ± 0.2	19.6 ± 0.4	6.2 ± 0.3	17.1 ± 0.4

The efficiency of the conversion of total OS of depolymerized sawdust into butyric acid at 35°C on the 14th day of experiment was from 19 to 22%; in this case, an increase in the formation of VFA was observed (butyric, propionic, and acetic) with a decrease in the initial concentration of OS. At 20°C, an inverse dependence of the yield of all three volatile fatty acids (VFAs) and ethanol vs. the initial concentration of OS was observed. At the 18th day of the experiment, the efficiency of the formation of butyric acid at 17, 13, and 11% was achieved at three concentration values (4.0, 6.4, and 8.0 g COD/L) of soluble OS.

The biocatalyst-mediated conversion of depolymerized lignin into VFA is given in Table 6.7.

The efficiency of the conversion of total OS of depolymerized lignin into butyric acid at 35°C at the 14th day of experiment was from 13 to 19% (Table 6.7); in this case, the inverse dependence of the formation of VFA (butyric, propionic, and acetic) vs. the initial concentration of OS was observed. The identical pattern of the yield of all three VFAs was observed at 20°C; however, ethanol did not form in both cases. At the 18th day of experiment at 20°C and a concentration of soluble OS corresponding to 4.3, 6.8, and 8.5 g COD/L, the efficiency of the formation of butyric acid of 13, 10, and 8%, respectively, was achieved (Table 6.6).

TABLE 6.7 Conversion of Depolymerized Lignin into Ethyl Alcohol and VFA Depending on the Concentration of Substrate, Temperature, and Time of Contact with an Acid Producing Biocatalyst.

Initial concentration of soluble OS, g COD/L	Time of contact with the substrate, days	Effectiveness of conversion into ethanol, %	Effectiveness of conversion into acetic acid, %	Effectiveness of conversion into propionic acid, %	Effectiveness of conversion into butyric acid, %
		Temperature of conversion 35°C			
8.5	14	0	6.1 ± 0.1	3.4 ± 0.1	13.3 ± 0.2
6.8	14	0	9.0 ± 0.2	4.7 ± 0.2	16.1 ± 0.3
4.3	14	0	12.5 ± 0.2	5.9 ± 0.3	19.0 ± 0.3
		Temperature of conversion 20°C			
8.5	18	0	5.1 ± 0.1	2.0 ± 0.1	7.6 ± 0.1
6.8	18	0	6.0 ± 0.2	2.8 ± 0.1	10.1 ± 0.2
4.3	18	0	9.9 ± 0.4	4.1 ± 0.2	13.0 ± 0.2

Thus, these studies showed that the most promising biomass to prepare VFA from agricultural wastes is pretreated straw. The maximum yields (up to 68%) of total VFA were obtained on it at 35°C and a period of 6 days under the action of an acid-producing biocatalyst. On depolymerized sawdust and lignin, the maxi mum yields of total VFA were 53 and 37%, respectively, at an analogous temperature and a period of the process of 14 days. These results were obtained due to preliminary selection in a UASB reactor of an acid-producing biocatalyst (first on glucose and then on serum), which allowed an increase in the acid-producing activity of the biocatalyst by 1.31 ± 0.07 times. The conversion of agricultural biomass can occur with high efficiency at both 30–35 and 20°C; however, in this case, a period of conversion and high yield of a side product, namely, acetic acid, should be assumed.

The results obtained, presented in more detail in [9], can be used in the organization of production of VFA and ethanol from waste products of agriculture and other renewable raw materials.

6.3.3   THE SEPARATION OF VFA AND ETHANOL FROM WATER SOLUTIONS—CONDENSATION TO ETHYLCARBOXYLATES

VFA and ethanol in biocatalytic process exist as low concentration water solutions [11]. The direct separation of products from large amount of water with distillation is energy-intensive process. Economic effect of recycling is largely determined by the efficiency of the separation of solutions and subsequent transformation into carboxylate for use mainly in the high-octane fuel compositions or compositions for modification of motor fuel based biodiesel. The most rational way to release acids is extraction with a water immiscible organic solvent. To this end the screening study a number of possible available extractants—aromatic hydrocarbons (benzene, toluene), esters and ethers (diethyl ether, dibutyl ether, diethyl carbonate, diethyl malonate), esters with additives tributyl or tributylamine (which are known reflect enhance the extraction efficiency), halogenated hydrocarbons (methylene chloride, chloroform, and carbon tetrachloride). Most of the reduced number of solvents exhibit a significant degree of flooding, so to completely remove the absorbed water requires the use of drying agents (salts, effectively absorbing water, and molecular sieves). It proved to be the best solvent CCl_4, as it provides sufficient isolation completeness of acids released easily from the residual water by azeotroping about 10% of the total extract and does not require special desiccant. It can also be carried out with lower alcohols esterification. Liberated during esterification water forms a separate phase that shifts the balance of the process in the product side.

The acid used as a homogeneous catalyst is removed from the aqueous phase, so the process does not require the neutralization step and the acid can be reused. Furthermore, it was found that the use of CCl_4 allows joint selection of VFA and ethanol. This is illustrated by the example of model binary solutions—ethanol – propionic acid and ethanol – butyric acid with 3 and 5% of the content of both components. Composition and completeness of the extraction were measured by NMR. The alcohol and acid are no longer detected spectrally in the aqueous phase after three times of extraction of 100-mL solution with portions of 10 mL of the extractant. After drying the extract and removing the extractant, extraction completeness was determined. It was $95 \pm 3\%$. Dried extracts without removing the extractant used for the condensation of the alcohol—VFA in the presence of acidic catalysts (amberlyst, sulfuric acid, and toluenesulfonic acid).

During the process, there is separation of the aqueous and organic phases, shifting the equilibrium toward carboxylate.

However, a more attractive option seems esterification of VFA with ethanol in formation conditions of supercritical fluid (SCF). SCF process performed on the example of butyric acid in the mode with the following parameters: temperature −350°C, pressure 300 bar, the processing time—10 min, the ratio of alcohol–acid—1:1. Yield of ethylbutyrate—45%.

Esters VFA which detect significant effect, increasing octane number in gasoline (3–6 pcs, Table 6.8), can also be seen as an easy analog of traditional biodiesel produced by alcoholysis of triglycerides.

TABLE 6.8   The Results of Investigations of Ethylpropionate (Ester 1) and Ethylbutyrate (Ester 2) in Fuel Compositions.

Name	Gasoline, mass%	Additive, mass%	Octane number	Increasing
		Ester 1		
Gasoline 1—Ester 1	90	10	79.5	3.5
	85	15	81.5	5.5
		Ester 2		
Gasoline 1—Ester 2	90	10	79.5	3.5
	85	15	81.0	5.0
		Ester 1		
Gasoline 2—Ester 1	90	10	93.9	1.9
	85	15	95.1	3.1
		Ester 2		
Gasoline 2—Ester2	90	10	93.5	1.5
	85	15	95.0	3.0

Note: Gasoline 1—octane number 76 (research method); Gasoline 2—octane number 92 (research method).

For diesel fuel, important parameter is the kinematic viscosity of which does not exceed certain limits. Viscosity control using esters of VFA as additives may be one of the directions of their use in the fuel industry.

The possibility of direct synthesis of fuel compositions with SCF method was tested by the example of the alcoholysis of rapeseed oil in a mixed media of acetone (for binding of glycerol)—butyric acid—the excess of the supercritical ethanol. Homogeneous mixture is obtained, which by gas chromatography–mass spectrometry in addition to the expected ethyl butyrate, ethyl esters of rapeseed oil acids, 2,2-dimethyl-4-hydroxy-methyl-1,3-dioxolane,   2-methyl-4-hydroxymethyl-1,3-dioxolane,   and mono-isopropyl and monoethylglycerol ether, dihydroxymethyldioxane were registered.

Detected products indicate that the triglyceride in a SCF-conditions undergoes complete transformation—alcoholysis ketalization—in the absence of catalyst. Availability of monoesters and acetale glycerol indicates the exchange redox-interaction between the alcohol and acetone to form acetaldehyde and iso-propyl alcohol that appeared in the products of their subsequent transformations. These processes appear to be specific for SCF conditions. In the future, they can serve as the basis for new technologies for processing of renewable raw materials.

6.4 CONCLUSIONS

Biodiesel and bioethanol are two main types of liquid biofuels produced in large industrial scale. Development of new approaches to their synthesis, removing a number of not fully resolved problems, is an urgent task.

Ethanol together with VFA—acetic, propionic, and butyric—produces as a result of biocatalytic processing plant of polysaccharide raw material, preferably waste agriculture and wood industry (straw, sawdust, etc.).

The process is implemented using specialized microbial associations (biocatalyst). A new biocatalyst is created as a catalytic system consisting of a consortium of free and immobilized cells acidogenic microorganisms that can withstand considerable loads on the organic matter in the flow reactor.

A kinetic study of biocatalytic systems and the optimal bioprocess parameters (temperature, pH of the medium, and the concentration of the original biosubstrate) was carried out.

New methods alcoholysis VFA were supposed to go to the carboxylate (easy analogue biodiesel) with a focus on the use of SCF technology and application of the results in the synthesis of new types of liquid biofuels.

KEYWORDS

- **methanogenesis**
- **selection**
- **biocatalyst**
- **biomass**
- **bioprocess**
- **biodiesel**
- **supercritical fluids**
- **carboxylates**

REFERENCES

1. Varfolomeef, S. D.; Kaluznii, S. V.; Medman, I. A. Chemical Bases of Producing Fuels Biotechnology. *Russ. Chem. Rev.* **1988**, *57*, 1201–1231 (In Russian).
2. Varfolomeef, S. D.; Efremenko, E. N.; Krylova, L. P. *Biofuels Russ. Chem. Rev.* **2010**, *79*, 544–564 (In Russian).
3. Usmanov, R. A.; Gabitov, R. R; Bektashev Sh. A.; Shamsutdinov, F. M.; Gumerov, F. M.; Gabitov, F.R, Zaripov, Z. I.; Gasisov, R. A.; Yarulin, S. S. Pilot Unit for Continuous Transesterification of Vegetable Oils in Supercritical Methanol and Ethanol Media. Supercritical Fluides. *Theory Pract.* **2011**, *6*(3), 45–61 (In Russian).
4. Kusnetsov B. N. Catalytic Chemistry of Plant Biomass. *Soros Educ. J. Chem.* **1996**, *12*, 47–57 (In Russian).
5. Skibida, I. P.; Aseeva, R. M.; Saharov, P. A.; Saharov, A. M. Fire Retardant, its Production Method, A Method of Flame-Retardant Materials and the Processing Method of Extinguishing the Hearth Burning. Patent RU # 2204547 C1 Published 20.05.2003 Bull # 14(2), 2003 (In Russian).
6. Varfolomeef, S. D.; Lomakin, S. M.; Gorshenev, V. N.; Saharov, P. A.; Saharov, A. M.; Demin, V.L. Flame Retardant, Process For its Preparation, Method of Fire-Retardant Treatment Materials and Method of Extinguishing the Fire Burning. Patent RU # 2425069. Published 10.02.2011. Bul. # 21, 2011 (In Russian).
7. Gladchenko, M. A. The Development of Biotechnological Methods of Utilization of Winary Production Wastes. PhD Thesis, Moscow, 2001 (In Russian).
8. Dubber, D.; Gray, N. F. Replacement of Chemical Oxygen Demand (COD) with Total Organic Carbon (TOC) for Monitoring Wastewater Treatment Performance to Minimize Disposal of Toxic Analytical Waste". *J. Environ. Sci. Health, Part A: Toxic Hazard. Subst. Environ. Eng.* **2010**, *45*(12), 1595–1600.
9. Klesov, A. A.; Rabinovich, M. L.; Sinitsin, A. P. Fermentative Hydrolysis of Cellulose. *Bioorg. Chem.* **1980**, *6*(8), 1225–1242 (In Russian).

10. Emelianov, V. E.; Skvortsova, V. N. *Motor Fuels: Anti-Knock Properties and Flammability*. Thechnics: Moscow, 2006; pp. 191 (In Russian).
11. Gladchenko, M. A; Gaydamaka, S. N.; Murygina, V. P.; Varfolomeef, S. D. The Optimization of the Conversion of Agricultural Waste Into Volatile Fatty Acids into Anaerobic Conditions. *Moscow State Univ. Vestnik Chem.* **2014,** *55*(4), 241–248.

CHAPTER 7

MECHANISM OF AMMONIA IMMOBILIZATION BY PEAT AND OBTAINING OF PEAT-BASED SORBENT

ALEXANDER R. TSYGANOV[1], ALEKSEY EM. TOMSON[1],
VICTOR P. STRIGUTSKIY[1], VICTORIYA S. PEHTEREVA[1],
TAMARA V. SOKOLOVA[1], SVETLANA B. SELYANINA[2*],
MARINA V. TRUFANOVA[2], and TAMARA IG. PONOMAREVA[2]

[1]*Institute of Nature Management of National Academy of Sciences of Belarus, 10, Francisk Scoriny St., Minsk 220114, Republic of Belarus*

[2]*Federal Center for Integrated Arctic Research of Russian Academy of Science, 23, Embankment of the Northern Dvina, Arkhangelsk 163000, Russia*

*Corresponding author. E-mail: smssb@yandex.ru

CONTENTS

ABSTRACT

Among the renewable phytogenous resources, peat is a unique natural formation, which combines properties of the peat-forming plants and products of their biodegradation. It is humic substances that determine the main properties of peat. It is considered that the sorption properties of humic substances are caused, only, by the presence of ion-exchange groups, primarily carboxyl. However, the amount of ammonia gas adsorbed by peat exceeds its ion-exchange capacity by several times. Mechanism of extra-equivalent adsorption was studied in this paper. This investigation enables to develop the theoretical foundations of directed peat modification and production of highly efficient sorbents for the purification of air on its base. The high-moor peat has been consistently disassembled into individual organic components, which have been studied by IR- and EPR-spectroscopy methods. Extra-equivalent ammonia sorption of peat due to energetically favorable formation of bridging hydrogen and the donor–acceptor bonds by its molecules with polyconjugated fragments was shown. Scientific base of obtaining of peat-based sorbents for the purification of air was developed.

7.1   INTRODUCTION

Complex utilization of renewable sources of organic matter is of great interest at the present time. Fields of manufactured or worked-out products' application are exceedingly diverse. One important aspect is to obtain sorbents based on plant porous matrices. In particular, bark, cardboard, soil, and peat are cheap adsorbing materials, which allow creating sufficiently effective and inexpensive filtering devices for air purification. Among the renewable plant resources, peat is considered to be a unique natural formation, which combines properties of the plants and its biodegradation products due to the specific hydrological conditions. Peat is characterized by great diversity and lability of organic compounds. Chemical group composition of peat is presented by the same groups of compounds as in plant tissues: extractives (also called peat bitumen), water-soluble, easy- and hard-hydrolyzed substances, and lignin. In addition, a number of specific products of biodegradation, called humic substances, are an integral part of peat organic matter. Humic substances (mainly humic acids) have a considerable influence on the basic properties of peat. It is considered in the literature, that the sorption properties of humic substances are caused, exclusively, by the presence of ion-exchange groups, primarily carboxyl [1, 2]. However, according to [3], the amount of ammonia gas adsorbed by peat exceeds its ion-exchange capacity by several times.

Therefore, investigation of mechanism of extra-equivalent adsorption allows us to develop the theoretical basics for directed modification of peat and thus producing of highly efficient sorbents for the purification of air on its base.

7.2 MATERIALS AND METHODOLOGY

The study was conducted on the samples of high-moor peat (cotton grass peat with a degree of decomposition of 40–45%), its metal-substituted forms—Ca, Cu, and Fe forms—obtained by substituting hydrogen ions of carboxyl groups with metal ions, as well as samples of consistently disassembled initial peat matrix, that is, organic matter was sequentially deprived of individual components of by the method of Instorf [4]. Sample of humic acids was extracted separately. Exchange capacity of peat and humic acids samples was determined by standard method, using calcium acetate [5]. The total nitrogen content in the initial peat sample was 0.9% (0.6 mEq/g), in humic acids—0.4% (0.3 mEq/g).

The synthetically fibrous cation-exchanger FIBAN K-4 (polypro-pylene-grafted-acrylic acid), manufactured by Institute of Physical Organic Chemistry of National Academy of Sciences of Belarus, was taken as the reference sample. Ion-exchange centers of FIBAN K-4 are carboxyl groups, the same in as peat, but there are no polyconjugation systems in it [6]. Sorption capacity of fiber was determinated according to GOST 20255.1-89. Initial fiber was treated with 0.1-N hydrochloric acid (in the ratio of 1 l of the solution to 1 g of fiber) for activation of ion-exchange centers.

Samples of high-moor peat (cotton grass peat with a degree of decomposition of 40% and ash content of 2.2%) and low-moor peat (sedge peat with a degree of decomposition of 20% and ash content of 5.8%) were selected as objects for developing environment-friendly technology of peat treatment, which allows to increase the absorption of ammonia and to improve the technological characteristics of the sorbent.

Modified sorbents were obtained by extrusion, by pushing the peat through a spinneret with 3 mm cross-section, followed by drying and separation. Peat was refluxed by 10^{-4} mol/l and 10^{-2} mol/l acids solutions with ratio of solid phase to the liquid 1:2.5 and 1:5 as a pretreatment. Determination of processing characteristics was carried out according to standard methods [4, 7].

Air-dry samples (the moisture content 11.0–13.6%) were ammoniated in dynamic conditions using equipment, reported in [8]. The amount of adsorbed ammonia was estimated by determining the total nitrogen content in the initial peat samples and ammoniated samples by the Kjeldahl method [9]. The strength of ammonia fixing in the samples was determined by drying at 105°C for 1 h.

The goal was solved by the methods IR- and EPR-spectroscopy methods. Method of EPR is highly informative because of paramagnetism of peat and humic acids, caused by energy gain during the formation of supramolecular associates due to the synergistic effect of the interaction of aromatic segments and hydrogen bonds [10], but not due to classic free radicals of a different nature, stabilized by rigid matrix, as it is considered to this day by a wide range of researchers [11, 12]. It must be emphasized that the unpaired electron in these associates is delocalized on a number of molecules, in contrast to the classical free radical in which it lies within the isolated molecular fragments. It determines both the unusual stability of the EPR signal and a high sensitivity of its parameters to the structure of polyconjugated systems.

Strictly speaking, delocalization of spin density, the characteristic of the low-dimensional structures, takes place in these paramagnetic associates, but not the migration of the unpaired electron as a particle [13]. That last feature gives to the unpaired electron function of "molecular reporter."

Infrared spectra have been acquired in a FTIR spectrophotometer "Specord-M80" using the KBr pellets technique. Methods of recording and processing of EPR spectra are shown in [14].

7.3   RESULTS AND DISCUSSION

The infrared spectra of original peat presented characteristic for high-moor peat bands on: 3400 cm^{-1}—ν OH…, 2930–2850 cm^{-1}—ν CH$_2$, CH$_3$, 1715 cm^{-1}—ν C=O of acids, 1610 cm^{-1}—ν C=C of polyconjugated aromatic systems, 1510 cm^{-1}—ν C=C of monoaromatic structures, 1040 cm^{-1}—ν C–OH of alcohols and carbohydrates. The same is for humic acids: 3400–3200–2600 cm^{-1}—ν OH…, ν NH…, ν OH… of acids dimers, 2920–2850 cm^{-1}—v CH$_2$, CH$_3$, 1720 cm^{-1}—v C=O of acids, 1630 cm^{-1}—v C=C of polyconjugated aromatic systems, 1250–1200 cm^{-1}— v C–O of acids and phenols.

The infrared spectra of both the ammonia-treated peat and humic acids presented a significant shrinking of band attributed to free carboxyl groups—1720 cm^{-1}, with the appearance of bands on: 3200 cm^{-1}—v NH..., 1410 cm^{-1}—v NH$_4^+$, 1600 cm^{-1} and 1400 cm^{-1}—v COO$^-$ and with sharp intensification of bands 3400, 1640, 600 cm^{-1} attributed to stretching and deformation vibrations of water OH groups. That indicates the formation of ammonium humates.

EPR-spectra parameters of investigated peat and humic acids samples (Table 7.1) are typical for high-moor peats and testify to the substantial absence of nitrogen in its aromatic polyconjugated systems. EPR signal of FIBAN K-4 sample is not registered because of the absence of polyconjugation systems.

Ammonia treatment leads to substantial intensification of EPR signal accompanied by its broadening and reducing of signal saturation with SHF power (increasing A/A_0 parameter), in the case of humic acids (see Table 7.1). The parameters of EPR spectra of ammoniated peat samples indicate the hyperfine interaction of unpaired electron with magnetic core of nitrogen [14].

Paying attention to alkaline properties of ammonia [16], this fact can, seemingly, be explained by production of semiquinone anion radicals [17, 18].

However, it is shown in [19] that treatment of humic acids with NH$_4$OH solution reduces the intensity of the EPR signal by times, that is, the effect of removing the hydrogen bonds by monovalent NH$_4^+$ cations [20] prevails over the alkaline nature of the solution. Moreover, the simulated EPR spectra of peat-treated samples by the means of the original signal and the signal of semiquinone radicals' superposition are not consistent with the experimental spectra. The parameters of the semiquinone radicals signal have been set previously for the cotton grass peat (ΔH = 5.5–6 G and g-factor = 2.0042–2.0043 [17]). In this case, it was expected that the ammonia treatment of the samples would not lead to a significant change of the spectrum width (the distance between the extreme points ΔH), and increase in the level of SHF power would cause its significant broadening [18]. It is interesting that the washing with distilled water does not lead to a significant reduction of the paramagnetic centers concentration, as

TABLE 7.1   Influence of Ammonia Adsorption on the EPR Spectra of Peat [15].

Sample	$\Delta H_{0.1\,mW}$, (G)	$\Delta H_{50\,mW}$, (G)	g-Factor	I (10^{17} spin/g)	A/A_0
Initial peat	3.8	3.9	2.0031	5.5	3.52
Peat treated with ammonia	5.4	5.6	2.0035	12	3.09
Peat treated with ammonia and washed	5.2	5.2	2.0034	10	3.15
Peat without bitumen	3.6	4.2	2.0035	5.2	3.60
Peat without bitumen, treated with ammonia	4.7	4.9	2.0037	10.5	3.12
Peat without bitumen and water-soluble substances	3.8	3.9	2.0031	9.2	3.25
Peat without bitumen and water-soluble substances, treated with ammonia	5.2	5.4	2.0034	13	2.74
Peat without bitumen, water-soluble and easy-hydrolyzed substances	3.3	2.8	2.0028	7.6	3.42
Peat without bitumen, water-soluble and easy-hydrolyzed substances, treated with ammonia	4.2	3.1	2.0034	9.8	2.46
Humic acids	2.9	3.6	2.0033	7.8	1.58
Humic acids, treated with ammonia	4.3	5.4	2.0036	13	2.69

Note: A/A_0—Amplitude ratio of signal, shot at 50 to 0.1 mW; ΔH—quantity that characterizes the width of the EPR signal. This value is measured in units of magnetic field—gauss (G); g-factor—a factor that connects a gyromagnetic ratio of the particles with the classical value of the gyromagnetic ratio. For classical particles, g-factor is 1; for free quantum particles with nonzero spin, these value is 2; for real particles experimentally certain value g-factor may be different from both 1 and 2 and is one of the characteristics of the particles; I—the concentration of free radicals, which characterizes the intensity of the EPR-signal. The I value is measured in relative units (spin/g).

well as return the parameters of EPR spectrum to the original sample (see Table 7.1). The intensity of the EPR signal of ammonia in the ammonia-treated peat decreased by two times, whereas the narrowing of the signal is only for 1 G, after 16 days. The intensity and parameters of the signal are practically identical with the same in the untreated sample, after a year of storage. What is more, the significant quantity of carboxyl groups stay in the substituted form. In our opinion, the behavior of the EPR spectra indicates that in peat treated with ammonia, polyconjugation systems enriched with nitrogen are also formed, along with the ammonium humates [14].

Restoring the EPR signal after long-term storage does not give a reason to state that the formation of the classical polyconjugated molecular fragments enriched with nitrogen is occurring. Taking into account the ability of ammonia to form donor–acceptor and hydrogen bonds [16], we can assume that the ammonia molecule, penetrating between the macromolecules, forms intramolecular cross-linking hydrogen and donor–acceptor bonds between them, for example, due to interactions with the hydroxyl, ketone, aldehyde and other oxygen-containing groups. The proposed explanation is confirmed by the appearance of band of hydrogen bonds NH at 3200 cm^{-1}. It is known that hydrogen bonds contribute to the "activation" of polyconjugation systems [21]. As shown in [22], it is hydrogen bonds that determine the high level of humic substances paramagnetism. Considering that the unpaired spin of paramagnetic centers, formed as a result of additional introducing of hydrogen bonds, is localized partially on the nitrogen atom, its hyperfine interaction with nitrogen magnetic core explains the difference between the parameters of the EPR-spectra of the treated and original peat samples. In fact, the effect of ammonia molecules on the humic acids' paramagnetism is similar to the effect of nonparamagnetic polyvalent metal cations [20]. In this case, the removal of metal ions results to the recovery of the humic acids EPR-spectra parameters back to the initial state.

This energy gain due to amplification of polyconjugation degree, resulting from the increase in the concentration of paramagnetic centers [23], explains the extra equivalent adsorption of ammonia (Fig. 7.1).

For the same reason, even after a rather harsh drying (105°C), the remaining ammonia content exceeds the exchange capacity (Fig. 7.1), despite the appearance of the free carboxyl groups band on 1720 cm^{-1}. Disappearance NH_4^+ band (1410 cm^{-1}), while maintaining the bands on 1400 and 1600 cm^{-1}, indicates the transition of the main share of the carboxyl groups to the amide form [24].

Experiments with metal-substituted forms of peat were performed to determine the possible role of the ash component of peat in ammonia sorption (Table 7.2). FTIR and EPR-spectroscopy data indicates replacing of hydrogen ion by metal ions in the main part of the functional groups (see Tables 7.1 and 7.2), which should prevent the formation of ammonium humates. Indeed, in the samples of metal-substituted forms of peat treated with ammonia, a clear appearance of NH_4^+ band (1410 cm^{-1}) is not observed. However, in this case, the sorption of ammonia also occurs, that is, Cu and Fe forms are slightly higher than for nonsubstituted forms (see Fig. 7.1).

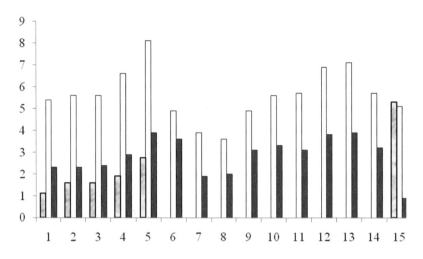

FIGURE 7.1   Sorption of ammonia by peat, its compounds and cationic forms.

1—initial peat; 2—peat without bitumen; 3—peat without bitumen and water-soluble substances; 4—peat without bitumen, water-soluble and easy-hydrolyzed substances; 5—humic acids of peat; 6—Ca form of initial peat; 7—Ca form of peat without bitumen; 8—Ca form of peat without bitumen and water-soluble substances; 9—Ca form of peat without bitumen, water-soluble and easy-hydrolyzed substances; 10—Cu form of initial peat; 11—Cu form of peat without bitumen; 12—Cu form of peat without bitumen and water-soluble substances; 13—Cu form of peat without bitumen, water-soluble and easy-hydrolyzed substances; 14—Fe form of initial peat; 15—FIBAN K-4.

It is considered in the literature that first NH_4^+ ions react with free functional groups and then enter the exchange reaction with cations of the substituted groups, during the saturation of the metal-substituted samples of peat with ammonia gas [1]. However, our FTIR- and EPR-spectroscopy data have not confirmed this point of view. Indeed, the treatment of metal-substituted forms of peat provides a further weakening of the band on 1720 cm⁻¹, but NH_4^+ band does not appear on 1410 cm⁻¹. The results of the EPR-spectroscopy indicated not a weakening, but rather strengthening of bonds of metal cations with organic matter. Increasing of the paramagnetic centers concentration in the ammonia treated Ca-form samples along with broadening of spectra and increasing of g-factor indicate additional cross-linking bonding between functional groups by Ca^{2+} ions [20].

TABLE 7.2 Ammonia Adsorption Influence on the EPR Spectra of Metal-Substituted Forms of Peat [15].

Sample	ΔH, G		g-Factor		I, 10^{17} spin/g	A/A$_0$
	0.1 mW	50 mW	0.1 mW	50 mW		
Ca form of initial peat	5.0	5.4	2.0037	2.0033	13.1	2.37
Ca form of peat, treated with ammonia	6.1	6.3	2.0041	2.0039	40.2	2.77
Ca form of peat without bitumen	4.6	4.4	2.0036	2.0027	11.3	2.76
Ca form of peat without bitumen, treated with ammonia	5.7	5.9	2.0038	2.0036	21.3	3.50
Ca form of peat without bitumen and water-soluble substances	5.2	5.5	2.0048	2.0035	17.8	2.52
Ca form of peat without bitumen and water-soluble substances, treated with ammonia	5.9	6.2	2.0037	2.0036	31.8	2.53
Ca form of peat without bitumen, water-soluble and easy-hydrolyzed substances	4.8	2.5	2.0035	2.0037	20.8	2.21
Ca form of peat without bitumen, water-soluble and easy-hydrolyzed substances, treated with ammonia	6.3	3.6/5.3/10	2.0037	2.0030	25.9	2.19
Cu form of initial peat	4.1	3.3	2.0030	–	0.16	4.50
Cu form of peat, treated with ammonia	Not available					
Cu form of peat without bitumen	3.8	3.5	2.0030	–	1.6	4.72
Cu form of peat without bitumen, treated with ammonia	4.5	3.4	2.0031	–	1.4	4.27
Cu form of peat without bitumen and water-soluble substances	3.8	3.5	2.0030	–	0.77	6.53
Cu form of peat without bitumen and water-soluble substances, treated with ammonia	Not available					
Cu form of peat without bitumen, water-soluble and easy-hydrolyzed substances	3.0	3.0	2.0028	2.0028	1.5	9.65

TABLE 7.2 (Continued)

Sample	ΔH, G		g-Factor		I, 10^{17} spin/g	A/A_0
	0.1 mW	50 mW	0.1 mW	50 mW		
Cu form of peat without bitumen, water-soluble and easy-hydrolyzed substances, treated with ammonia	2.7	1.8	2.0029	–	0.085	5.34
Fe form of initial peat	4.4	4.2	2.0031	–	1.8	4.85
Fe form of peat, treated with ammonia	4.7	5.0	2.0031	–	1.3	4.83

Note: A/A_0—amplitude ratio of signal, shot at 50 and 0.1 mW; ΔH—quantity that characterizes the width of the EPR-signal. This value is measured in units of magnetic field— – gauss (G); g-factor—a factor that connects a gyromagnetic ratio of the particles with the classical value of the gyromagnetic ratio. For classical particles, g-factor is 1; for free quantum particles with nonzero spin, these value is 2; for real particles experimentally certain value g-factor may be different from both 1 and 2 and is one of the characteristics of the particles; I—the concentration of free radicals, which characterizes the intensity of the EPR-signal. The I value is measured in relative units (spin/g).

The well-known fact of incomplete use of valences by a significant part of polyvalent metal ions adsorbed by humic substances should be taken into account for explanation of obtained results [25]. The part of Ca^{2+} ions is bonded with ion-exchange groups only by one valence not forming cross-linking bondings between the molecular fragments. Consequently, they do contribute not to the increasing, but to the decreasing of the concentration of paramagnetic centers, while maintaining signal parameters [20]. Cross-linking hydrogen and donor–acceptor bonds, mentioned above, formed by ammonia molecules, contribute to the convergence of molecular fragments of peat organic matter, and thus create steric conditions for the full implementation of valences by the Ca^{2+} ions. Another factor contributing to the interaction of the free valences of Ca^{2+} ions with functional groups is increasing of the pH of the samples after ammonia treatment of peat [1]. It is obvious that the interaction of monovalent ammonium cations with the free carboxyl groups and their exchange with divalent calcium cations would have the opposite effect.

Treatment with ammonia of Cu- and Fe-substituted forms of peat leads to decreasing of the EPR-signal intensity of organic paramagnetic centers that is explained by more comprehensive implementation of valences by cations and is a forcible argument against the formation of semiquinone radicals. Obviously, there will be a reduction of the effective (average) distance between the paramagnetic ions and organic paramagnetic centers. In a view of strong dependence of their interaction from the distance [26–28], slight changes in the topology of relative positions of ions and organic paramagnetic centers induce further reducing of signal of latter in Cu- and Fe-substituted samples during ammonia sorption (Table 7.2), although the ion content is not changed. It should be noted that the narrowing of the EPR spectrum of Cu^{2+} ions during ammonia sorption is a direct evidence of the energy amplification of their interaction with the carboxyl groups [29].

Substantial increasing of signal intensity in the low-field (g-factor = 4) is observed for the Fe-substituted sample, due to the Fe^{3+}-compounds with a purely ionic bonding [28]. Due to the peculiarities of its topology, the first have a strong influence on the EPR spectrum of organic matter in comparison with the iron compounds that are responsible for a much more intensive broad spectrum in the range of g-factor = 2 [27]. Obviously, the substitution of the Cu^{2+} and Fe^{3+} paramagnetic ions for ammonium ions should lead to increasing of signal intensity of organic paramagnetic centers, and broadening and symmetrization of the Cu^{2+} ions spectrum, as well as the disappearance of the Fe^{3+} ions signal in low fields (g-factor = 4). In our opinion, peculiarities of the supramolecular structure transformation of the peat organic matter under sorption of Cu^{2+} and Fe^{3+}, consisting of the convergence of its molecular fragments, are so favorable for its physicochemical interaction with ammonia that it compensates the "loss" of ion-exchange groups.

Thus, it was found that the extra equivalent sorption of ammonia by peat is defined by energetically favorable formation of cross-linking hydrogen and donor–acceptor bonds with polyconjugation fragments by its molecules. Treatment of metal-substituted forms of peat with ammonia gas does not lead to the substitution of the metal cations associated with the functional groups by ammonium ions, but, conversely, lead to amplification of bonding. Transformation of the supramolecular structure under the sorption of Cu^{2+} and Fe^{3+} ions contributes to the formation of the bonds, mentioned above, which compensates the "loss" of ion-exchange groups.

Assignment of the essential role of the peat polymer matrix, formed by aromatic polyconjugation systems, served as the basis for the development of environment-friendly technology of peat treatment by weak acids (phosphoric, citric, and oxalic acid), which does not destroy the structure of peat organic matter. This technology allows to increase the ammonia sorption and to improve the technological characteristics of the sorbent.

The concentration of sorbed ammonia substantially exceeds the ion-exchange capacity (1.2 mEq/g for cotton grass peat, 0.6 mEq/g for sedge peat), as it seen from the data in Figure 7.2. Soft acid treatment of peat increases the uptake of ammonia at 6.4–39.7%. Moreover, sorption activity grows up with increasing of acid concentration and the fluid module.

At first sight, these results seem to be quite trivial because the technology of treatment of organogenous ammonia sorbents with mineral acids is well known for increasing of the content of ion-exchange groups [3, 31].

However, the calculations show that the concentration of introduced functional groups substantially lower of amplification effect of sorption activity. Let us consider, for example, the situation occurs at peat treatment with oxalic acid. It is obvious that the highest concentration of dopant carboxyl groups (supposing there is full implementation of the acid molecules in the organic matrix of peat) is provided by using a 10^{-2} mol/l solution with a modulus of 1:5, and the maximum possible amount of introduced

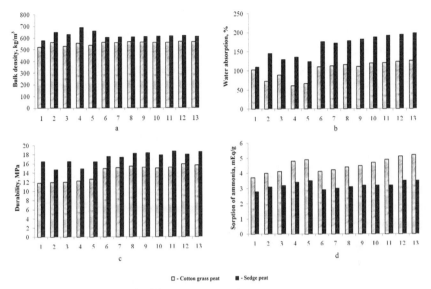

FIGURE 7.2   Influence of acids on peat properties.

a—bulk density, b—water absorption, c—durability, d—sorption of ammonia: 1—initial peat; peat modified by: 2—10^{-4} N solution of orthophosphoric acid with a modulus 1:2.5; 3—10^{-4} N solution of orthophosphoric acid with a modulus 1:5; 4—10^{-2} N solution of orthophosphoric acid with a modulus 1:2.5; 5—10^{-2} N solution of orthophosphoric acid with a modulus 1:5; 6—10^{-4} N solution of citric acid with a modulus 1:2.5; 7—10^{-4} N solution of citric acid with a modulus 1:5; 8—10^{-2} N solution of citric acid with a modulus 1:2.5; 9—10^{-2} N solution of citric acid with a modulus 1:5; 10—10^{-4} N solution of oxalic acid with a modulus 1:2.5; 11—10^{-4} N solution of oxalic acid with a modulus 1:5; 12—10^{-2} N solution of oxalic acid with a modulus 1:2.5; 13—10^{-2} N solution of oxalic acid with a modulus 1:5.

acid molecules not exceeding $5 \cdot 10^{-2}$ mmol/g. Oxalic acid is a dibasic acid, respectively, exchange capacity increases, in this case, to 10^{-1} mEq/g. The strengthening of ammonia sorption by cotton grass peat reaches 1.5 mEq/g (Fig. 7.2), that is, 15 times higher. An even greater difference occurs after treatment of cotton grass peat with weak acid solution (10^{-4} mol/l) with a modulus of 1:2.5. The exchange capacity increases not more than to $5 \cdot 10^{-4}$ mEq/g, the observed increase in the concentration of sorbed ammonia is 1 mEq/g (see Fig. 7.2), that is, more than 2000 times. For the entire set of studied samples, the strengthening of sorption activity is 5–2000 times higher than the concentration of dopant ion-exchange groups.

The effect of treatment with the acids, mentioned above, on EPR-spectra of the sorbents was investigated to determinate mechanisms of increase in the sorption activity. The EPR-spectra parameters of the initial peat samples (Tables 7.3 and 7.4) are typical of high- and low-moor peat [14]. Low intensity of the EPR signal of the low-moor peat is due to its lower degree of decomposition. The large width and a lesser degree of saturation signal with SHF-power (the high value of the parameter A/A_0) are due to the significant nitrogen content in the polyconjugation systems [14].

The effect of treatment with the acids, mentioned above, on EPR spectra of the sorbents was investigated to determinate mechanisms of increase in the sorption activity. The EPR spectra parameters of the initial peat samples (Tables 7.3 and 7.4) are typical for high- and low-moor peat [14]. Low intensity of the EPR signal of the low-moor peat is explained by its lower degree of decomposition. The large width and a less degree of saturation signal with SHF power (the high value of the parameter A/A_0) are due to the significant nitrogen content in the polyconjugation systems [14].

Besides a narrow signal of organic paramagnetic centers, the blurred spectrum of Fe^{3+} compounds is registered for the sedge peat that is also typical of low-moor peat [14]. The relatively narrow signal at low fields ($\Delta H = 100$ G, g-factor = 4.3) is due to iron compounds with purely ionic bonding (Fe^{3+} ions, associated with the carboxyl groups of the peat organic matter), while the broad signal ($\Delta H = 600–900$ G) at g-factor = 2.2 is due to iron compounds with covalent bonding, cluster of ions (Fe^{3+} coordination bondings with both organic and mineral components).

The presence of magnetic iron compounds is an additional factor explaining the weak organic paramagnetic centers signal saturation with SHF power (the high value of the parameter A/A_0). It should be noticed that low content of Fe^{3+} ions, associated with the carboxyl groups, is also registered when a cotton grass peat is used as the sample. Apparently, it explains the relatively weak saturation of the EPR signal with SHF power for high-moor peat.

The ammonia sorption results in the significant increase in the intensity of the EPR signal, consistent with the above results, and the transformation of its parameters, explained by the interaction of molecules of ammonia with polyconjugation systems, for all the samples studied.

Treatment with the acids, mentioned above, leads to elevation of g-factor (Δg is 0.0004 ± 0.0002 for the high-moor peat and 0.0003 ± 0.0001 for the low-moor peat) that may be explained by the partial localization of the unpaired π-electron on the oxygen atoms of the acid residue.

TABLE 7.3 Influence of Acids on EPR Spectra of Cotton Grass Peat Samples [32].

Modifying agent	Concentration, mol/l	Module	Ammonia treatment	A_{tt}	ΔH, G 0.1 mW	ΔH, G 50 mW	g-Factor	$I, 10^{17}$ spin/g	A/A_0	Fe^{3+} g-Factor = 4.3 ΔH, G	Fe^{3+} g-Factor = 4.3 $I_{rel}, 10^6$	Fe^{3+} g-Factor = 2.2 ΔH, G $I_{rel}, 10^8$
No agent	–	–	–	46	4.0	4.0	2.0032	6.0	3.57	80	0.4	Undetectable
No agent	–	–	+	46.5	5.6	7.3	2.0039	10.8	2.99	80	0.4	Undetectable
Orthophosphoric acid	10^{-4}	1:2.5	–	43.5	4.0	4.8	2.0039	6.1	2.74	80	0.4	Undetectable
Orthophosphoric acid	10^{-4}	1:2.5	+	42	5.8	7.3	2.0036	13.2	2.88	80	0.3	Undetectable
Orthophosphoric acid	10^{-4}	1:5	–	43.5	3.7	4.8	2.0032	6.0	3.20	80	0.4	Undetectable
Orthophosphoric acid	10^{-4}	1:5	+	43.5	5.6	6.6	2.0036	12.9	3.43	80	0.3	Undetectable
Orthophosphoric acid	10^{-2}	1:2.5	–	39	3.8	5.6	2.0033	6.2	2.50	80	0.3	Undetectable
Orthophosphoric acid	10^{-2}	1:2.5	+	46.5	5.6	6.6	2.0036	11.9	3.22	80	0.4	Undetectable
Orthophosphoric acid	10^{-2}	1:5	–	42	4.0	4.8	2.0037	7.7	2.96	80	0.3	Undetectable
Orthophosphoric acid	10^{-2}	1:5	+	44	6.3	7.2	2.0036	17.1	3.34	80	0.3	Undetectable
Citric acid	10^{-2}	1:2.5	–	43.5	4.0	4.3	2.0037	7.1	3.33	80	0.3	Undetectable
Citric acid	10^{-2}	1:2.5	+	44	5.5	6.4	2.0038	11.2	2.16	80	0.3	Undetectable
Oxalic acid	10^{-4}	1:2.5	–	45	3.9	5.0	2.0039	7.8	2.84	80	0.3	Undetectable
Oxalic acid	10^{-4}	1:2.5	+	46	5.3	5.4	2.0033	15.2	2.81	80	0.4	Undetectable

Note: A_{tt}—indicated value of SHF-channel attenuator; A/A_0—amplitude ratio of signal, shot at 50 and 0.1 mW. ΔH—quantity that characterizes the width of the EPR-signal. This value is measured in units of magnetic field—gauss (G); g-factor—a factor that connects a gyromagnetic ratio of the particles with the classical value of the gyromagnetic ratio. For classical particles, g-factor is 1; for free quantum particles with nonzero spin, these value is 2; for real particles experimentally certain value g-factor may be different from both 1 and 2, and is one of the characteristics of the particles; I—the concentration of free radicals, which characterizes the intensity of the EPR signal. The I value is measured in relative units (spin/g); I_{rel}—the relative concentration of free radicals.

TABLE 7.4 Influence of Acids on EPR Spectra of Sedge Peat Samples [32].

Modifying agent	Concentration, mol/l	Module	Ammonia treatment	A_{rr}	ΔH, G 0.1 mW	ΔH, G 50 mW	g-Factor	$I, 10^{17}$ spin/g	A/A_0	Fe^{3+} g-Factor = 4.3 ΔH, G	g-Factor = 4.3 $I_{rel}, 10^6$	g-Factor = 2.2 ΔH, G	g-Factor = 2.2 $I_{rel}, 10^8$
No agent	–	–	–	46	6.8	7.0	2.0035	3.1	4.87	80	0.4	950	1.15
No agent	–	–	+	46.5	7.0	8.5	2.0036	6.1	5.63	110	1.9	900	1.13
Orthophosphoric acid	10^{-4}	1:2.5	–	42.5	6.6	7.2	2.0040	2.8	5.08	110	1.7	810	0.83
Orthophosphoric acid	10^{-4}	1:2.5	+	46.5	7.2	8.5	2.0040	5.5	5.63	110	1.9	690	0.82
Orthophosphoric acid	10^{-4}	1:5	–	45.5	6.6	7.2	2.0037	2.8	5.32	110	2.3	660	0.73
Orthophosphoric acid	10^{-4}	1:5	+	47	7.1	7.4	2.0040	5.6	5.45	110	2.6	710	0.68
Orthophosphoric acid	10^{-2}	1:2.5	–	39.5	6.9	7.3	2.0038	3.2	4.37	110	2.4	680	0.70
Orthophosphoric acid	10^{-2}	1:2.5	+	46.5	7.0	8.5	2.0036	6.1	5.63	110	2.4	660	0.75
Orthophosphoric acid	10^{-2}	1:5	–	42	6.2	6.3	2.0038	2.7	5.83	110	1.9	670	0.75
Orthophosphoric acid	10^{-2}	1:5	+	46	7.1	9.0	2.0041	5.5	6.72	110	2.1	690	0.73
Citric acid	10^{-2}	1:2.5	–	40	6.4	8.0	2.0039	5.1	4.59	110	2.2	635	0.68
Citric acid	10^{-2}	1:2.5	+	47	7.2	7.6	2.0040	9.4	5.60	110	2.7	640	0.67
Oxalic acid	10^{-4}	1:2.5	–	43	5.8	6.3	2.0036	5.7	4.23	110	1.6	700	0.74
Oxalic acid	10^{-4}	1:2.5	+	47	7.1	7.8	2.0038	6.6	5.62	110	1.7	715	0.95

Note: For designations see Figure 7.3.

Intensification of the EPR signal both for initial and ammoniated samples of cotton grass peat is also observed. The latter statement is consistent with the increase in sorption capacity. In the case of sedge peat, weakening and narrowing of the spectrum of the covalent-bonded iron, followed by amplification of the signal of Fe^{3+} with purely ionic bonding (g-factor = 4.3), are also observed (Table 7.3).

When working with the acid-treated peat samples, there is a slight but proved reduction of quality factor of measuring resonator, which is manifested in the necessity of ascension of SHF-power delivered into the resonator for storing the current of SHF-detector (reduction of A_{tt} parameter). This reduction proves the introduction of acids in the peat matrix, and is explained by dielectric loss, contributed by its polar molecules. Reduction of quality factor is stronger when peat is treated with acid solution with a concentration of 10^{-2} mol/l. Following ammonia treatment tends to return the quality factor of measuring resonator to the initial state; it is logically explained by the reaction of neutralization.

To understand a visual effect of weak acids of low concentrations on the EPR spectra of peat, it should be taken into account that used acids are di- (oxalic) and tribasic (orthophosphoric and citric). Therefore, it is possible to draw an analogy with the effect of not paramagnetic polyvalent metal cations [20]. It can be assumed that they form cross-linking hydrogen and donor–acceptor bonds between polyconjugation fragments. Whereas, in this case, the unpaired π-spins are partially localized on the oxygen atoms of the acid groups, then the g-factor of signal of organic paramagnetic centers is increasing, as well as broadening with SHF-power level elevation takes place. The proposed analogy is supported by the greatest improvement of sorption activity and amplification of EPR-signal while using dibasic oxalic acid. The reason is that the divalent metal cations more efficiently modify polyconjugation systems of humic acids in comparison with trivalent [20].

However, there are fundamental differences compared to the sorption of nonparamagnetic polyvalent cations by humic substances. First, the influence of the nonparamagnetic polyvalent cations on the EPR spectra is evident at a concentration not lower than 1 mEq/g [31]. The effect of the acids is already registered at concentration $2.5 \cdot 10^{-4}$ mEq/g. In addition, the metal-substituted forms are characterized by the presence of two types of paramagnetic centers: those with metal ion in its structure, and initial paramagnetic centers [19]. Thus, the transformation of the EPR signal parameters is accompanied by a significant enhancement of its intensity for them. In our case, it is insignificant.

For the explanation of identified differences ideas of "microcoordination" and "macrocoordination" in the interaction of metal ions with humic acids developed in [33] should be taken into account The first refers to the process of binding of certain functional groups with cations with formation of coordination centers; the second refers to the binding of cations by macromolecular aggregates with formation of macromolecular complexes. It is obvious that the sequence of sorption space filling is carried out according to the principle of the energy gain, when the sorbent is interacting with the metal ions. In the case of carboxyl cation-exchanger, what the peat essentially is, interaction of metal ions directly with carboxyl groups is energetically efficient, and, that is, sorption is proceeding by mechanism of "microcoordination". In the experiments described in [19, 31], the concentration of metal ions did not exceed the carboxyl group content, and, consequently, the mechanism of "microcoordination" only was implemented. Therefore, the localization of the unpaired spin on metal ions took place only in the formed coordination centers, that explains the registration of two EPR signals caused by both localized on the embedded metal ion and initial unpaired spins.

Obviously, the chemical nature of used acids does not allow their interaction directly with the carboxyl groups, that is, they can't be embedded by the mechanism of "microcoordination". The interaction with the aggregates formed by polyconjugated fragments of macromolecules is only possible for them, due to the formation of hydrogen and donor–acceptor bonds, that is, acids sorption is proceeded by the mechanism of "macrocoordination". In this regard, nearly all unpaired π-electrons are sensitive to the presence of acid residues, despite the low content of the latter.

Considering that ammonia forms the ammonium cation in the aqueous medium and is able to form hydrogen and donor–acceptor bonds, ammonia sorption by peat organic matrix and humic acids involves both mechanisms: of the "microcoordination" and of the "macrocoordination".

Efficiency of "macrocoordination" mechanism can explain a significant effect of low concentrations of weak acids on the organomineral complex's structure of sedge peat samples (Table 7.3). Decrease of registered EPR-signal intensity of the organic matrix of the latter during the phosphoric acid treatment is consistent with amplification of the ion-bonded iron signal (g-factor = 4.3). Ions of Fe^{3+}, responsible for this signal, due to the peculiarities of its topology in relation to organic para-magnetic centers, have a stronger "quenching" effect on their EPR-signal in comparison with the iron compounds responsible for a more intense spectrum at g-factor = 2.2 [27].

Concept of supramolecular associates' transformation, formed by poly-conjugation systems, logically explains the effect of doping by weak acids with low concentrations on the technological parameters of the sorbent, such as increasing of the bulk density, improving of the durability, and water–physical characteristics (Fig. 7.2). This concept explains the positive effect of the densification of concentration of the solution both on the sorption activity of the produced sorbent and on its technical characteristics. The greatest influence of dibasic oxalic acid on the latter, along with increasing of its sorption activity and the EPR-signal, mentioned above, also supports the proposed concept.

At the same time, our attention is attracted by the experimental data obtained after orthophosphoric acid treatment of peat, with a formation of hemihydrates [16] that are used, in particular, as dehumidifiers of air [34]. Excess water molecules react with H_2O molecules of hemihydrates, thereby forming a hydrate layers [35]. High moisture content explains lower water absorption in comparison with the sorbents produced by treatment with citric and oxalic acids (in case of peat decreasing of water absorption in comparison with initial samples is even observed), as well as high bulk density of sorbents based on sedge peat. The presence of hydrate layers explains the absence of intensity strengthening and ampli-fication of the EPR signal [31], and attenuation of these characteristics in comparison with initial sample, in the case of sedge peat. The lower efficiency of citric acid for enhancement of the sorption activity can be explained by the fact that it is a weaker acid in comparison with phos-phoric and oxalic acids.

It can be stated that more complicated pattern of acid influence on the sorption, spectral, and technological characteristics of sorbents based on sedge peat, compared to those obtained on the base of cotton grass peat, is first caused by the transformation of organic complexes.

The results of investigation could be the scientific base of producing and application of environment-friendly peat-based sorbents for the puri-fication of gas emissions from toxic pollutants, primarily from ammonia. Industrial tests, carried out on poultry farms of Minsk region, showed the economic feasibility of the peat sorbents use for the purification of air, due to a decrease in deaths of birds with increase in average daily gain, as well as lower energy costs, benefits of environmental tax and saving production costs (medicines, veterinary attendance, etc.).

7.4 CONCLUSIONS

Extra equivalent sorption of ammonia by peat is based on energetically favorable formation of cross-linking hydrogen and donor–acceptor bonding by peat molecules with polyconjugation fragments. Treatment with ammonia gas of the metal substituted forms of peat does not lead to the replacement of metal cations bonded with the functional groups–ammonium ions, and, conversely, strengthen their bonding with the organic matter. Transformation of the supramolecular structure during the sorption of Cu^{2+} and Fe^{3+} facilitates formation of the bonds, mentioned above, and compensates for the "loss" of ion-exchange groups.

Transformation of polyconjugation systems and rearrangement of organomineral complexes structure aligned with it provide the mechanisms of enhancement of sorption activity and improvement of technological properties of peat-based sorbents during its modification with low concentrations of weak acids. Interaction of modifying acids with organic matrix of peat occurs by the mechanism of "macrocoordination."

The scientific fundamentals of obtaining of peat-based sorption materials for solving the problem of cleaning of air environments were suggested. Industrial tests proved the feasibility of peat sorbents use for removal of ammonia on agricultural enterprises.

ACKNOWLEDGMENTS

This research was funded from RFBR under the project agreement Nos. 12-03-90018-Bel_a, 14-05-90011-Bel_a, from FBR of Belarus under the project agreement No. 12P-147, X 143-233 and from FASO of Russia under the project agreement No. 0410-2014-0029 "Physico-chemical basis for studying of main regularities of the fundamental cycle "structure—functional nature—properties" of natural polymer matrices".

KEYWORDS

- peat bitumen
- humin
- extra equivalent sorption
- EPR-spectroscopy

REFERENCES

1. Tishkovich, A. V. *Theory and Practice of Ammoniation of Peat.* Science and Technique: Minsk, Belarus, 1972; p 169 (In Russian).
2. Gorovaya, A. I.; Orlov, D. S.; Stcherbenko, O. V. *Humic Substances. Structure, Functions, Mechanisms of Action, Protection Properties, Ecological Role.* Scientific Thought: Kiev, Ukraine, 1995; p 304 (In Russian).
3. Zhukov, V. K.; Tomson, A. E.; Samsonova, A. S. *et al.* New Aspects of the Formulation and Use of Adsorption Materials Based on Peat. *Nat. Manage.* **2002,** *8,* 167 (In Russian).
4. Lischtvan, I. I.; Korol, N. T. *Main Peat Properties and Methods of Its Determining.* Science and Technique: Belarus, 1975; p 320 (In Russian).
5. Syskov, K. M.; Kuharenko, T. A. Definition of the Constitutional Groups in Coals and Their Constituents by Sorption Method. *Fact. Lab.* **1947,** *1,* 25–27 (In Russian).
6. Tomson, A. E.; Shunkevich, A. A.; Velikaya, E. N. *About Nature of Extra Equivalent Sorption of Ammonia by Peat, Colloid-2003.* Proceedings on II International Conference, Minsk, Belarus, 2003, 106 (In Russian).
7. Golovach, A. A.; Kuptel, G. A.; Terentyev, A. A. *Laboratory Workshop on the Physics and Chemistry of the Peat. Part 1. Physical and Technical Properties of Peat.* Belarusian Polytechnic Institute: Minsk, Belarus, 1980; p 69 (In Russian).
8. Tomson, A. E.; Sokolova. T. V.; Navosha, Yu. Yu. Evaluating the Potential of Peat as an Ammonia Gas Sorbent. *Nat. Manage.* **2001,** *7,* 110–112 (In Russian).
9. GOST 2671585 (1987). *Organic Fertilizers. Methods of Determination of Total Nitrogen* (In Russian).
10. Jezierski, A.; Czechowski, F.; Jerzykiwicz, M.; Drozd, J. EPR Investigations of Humic Acids Structure from Compost, Soil, Peat and Soft Brown Coal upon Oxidation and Metal Uptake. *Appl. Magnet. Resonan.* **2000,** *18*(1), 35–52 (In Russian).
11. Chukov, S. N. *Structural and Functional Parameters of Soil Organic Matter under Conditions of Anthropogenic Impact.* Chukov SN, Saint-Petersburg, 2001; p 216 (In Russian).
12. Lischtvan, I. I.; Strigutskiy, V. P. *Associative Nature of Paramagnetism of Humic Acids.* Colloid Chemistry in Solving of Ecological Problems, Proceedings on International Conference, Minsk, Belarus, 1994; 136–139 (In Russian).

13. Erchak, D. P.; Efimov, V. G.; Stelmakh, V. F. EPR of the Low Dimensions Systems. Phys. Status Solidi. **1997**, *203*(2), 529–548 (In Russian).

14. Strigutskiy, V. P.; Bambalov, N. N.; Prohorov, S. G.; Smirnova, V. V. Similarity of the Structure of the Aromatic Ring of the Initial Complex of Humic Acid Preparations. *Solid Fuel Chem.* 1996, *6*, 29 (In Russian).

15. Tsyganov, A. R; Tomson, A. E.; Bogolytsyn, K.G.; Sokolova. T. V.; Strigutskiy, V. P., Pehtereva, V.S.; Selyanina, S.B.; Parfenova, L.N.; Trufanova, M.V. On the role of the polymer matrix of the peat in the sorption of ammonia. *Fundamental research.* **2013**, **4**, *2,* 345-350 (In Russian).

16. Encyclopedic Dictionary of Chemistry. Soviet Encyclopedia: Moscow, 1983; p 417 (In Russian).

17. Navosha, Yu. Yu.; Strigutskiy, V. P.; Lyogonkiy, B. I. On the Contribution of Semiquinone Radical-ions in Humic Acids Paramagnetism. *Solid Fuel Chem.* **1982**, *2*, 24–26 (In Russian).

18. Lyogonkiy, B. I.; Strigutskiy, V. P.; Aleksanyan, R. Z.; Bel'kevich, P. I.; Navosha, Yu. Yu.; Mamedov, B. A.; Ragimov, A. V. On the Contribution of Semiquinone Radical-ion Paramagnetism States in Redox Polymers with Conjugated Bonds. *Rep. Sci. Acad. BSSR.* **1981**, *25*(9), 825–827 (In Russian).

19. Vasilyev, N.G.; Suleymanov, S. P. Effect of Transition Metal Ions in the EPR Spectra of Humic Acids. *Solid Fuel Chem.* **1986**, *3,* 26–31 (In Russian).

20. Lyogonkiy, B. I.; Lischtvan, I. I.; Lyubchenko, L. S.; Navosha, Yu. Yu.; Osika, V. A.; Strigutskiy, V. P. Role of Non-paramagnetic Metal Ions in the Formation of Paramagnetism of Polyconjugated Structures. *Rep. Sci. Acad. BSSR.* **1986**, *288*(6), 1411–1415 (In Russian). Chemical Encyclopedia. **1**, Soviet Encyclopedia: Moscow, 1988; 274 p(In Russian).

21. Terentyev, V. A. *Thermodynamics of Hydrogen Bonding.* University of Saratov: Saratov, Russia, 1983; p 258 (In Russian).

22. Dudarchik, V. M.; Prohorov, S. G.; Smychnik, T. P.; Strigutskiy, V. P.; Terentyev, A. A. On the Role of Hydrogen Bonds in the Formation of Humic Acids Paramagnetism. *J. Coll.* **1997**, *59*(3), 313–316 (In Russian).

23. Lyubchenko, L. S.; Misin, V. M.; Cherepanova, E. S.; Cherkashin, M. I. Investigation of the Properties of Diphenyl Acetylenes Substituted Polymers by Electron Paramagnetic Resonance. III. The Mechanism of Paramagnetic Centers in Polyconjugated Systems. *J. Phys. Chem.* **1995**, *59*(12), 3085 (In Russian).

24. Bellami, L. *Infrared Spectra of Complex Molecules.* World: Moscow, 1963; p 285 (In Russian).

25. Lischtvan, I. I.; Kruglitskiy N. N.; Tretinnik, V. Yu. *Physico-Chemical Mechanics of Humic Substances.* Science and Technique: Minsk, Belarus. 1976; p 264 (In Russian).

26. Stelmah, V. F.; Strigutskiy, L. V. Peculiarities of EPR-spectroscopy of System of Centers with Different Relaxation Times. *J. Appl. Spectr.* **1998**, **65**, *2,* 224–229 (In Russian).

27. Strigutskiy, L. V. *Influence of Centers with Short Relaxation Times on the EPR-spectra of Carbon Materials: Doctoral Thesis.* State University of Belarus: Minsk, Belarus. 1999; p 21 (In Russian).

28. Ingram, D. Electronic Magnetic Resonance in Biology. World: Moscow, 1972; p 296 (In Russian).

29. Vishnevskaya, G. P.; Molochkov, A. S.; Safin, R. Sh. EPR of Ion Exchangers. Science: Moscow; 1992; p 165 (In Russian).

30. Shmankova, N. A.; Orehova, S. E.; Hmylko, L.I.; Ashuyko, V. A. Patent # 7747 Method of Gas Purification from Ammonia. Published: 28.02.2006 IPC: B01J 20/30, B01D 53/58, 2006.

31. Navosha, Yu. Yu.; Parmon, S. V.; Prohorov S. G. *Investigation of the Effect of Non-Paramagnetic Metal Ions on the Paramagnetic Properties of Humic Acids*. Organic Matter of Peat: Theses of International Symposium. Academy of Science of Belarus: Minsk, Belarus, 1995; pp 17–18 (In Russian).

32. Tsyganov, A. R; Tomson, A. E.; Bogolytsyn, K.G.; Sokolova. T. V.; Strigutskiy, V. P., Pehtereva, V.S.; Selyanina, S.B.; Parfenova, L.N.; Trufanova, M.V. Preparation of sorption materials based on peat. Chem. of plant raw materials. 2014, 4, 295–302 (In Russian).

33. Zhorobekova, Sh. Zh. Macroligand Properties of Humic Acids. Frunze, Uzbekistan, Ilim, 1987; p 194 (In Russian).

34. Rabinovich, V. A.; Havin, V. Ya *A Brief Chemical Handbook*. Chemistry: Leningrad, 1977; p 135 (In Russian).

35. Gamayunov, N. I.; Gamayunov, S. N. Sorption in Hydrophilic Materials, Tver State Technical University, Tver, 1997; p 160 (In Russian).

CHAPTER 8

PROBLEM OF MODIFICATION OF TECHNICAL LIGNINS USING ACYLATION METHOD

ANDREY V. PROTOPOPOV*, DANIL D. EFRYUSHIN, and VADIM V. KONSHIN

Polzunov Altai State Technical University 46, Lenin Av., Barnaul 656038, Russia

Corresponding author. E-mail: a_protopopov@mail.ru, vadandral@mail.ru

CONTENTS

ABSTRACT

This chapter studied the possibility of chemical modification of sulfate lignin carboxic acids. The kinetics of the reaction of acylation and the thermodynamic parameters of the acylation reaction of the activated complex and the overall activation energy were also studied. Research on absorption of methylene blue, phenol, and ion of metal is also discussed.

8.1 INTRODUCTION

The problem of the industrial waste utilization is one of the most relevant tasks of the rational nature management. Accumulated industrial waste takes up a lot of areas; it is a source of the environmental pollution and causes deterioration of conditions for human living and habitats of organisms. However, some carbon-containing wastes can be considered as secondary technogenic raw material resources.

The lignin being a wood component is the most difficult recycle material resulting from its chemical processing of pulp and paper, as well as hydrolysis plants. On the other hand, the lignin can act as potential raw material resource for synthesis of new products.

It should be emphasized that there are no comprehensive technologies to utilize technical lignins at the present time, although the scientific literature review over the years shows a rising interest of researchers for this raw material resource.

Methods of lignin recycling in the initial (unmodified) form are mainly based on using its dispersing, adhesive and surface-active properties.

The most large-scale direction of using of technical lignins is a synthesis of carbon sorbent agents. At the present time, new methods of thermal recycling of the lignin with porous carbon materials having controlled pore size and requested surface functional groups as well as high strength properties are developed.

The chemical modification of lignins provides an opportunity to derive new products for use in different areas: nitrolignin for manufacturing rust converter solutions, chlorlignin for boring, sulphated lignin for color dispersing, in biodegradable composite materials, and others. It is possible to derive aromatic aldehydes, organic acids, and liquid fuels using technical recycling of lignins.

Researchers in the area of recycling of industrial timber wastes, primarily technical lignins and lignin-containing materials (lignosulphonic acids), are up against an important task: search for modification methods of technical lignins which minimize the energy cost, reagents, and negative impact on human beings and environment as result, and due to this, the recycling of industrial timber wastes will be large-tonnage.

The chemical modification of the lignin provides an opportunity to derive new products for use in different areas. Due to high activity of hydroxyl groups, it is possible to include the lignin-containing materials of different functional groups. The performance of such chemical modification provides an opportunity to derive and derivate technical lignins with determined properties.

8.2 MATERIALS AND METHODOLOGY

8.2.1 LIGNIN ACYLATION

The previously prepared acylating mixture was added to the lignin quantity with weight of 0.5 g weighted accuracy to 0.002 g. The acylation was performed varying the synthesis time in the limit of 1–5 h at the temperature between 25 and 55°C. The product was isolated from a reaction mixture by adding the precipitator. The products were carefully washed free from present, not reacted acids on the Schott filter and dried to the constant mass in the air.

8.2.2 ACYLATING MIXTURE

The mixture preparation was performed as follows: the acid with quantity of 3.5 mol equivalent to lignin-weighted quantity was dissolved in 15 ml of toluene, and then the thionyl chloride with equivalent quantity of 1 ml and the catalyst (H_2SO_4) with quantity of 0.1 ml were added. A derived mixture was incubated for 30 min at chosen synthesis temperature for interaction of thionyl chloride and acid.

8.2.3   DETERMINATION OF COMBINED HYDROXYL GROUPS IN LIGNIN

The tested substance with quantity of 0.35 ± 0.01 g (accuracy to 0.002 g) was put in the conical flask with volume of 100 ml and wetted by 10 ml of ethanol. Then, the flask was capped and put in the thermostat with temperature of 56–60 °C for 15 min for the best sample expanding. After this time, the flask was added with 10 ml of KOH solution (0.5 mol/l) and allowed fully to saponify for 24 h at temperature of 55–60°C. Then, the flask content was titrated against the HCl solution (0.5 mol/l) by potentiometry; the reference electrode is silver-chloride, indicating electrode as glass.

The content of combined amino acid is calculated according to the following formula:

$$C,\% = \frac{\left(N_1 \cdot V_1 - N_2 \cdot V_2\right)}{1000 \cdot q} \cdot M \cdot 100\%,$$

where C is the content of combined acid, %; N_1 is the normality of NaOH solution; V_1 is the volume of NaOH solution; V_2 is the volume of chlorohydric acid sent for titration; N_2 is the normality of chlorohydric acid; M is the equivalent acid mass; q is the weighted quantity mass.

8.2.4   DETERMINATION OF ACID HYDROXYL GROUPS IN LIGNIN

The dry delivery flask with volume of 25 ml is added with a lignin weighted quantity (40–60 ml) and exactly 5 ml of 0.1 N LiOH, as well as 2 ml of 96% ethanol. Then, the closed flask is put in water bath heated to 85 °C for 3 min. The hot solution or suspension is added with 1 ml of 10% barium chloride solution, the flask is capped and allowed to cool for 15 min; thereafter, the mixture is made up to volume using boiled distilled water, mixed and poured quickly into the flask which is capped and centrifuged for 3–5 min. Then, the clear centrate with volume of 20 ml is pipetted and poured into the conical flask which contains exactly 5 ml of 0.1 N HCl. It is added with the indicator, and the acid excess is titrated by 0.1 N LiOH from the microburet. Under the same conditions, a blank test is performed without lignin weighted quantity.

Content calculation of OH groups, %

$$[OH] = \frac{(a_0 - a) \cdot f \cdot 1.25 \cdot 1.7 \cdot 100}{A},$$

where a and a_o are the volumes of 0.1 N LiOH expanded against titration during working and blank tests, respectively, ml; f is the factor for titer of 0.1 N LiOH; 1.25 is the conversation factor for the whole volume; 1.7 is the mass of OH groups equivalent to 1 ml of 0.1 N LiOH, mg; A is the lignin weighted quantity, mg.

8.2.5 DETERMINATION OF STRONGLY ACID (CARBOXYL) GROUPS IN LIGNIN

A total of 40–60 mg lignin is put into the delivery flask with volume of 25 ml, it is added with 20 ml of 0.4 N $(CH_3COO)_2Ca$ solution and heated loosely capped for 0.5 h at 85°C. Then, the flask is closed and allowed to cool for 15 min. The mixture is made up to volume using distilled water, mixed and filtered through a dry ashless paper filter into the dry flask with volume of 50 ml. A volume of 20 ml is pipetted from the filtrate, and a free acetic acid is titrated therein by 0.05 N LiOH over phenolphthalein. Under the same conditions (including filtration), a blank test is performed without lignin weighted quantity. Content calculation of OH groups, %:

$$[OH] = \frac{(a_0 - a) \cdot f \cdot 1.25 \cdot 0.85 \cdot 100}{A},$$

where a and a_o are the volumes of 0.05 N LiOH expanded against titration during working and blank tests, respectively, ml; f is the factor for titer of 0.05 N LiOH; 0.85 is the mass of OH groups equivalent to 1 ml of 0.05 N LiOH, mg; 1.25 is the conversation factor for the whole volume; A is the lignin weighted quantity, mg.

8.2.6 SPECTROMETRY

Infra-red spectra were recorded with IRS-40 using solid phase suspension in KBr. ^{13}C NMR was performed on device BRUKER AVANCE III

300WB. The ^{13}C spectra of high resolution in the solid body were registered on the frequency of 75 MHz using standard method of cross-polarization with proton reduction and magic angle spinning.

8.2.7 RESEARCH OF METAL ION ADSORPTION

Adsorbent weighted quantities with mass of 0.01 g were put into flasks, poured by 50 ml of the tested salt solution in different concentrations and allowed to stand for 24 h. The equilibrium concentration of metal ions was determined by the colorimetric method according to GOST (Code of Regulations) 4388–72 for Cu^{2+} ions, GOST 18,293–72 for Pb^{2+} ions, and GOST 23,862.31–79 for Th^{4+} ions.

8.2.8 PROCEDURE FOR DETERMINATION OF METHYLENE BLUE ADSORPTION BY ACYL PRODUCTS OF TECHNICAL LIGNINS

The methylene blue solution is prepared by weighting 1.5 g of methylene blue and by dissolving in the flask per 1 l up to volume. Comparison solutions are made for developing of a calibration curve. For this purpose, 10 delivery flasks with volume of 50 ml are added with 0.5, 1.0, 1.5, 2.0, 3.0, 4.0, 5.0, 6.0, 7.0, and 8.0 ml of methylene blue solution; each solution is made up to volume using water. 1 l of each obtained solution contains 15, 30, 45, 60, 90, 120, 150, 180, 210, and 240 mg/l of methylene blue, respectively. The optical density D of the mixed comparison solutions was messed on a photoelectric colorimeter using blue color filter with wavelength (λ) of 400 nm in cuvettes with distance of 10 mm between active faces. Distilled water was used as control solution. The calibration curve of dependence between optical density and comparison solution concentration was plotted according to the findings.

The lignin weighted quantity (0.1 g) predried at temperature of 100°C for 2 h was put into the delivery flask with volume of 50 ml; it was added with 25 ml of methylene blue solution and was shaken on a vibrating mixing device for 20 min. After shaking, the flasks were allowed to stand for 5 min; then 5 ml of the clarified solution were pipetted, and its optical density was estimated on the photoelectric colorimeter.

The lignin adsorption activity due to the methylene blue, in milligrams of methylene blue per 1 g of lignin, is calculated according to the following formula:

$$X = \frac{(C_1 - C_2 K) \times 0.025}{m}$$

where C_1 is the concentration of the original coloring solution, mg/dm^3; C_2 is the concentration of the solution after contact with the lignin preparation, mg/dm^3; K is the dilution factor of the solution, sampled for analysis, after contact with the lignin preparation, $K = 10$; m is the mass of the lignin preparation weighted quantity, g; 0.025 is the volume of the methylene blue solution, sampled for clarifying, dm^3.

8.2.9   PROCEDURE FOR DETERMINATION OF PHENOL ADSORPTION BY ACYL PRODUCTS OF TECHNICAL LIGNINS

A total of 500 cm^3 tested water was sampled by the measuring cylinder and was put into the flask for stripping. If the sample contained available chlorine, an equivalent quantity of the sodium thiosulphate solution was poured into the flask and allowed to stand for 5 min. Then, 5 cm^3 of 10% copper sulphate solution and 10 cm^3 of 10% sulphuric acid solution were added. The flask equipped with liquid trap and cooler was placed on the hot plate. The flask was wrapped in glass-wool blanket to reduce the heat transfer. An outlet hose of the cooler was put in the flask with volume of 500 cm^3 which contained 10 cm^3 of 0.05N NaOH solution. The flask was heated to even boiling. The sample stripping was performed for 3 h until the stripping volume was 460 cm^3. The stripping product was transferred to the separation funnel with volume of 1000 cm^3. The flask was rinsed with 30–40 cm^3 of distilled water which was put in the same funnel. A total of 10 cm^3 buffer solution, 3 cm^3 of 2% 4-aminoantipyrine solution, and 3 cm^3 of 8% potassium hexacyanoferrate (III) solution were added; the sample was shaken after addition of each solution.

The sample was allowed to stand for 10–15 min, and then, it was extracted twice using 20 cm^3 of chloroform for the first extraction and 10 cm^3 of chloroform for the second. The first extraction was performed for 2 min and the second for 1 min. After phase disengagement, the chloroform extracts were filtered through a glass filter into the delivery flask with a volume of 25 cm^3 and made up to the volume using chloroform. The optical density of the extract was measured on the spectrophotometer with a wavelength of 460 nm or the photometer with a wavelength of 460–490 nm in cuvettes with optical path length of 50 mm.

The blank test was performed together by sampling 500 cm^3 of fresh boiled distilled water. The optical density of the blank test was subtracted from the optical density of the samples.

The mass concentration of volatile phenols contained by the tested water sample X is placed on the calibration curve.

8.2.10   THERMOGRAPHIC ANALYSIS OF TECHNICAL LIGNINS AND PRODUCTS OF THEIR CHEMICAL MODIFICATION

The thermographic analysis is based on the enthalpy measurement of the tested material in the heating process. However, T and DTA (Differential termical analysis) curves are recorded for one sample at the same time. The T curve is auxiliary. When imaging of appropriating points of the main curve on it, it is possible to estimate the temperature of phase changes and chemical reactions of the tested material.

The thermography device, thermograph Thermoscan-2, is provided with an inertialess registering digital unit for continuous recording of heating curves.

The weighted quantity volume is 0.2 g. For analysis of lignin acyl products, Al_2O_3 was used as reference standard.

After deconvolution of the thermogram, it was concluded indicating the temperature of material decomposition, moisture, and impurity content.

8.3   RESULTS AND DISCUSSION

Within the research performed, the lignin was treated with acyl mixture: carboxylic acid–toluene–thionyl chloride. The acylation process can be divided into the following stages (Fig. 8.1):

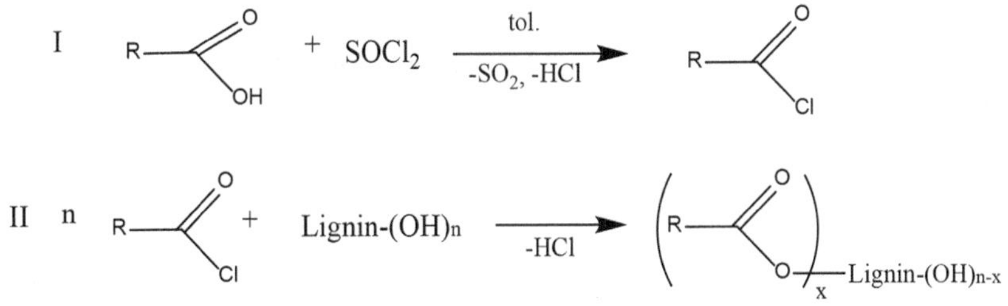

FIGURE 8.1   Scheme of acylation reactions.

- dissolution of carboxylic acid in toluene with thorough mixing;
- adding of thionyl chloride to the mixture and standing at the fixed temperature till full interaction (I); and
- treating of lignin with acyl mixture (II).

After this reaction, the acylated lignin was transferred to the Schott filter, dried from toluene, washed up to the neutral medium with an appropriate solvent for unreacted carboxylic acids, dried, and made up to constant mass. The derived products were checked for the number of combined acids.

The method of potentiometric titration was used to determine the combined acid. According to the data of chemical analysis for combined acid content, the number of combined hydroxyl groups in lignin was identified, and the conversion was calculated for each acid.

The aromatic amino acids are biologically active substances, and the lignin modified using such acids will be a preparation with pharmacological action. We have done the research of lignin acylation using aromatic amino acids (Table 8.1) [1].

The presented data show that the increase of temperature and synthesis duration entails rising of number of combined hydroxyl groups in acylated lignin. The number of combined hydroxyl groups does not exceed the maximum number of OH groups in sulphated lignin stated according to the experiment. It is evidence that aliphatic groups of lignin are acylated as the most chemically reactive only.

In the process of analysis performance for content of aliphatic and aromatic hydroxyl groups, it was stated that the total number of hydroxyl groups in sulphated lignin is 11.779; in addition, the average number of initial (aliphatic) OH groups is 9.122, and aromatic (acid) hydroxyl groups are 2.657.

TABLE 8.1 Conversion in Lignin Modified by Aromatic Amino Acids.

Acid in modified lignin	Time of synthesis performance, h	Temperature of synthesis,°C				
		25	35	45	55	65
		Conversion degree of lignin OH group				
I	1	0.44	0.55	0.66	0.73	1.05
	2	0.83	1.24	1.39	2.03	2.12
	3	0.88	1.35	2.61	2.88	2.76
	5	2.49	2.89	2.93	3.15	3.18
II	1	–	0.53	2.07	3.37	3.61
	2	–	1.43	2.22	3.76	4.14
	3	–	2.22	2.68	3.84	4.49
	5	–	2.53	3.85	4.56	4.87
III	1	–	1.76	2.14	3.49	5.22
	2	–	1.80	3.16	4.69	6.90
	3	–	2.45	3.69	6.69	7.51
	5	–	2.46	4.06	6.83	8.35

Note: I—p-amino benzoic acid, II—o-amino benzoic acid, and III—m-amino benzoic acid.

The acylation reaction of this system initiates in the heterogeneous medium, proceeds on the surface of the lignin macromolecule and is topochemical, because of that, the kinetic data processing was performed according to the Kolmogorov–Erofeev equation. The reaction constant was determined based on the kinetic dependence (correlation factor 0.94–0.97).

The reaction constant measured in such way has a nondimensional value. The Sakovich method was used to transfer the received values to the rates with dimension of s^{-1}.

The thermodynamic parameters of activated complex of acylation reaction were determined based on the Eyring equation when constructing a graph in coordinates of $\ln K\hbar/Tk_b$ from $1/T$, the energy and the activation entropy were estimated using this equation.

Values of the thermodynamic parameters of activated complex of sulphated lignin acylation reaction and kinetic dependency are listed in Table 8.2.

The value of the thermodynamic parameters for lignin acylation reaction using p-amino benzoic acid for heat reaction is $\Delta H^{\neq} = -38.7$ kJ/mol, the activation entropy is $\Delta S^{\neq} = -449.7$ J/(mol·K); for lignin acylation reaction using m-amino benzoic acid, $\Delta H^{\neq} = -12.7$ kJ/mol, $\Delta S^{\neq} = 366.0$ J/(mol·K); and for lignin acylation reaction using o-amino benzoic acid, $\Delta H^{\neq} = -60.5$ kJ/mol, $\Delta S^{\neq} = -506.9$ J/(mol·K). According to the found parameters, the Gibbs free energy of activation ($\Delta G^{\neq}$) of the transient formation for lignin acylation reaction using was calculated; its value is 120.3 kJ/mol for p-amino benzoic acid, 114.5 kJ/mol for m-amino benzoic acid, and 160.6 kJ/mol for o-amino benzoic acid. The received value of the activation enthalpy is pointed at the energy cost necessary for diffusing of acylating agent to lignin hydroxyls and for transient formation. According to the received value of activation entropy, it is possible to suppose that a fast fracture of the activated complex is derived with passing into reaction products.

TABLE 8.2 Thermodynamic and Kinetic Parameters of Lignin Acylation Reaction Using Aromatic Amino Acids.

Name of aromatic amino acid	Reaction constant, $K \cdot 10^4$, s^{-1}					Thermodynamic parameters of activated complex		
	25	35	45	55	65	ΔH, kJ/ mol	ΔS, J/ (mol·K)	ΔG, kJ/mol
pABA	1.88	4.01	4.57	4.97	5.34	−38.7	−449.7	120.3
mABA	–	8.13	8.34	13.07	22.91	−12.7	−366.0	114.5
oABA	–	3.96	7.22	26.59	168.87	−60.5	−506.9	160.6

Note: pABA—p-amino benzoic acid, mABA—m-amino benzoic acid, and oABA—o-amino benzoic acid.

The analysis of acylated lignin using infra-red spectroscopy (Fig. 8.2) showed absorption bands in the area of 3600–3400 cm^{-1}, typical for absorption bands of stretch vibrations for NH$_2$ and OH groups; in addition, these absorption bands are moving to the most high-frequency area toward absorption bands of pour lignin, it gives evidence of the amino predominance of the added acids. The absorption band in the area of 1730–1750 cm^{-1} is typical for of the CO groups as part of esters and proves the formation of lignin amino benzoates high in o-, m-, or p-amino benzoic acid in combined form. In addition, its intensity increases with the rise in temperature and synthesis duration. The presence of absorption bands responsible for vibration of the aromatic ring is based on lignin structural units, which increases the intensity due to adding of aromatic acid.

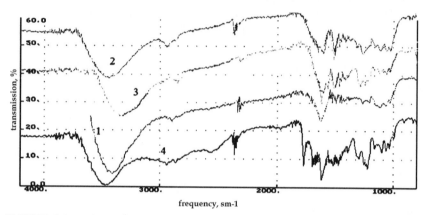

frequency, sm-1

FIGURE 8.2 Infra-red spectrum of sulphated lignin and acylated lignin products by aromatic amino acids: 1—sulphated lignin; 2—lignin acylated by *n*-amino benzoic acid; 3—products of acylation of lignin and *m*-amino benzoic acid; 4—acylated by *o* amino benzoic acid.

During acylating mixture research at different stages of its preparation, the infra-red spectrums were derived, which show the presence of absorption bands in the area of 3000 sm^{-1} for pure toluene that is equivalent to aromatic ring vibrations and also absorption bands in the area of 1610–1450 sm^{-1}, which are responsible for skeletal vibrations of carbon atoms in aromatic ring. Mixing of toluene with aromatic amino acid–specific bands appearing on spectrum in the area of 3600–3400 sm^{-1} indicates the presence of amino and OH groups of aromatic amino acids in mixture. The absorption band in the area of 3000 sm^{-1} is characterized with bigger intensity in comparison with pure toluene that relates to increase of concentration of molecules, which include aromatic rings. By comparison of infra-red spectrums of mixture of toluene with aromatic amino acid and ready acylating mixture, a presence of specific bands is noted in the latter case in the area of 1780–1700 sm^{-1}, which is equivalent to the presence of acid chloride of aromatic amino acid in the mixture. Therefore, the conceived mechanism of carrying out reaction is confirmed and carried out by researches of acylating mixture on different stages of its preparation with infra-red spectroscopy method (Fig. 8.3).

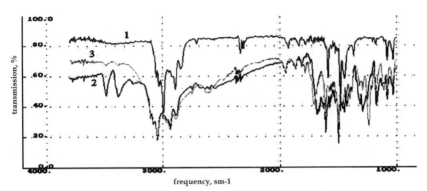

FIGURE 8.3 Infra-red spectrum of acylating mixture on different stages of its preparation: 1—toluene; 2—mixture of toluene and *n*-amino benzoic acid; 3—mixture of toluene, *n*-amino benzoic acid and thionyl chloride after holding at intensive mixing during 30 min.

During carried-out experiments, the esters of lignin with aromatic amino acids, *n*-amino benzoic acid, *m*-aromatic amino acid, and *o* aromatic amino acid, were derived, which was proved by infra-red spectroscopy method.

The content of combined aromatic amino acids in acylated products amounts from 0.44 to 9.63% of product mass. Chemical yield amounts from 90 to 150%, which is acceptable for such reaction type. The conceived mechanism of reaction behavior of lignin acylation by aromatic amino acid is confirmed by infra-red spectroscopy method.

Researches of sulphated lignin acylation by terephthalic acid were also carried out (Table 8.3) [2]. Derived product will be promising by using at derivation of phenolformaldehyde resins analogs.

TABLE 8.3 Quantity of Reacted Hydroxyl Groups in Lignin, Modified by Terephthalic Acid.

Time, h	Temperature of synthesis,°C			
	35°	45°	55°	65°
	Conversion degree of lignin OH group			
3	4.48	4.99	5.26	5.91
5	4.65	5.77	6.03	6.58

By reference to attained results, we can say that the number of reacted hydroxyl groups of lignin increases together with the increase in temperature and time of synthetics. For determination of total number of hydroxyl groups in lignin, the analysis of number of hydroxyl groups in aliphatic chain and in lignin aromatic ring was carried out. It was found that total amount of hydroxyl groups in sulphated lignin is 11.8%, and in this case, total amount of primary (aliphatic) OH groups is 9.1% at average, and the amount of aromatic (sour) hydroxyl groups is 2.7%.

Reactions proceeded in lingnin occur on the surface and relate to topo-chemical ones. To perform calculations of kinetics of these processes, the Erofeev–Kolmogorov equation is widely used. Based on chemical analysis of total number of hydroxyl groups, the conversion degree was calculated, and afterwards, kinetic anamorphoses were constructed.

After proceeding of received results according to Erofeev–Kolmogorov of constructed anamorphosis, constants of chemical reaction speed were defined, which amounted 6.63×10^{-6} s^{-1} at 35 degrees, 7.15×10^{-6} s^{-1} at 45 degrees, 8.03×10^{-6} s^{-1} at 55 degrees, and 8.13×10^{-6} s^{-1} at 65 degrees. Derived results confirm insignificant deviation of reaction speed at 55 and 65 degrees. By reference to attained speed constants values, we can make a conclusion on reaction proceeding in diffusive area. Derived products of chemical modification of lignin were analyzed with infra-red spectros-copy (Fig. 8.4). Absorption bands in the area of 3600 sm^{-1} are typical for valence vibrations of OH groups. At increase of C, the intensity increases absorption band of hydroxyl groups, vibrations of carboxylic acids 3500–3600 sm^{-1}, and also, there is an increase of intensity of absorption band of ester groups 1750–1730 sm^{-1} and of absorption bands responsible for vibrations of aromatic ring C–C bond 1560–1460 sm^{-1}.

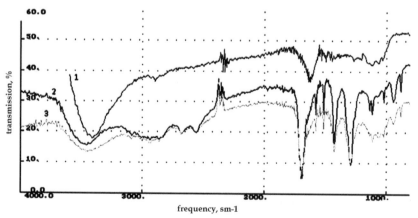

FIGURE 8.4   Infra-red spectrums of chemically modified lignin: 1—initial lignin; 2—product derived at 3 h 45°C; 3—product derived at 5 h 55°C.

Besides lignin products replaced by aromatic acids, products of inter-action with aliphatic acids are of big interest. Modified by aliphatic acids, lignin can be used in particular as adsorbent. Researches of lignin inter-action with different aliphatic acids were performed and correlation of reactivity with hydrocarbon radical length was considered. The optimality conditions of synthesis were discovered for all derived products [3], namely duration and temperature. The results show that maximum conver-sion degree of lignin OH group into the acyl ones is attained at a synthesis duration of 4 h in a temperature of 40°C (Table 8.4).

TABLE 8.4 Analysis Results of Acylated Products of Sulphated and Hydrolytic Lignin.

Carboxylic acid used for acylation	Conversion degree, α	
	Sulphated lignin	Hydrolytic lignin
Acetic acid	0.30	0.11
Valeric acid	0.66	0.13
Myristic acid	0.76	0.27
Palmitic acid	0.28	0.07
Stearic acid	0.12	0.04

Results of chemical analysis demonstrate higher reactivity of sulphated lignin in comparison with hydrolytic one that can be caused by the fact that hydrolytic lignin is more condensed than sulphated one.

With extension of with hydrocarbon radical of used carboxylic acid (from acetic acid to myristic one), the number of reacted OH groups of technical lignins increases too. The number of reacted OH groups is reduced at appending of carboxylic acids with large-sized hydrocarbon radical (e.g., palmitic and stearic acids) into technical-lignin macromol-ecule due to steric factor.

To research the behavior of derived acylated products in a wide temper-ature range, thermograms of initial lignins and products of their interac-tion with the system "carboxylic acid–thionyl chloride–toluene–sulphuric acid" were taken.

Thermograms' analysis (Figs. 8.5–8.7) enables to reveal the following areas of temperature changes:

At 100°C—it corresponds with process of free water removal; in the range of 100–175°C, there is a removal of combined water in material and vitrification of lignin; at temperature 140–200°C, the process of lignin

vitrification occurs; in the temperature range of 200–350°C, there is a
lignin decomposition. The biggest ΔT change is characteristic for the area
to 160°C. In this area, two processes are possible, which correspond with
material drying and lignin vitrification process.

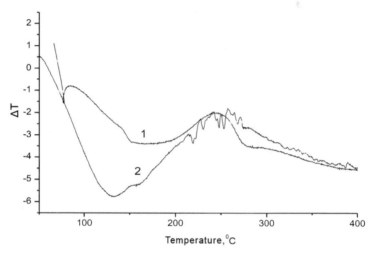

FIGURE 8.5 Thermograms of technical lignins: 1—sulphated lignin; 2—hydrolytic
lignin.

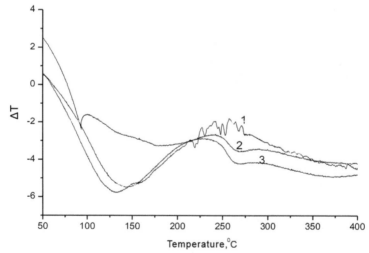

FIGURE 8.6 Thermograms of initial (1) and acylated hydrolytic lignin: 2—by myristic
acid (conversion degree CD = 0.27), and 3—by acetic acid (CD = 0.40).

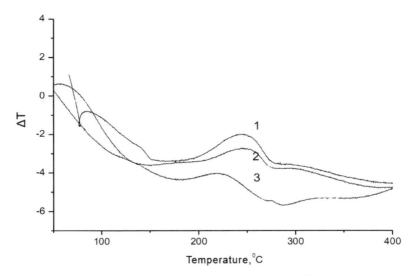

FIGURE 8.7 Thermograms of initial (1) and acylated sulphated lignin: 2—by acetic acid (CD = 0.30), and 3—by myristic acid (CD = 0.76).

By analyzing thermograms of pure technical lignins (see Fig. 8.5), we can draw a conclusion of higher thermostability of hydrolytic lignin in comparison with sulphated one due to higher condensation and bigger number of hydrogen bonds.

Hydrophilic products contain ester groups with acetic acid (see Figs. 8.6 and 8.7) and also initial lignins lose their water quite smoothly, without abrupt changes of heat absorption.

In products containing hydrophobic remainder of carboxylic acid (e.g., product of interaction of hydrolytic lignin with myristic acid) (see Figs. 8.6 and 8.7), the change from combined moisture loss to loss of free moisture is clearly observed at temperature 98–100°C. That can be explained with that, that water molecules are isolated by large acyl radical of myristic acid. Moreover, higher temperature (right up to 175°C) is necessary for moisture extraction from these products in comparison with initial lignin.

In the temperature range of 200–350°C, active decomposition of lignin and products of its acylation is accompanied by formation of the volatile compounds which occurs as a result of decomposition of bonds containing oxygen. According to literature data, the following bonds in structural lignin units can be destroyed in this temperature range at lignin pyrolysis: at 250°C, C_β–C_γ c by γ- alcohol group, at 300°C, C_{ap}–C_α α-benzyl group and also bonds between structural units: at 170–220°C, alkyl aryl ester bond C_β–OC_{ap}, at 250°C, C_α alkyl aryl ester bond –OC_{ap}. Moreover, the decarboxylation process can occur in such conditions.

Acylated lignins with high conversion degree of OH groups are notable for more "loose" molecular structure as acyl remainder of carboxylic acid penetrates into material structure deeper (and maybe destroy it), and as a result of it, there is a shift of decomposition temperature of acylated lignin toward lower temperature for 15–20°C, that makes the product less thermostabile.

One of typical specimens of cationic dyes is methylene blue, which can be sorbed by lignin from water solution both as monomeric form and as dimeric complexes (Table 8.5) [4, 5].

TABLE 8.5   Results of Analysis of Sorption Methylene Blue Cations.

Index	Products based on hydrolytic lignin (HL)			Products based on sulphated lignin (SL)		
	HL	HL-AA	HL-PA	SL	SL-AA	SL-PA
Adsorption activity by methylene blue, mg/g	132	137	120	132	155	118
Specific surface, m^2/g	49.6	51.4	45.2	49.6	58.4	44.2

Note: HL—hydrolytic lignin, HL-AA—acetylated hydrolytic lignin; HL-PA—hydrolytic lignin acylated by palmitic acid; SL—sulphated lignin; SL-AA—acetylated sulphated lignin; SL-PA—sulphated lignin acylated by palmitic acid.

Based on obtained results, it is evident that sulphated and hydrolytic lignin and products based on them possess a relative high adsorption activity by methylene blue.

The most effective sorbents are products HL-AA and SL-AA. Reduction of adsorption activity modified products from acetic acid to palmitic, one relates to hydrophobicity of synthesized lignin derivative with higher carboxylic acids, and as a result of it, some sectors become inert in relation to sorption and access for large organic cations deep into the material become very difficult.

Main and the most dangerous toxic substances, which pollute the environment, are considered phenol and its derivates, which promote malfunction of nervous system. The most effective sorbents by phenol extraction from water solutions are carbonic adsorbents. Results of graphical determination of adsorption capacity by Freundlich isotherm analysis are given in Table 8.6 [6].

TABLE 8.6 Adsorption Activity of Products of Chemical Modification of Technical Lignins by Phenol.

Index	Products based on hydrolytic lignin, HL				Products based on sulphated lignin, SL			
	HL	HL-AA	HL-MA	HL-PA	SL	SL-AA	SL-MA	SL-PA
Adsorption activity, mkg/g	27.9	1198.4	110.4	6.5	45.4	4742.8	87.6	82

Note: HL—hydrolytic lignin, HL-AA—acetylated hydrolytic lignin; HL-MA—hydrolytic lignin acylated by myristic acid; HL-PA—hydrolytic lignin acylated by palmitic acid; SL—sulphated lignin; SL-AA—acetylated sulphated lignin; SL-MA—sulphated lignin acylated by myristic acid; and SL-PA—sulphated lignin acylated by palmitic acid.

Data in Table 8.6 indicate that the most effective sorbent is product SL-AA. Similarly to adsorption of methylene blue, adsorption activity of modified products reduces from acetic acid to palmitic one, which can be related to steric factors and increase of hydrophobicity among acyl radicals:

$$H_3C—C—\underset{O}{\overset{\parallel}{}} \quad < \quad H_3C\cdot(H_2C)_{12}—C—\underset{O}{\overset{\parallel}{}} \quad < \quad H_3C\cdot(H_2C)_{14}—C—\underset{O}{\overset{\parallel}{}}$$

8.4 CONCLUSION

The necessity of waste regeneration or utilization increases together with increase of development tempo of chemical industry enterprises to reduce environmental damage. One of the most important aspects of rational nature management policy is processing of wood industry waste. Frequently, reagent regeneration is not duly conducted, which are the part of manufacture of different wood articles, and the same with waste utilization of this sector of national economy. It is essential to note that adoption of energy-saving technologies and different methods and means for return of wood waste into manufacture of different materials and articles would enable to reduce manufactured products cost and to reduce emissions and negative impact on the environment.

We have demonstrated the possibility of derivation of chemically modified lignin products, which contain combined carboxylic acids with application of acylating mixture "Carboxylic acid–toluene–thionyl chloride". For derived products, with the increase in temperature and synthesis duration, conversion degree increases too. Kinetic regularity of acylation reaction was studied for some derived products. Speed constants of acylation reaction of lignin by different aromatic amino acids were determined. According to reactivity, amino acids are ranged as follows: oABA > pABA > mABA

Performed researches for adsorption for lignin, modified by aliphatic acids showed a relative good adsorption activity, which is commensurable and in some cases bigger than one of known adsorbents. That gives an opportunity to use derived products as adsorbents of pharmacological application.

KEYWORDS

- acylation
- kinetics
- adsorbents
- absorption
- thermographic analysis

REFERENCES

1. Protopopov, A. V.; Klevtsova, M. V.; Konshin, V. V.; Efryushin, D. D. Chemical Modification of Lignins by Amino Acids. *Chem. Nat. Compounds*. **2015**, *51*, 934–936 (DOI 10.1007/s10600-015-1451-0).
2. Protopopov, A. V.; Klevtsova, M. V.; Bobrovskaya, S. A.; Voroshilova, A. V. Modification of Lignin by Aromatic Amino Acids. In *Biotechnology and Society in XXI Century: Collection of Articles*, Publishing office of Altai University: Barnaul, 2015; pp. 258–262 (In Russian).
3. Protopopov, A. V.; Bobrovskaya, S.A.; Voroshilova, A.V., Derivation of Products of Lingnin Modified by Terephthalic Acid. *Int. Sci. Res. Mag.*. **2015**, *43*, 119 (In Russian).

4. Efryushin, D.,D.; Konshin, V. V.; Protopopov, A. V.; Beushev, A. A. Modification of Technical Lignins by Carboxylic Acids. *Chem. Nat. Compounds*. **2015**, *51*, 1007–1008 (DOI 10.1007/s10600-015-1481-7).

5. Efryushin, D. D.; Konshin, V. V.; Evseyeva, T. P.; Shabalina, A. S.; Poteshkina, O. O. In *Researches of Methylene Blue Adsorption with Chemically Modified Technical Lignins Products*, Proceedings of XVI All-Russian Research and Training Conference for Students and Young Scientists with International Participation "Chemistry and Chemical Technology in XXI Century", Publishing office of Tomsk Polytechnic University: Tomsk, 2015, pp. 273–274 (In Russian).

6. Konshin, V. V.; Shabalina, A. S.; Evseyeva, T. P.; Efryushin, D. D. In *Adsorbents for Sewage Treatment on Basis of Chemically Modified Technical Lignins Products*, Proceedings of VI International Conference "Physicochemistry of Plant Polymers", Institute for ecological problems of the Nord Ural Department of the Russian Academy of Science: Arkhangelsk, 2015, pp. 109–110 (In Russian).

PART II
Biological Activity of Plant Substances

CHAPTER 9

COMPOUNDS OF PLANT ORIGIN AS AMP-ACTIVATED PROTEIN KINASE ACTIVATORS

DARIA S. NOVIKOVA*, GLEB S. IVANOV,
ALEXANDER V. GARABADZHIU, and
VIACHESLAV G. TRIBULOVICH

Saint Petersburg State Institute of Technology (Technical University), 26, Moskovsky Pr., Saint Petersburg 190013, Russia

*Corresponding author. E-mail: dc.novikova@gmail.com

CONTENTS

ABSTRACT

Adenosine monophosphate-activated protein kinase (AMP-activated protein kinase, AMPK) has become an attractive target for the treatment of diseases, associated with metabolic disorders over the last decade. The therapeutic significance of the AMPK activation is evident in diabetes, obesity, metabolic syndrome, cardiovascular diseases, and even in cancer. AMPK activators of plant origin and their synthetic modifications are considered; presumed mechanisms of their action and therapeutic effects are noted in this chapter.

9.1 INTRODUCTION

AMP-activated protein kinase (AMPK) was discovered more than 20 years ago, but only during the recent years, researchers have got a better understanding of its interactions with multiple signaling cascades. AMPK is considered as the main regulator of energy metabolism; it maintains energy homeostasis, both at cellular and whole-body levels [1]. In stressful conditions, when an increase in the adenosine monophosphate (AMP) level and decrease in the adenosine triphosphate (ATP) level is observed, the activation of AMPK leads to a rapid switching to reserve resources. If there is an energy deficiency, AMPK discontinues processes accompanied by energy consumption and promotes processes aimed at energy production. This switching from energy-consuming to energy-producing reactions enables to maintain the energy status of the body under stressful conditions [2].

The unique functions of AMPK allowed to suggest that the kinase targeting by pharmacological agents may be useful for the treatment of diseases associated with disturbed energy balance. Indeed, the therapeutic significance of the AMPK activation has been shown for such diseases as diabetes type 2, obesity, metabolic syndrome, as well as for a number of cardiovascular diseases, and cancer [3]. New evidences on the effectiveness of the AMPK targeting have been revealed for atherosclerosis and aging [4, 5].

Small molecule plant compounds are able to regulate the activity of proteins involved in the most important signaling pathways [6]. Therefore, the search and study on active substances of natural origin are of a great current interest. The aim of this chapter is to briefly summarize the data on plant compounds, which are able to activate AMPK. Many of the reviewed compounds are used in traditional medicines for different purposes and can be potentially reoriented for the treatment of diseases associated with the dysfunction of AMPK.

9.1.1 STRUCTURE AND FUNCTIONS OF AMPK

AMPK is a complex of three subunits: the catalytic α-, the regulatory β-, and γ-subunits [3]. The catalytic α-subunit includes the kinase domain, mediating the catalytic activity of AMPK, the autoinhibitory domain, and a domain responsible for binding to two other subunits. The β-subunit acts as the core of the kinase complex; in addition, it is also responsible for the interaction with carbohydrates due to the presence of the glycogen-binding domain. Two AMP-binding sites within the γ-subunit, capable of competitive binding with adenine nucleotides, act as a sensor of the energy state [7–9].

The AMPK activity can be regulated in several ways. The interaction of AMP with the AMP-binding sites causes a number of conformational changes resulting in the so-called allosteric activation [10]. In turn, the phosphorylation of the main activation site, a conserved threonine residue (Thr-172), which belongs to the kinase domain, contributes to a significant increase in the kinase activity [11]. To date, there are only three kinases that are able to activate AMPK: LKB1 (liver kinase B1), CaMKKβ (Ca^{2+}/calmodulin-dependent protein kinase kinase β), and transforming growth factor β-activated kinase-1 (TAK1) [12–14]. It is known that the AMPK activation can be achieved by blocking the autoinhibitory domain [15]. Moreover, AMPK has alternative sites, the binding to which results to AMP-like effects [16].

The main function of AMPK is to maintain the energy metabolism. This function is carried out by AMPK via the regulation of the main processes in the organism. In particular, AMPK is involved in carbohydrate, lipid, and protein metabolism and also performs a number of nonmetabolic functions such as the regulation of cell growth and proliferation, apoptosis, and autophagy [17]. The main effects of the AMPK activation, representing a therapeutic interest, are the increase in glucose uptake, stimulation of glycolysis, and inhibition of hepatic gluconeogenesis, which are of particularly importance for diabetes type 2 [18–20]. In addition, activated AMPK is capable of inhibiting the enzymes of fatty acid and sterol biosynthesis, while stimulating the oxidation of fatty acids, which can be useful for the control of obesity [21–23]. More detailed information on numerous functions of AMPK can be found in this chapter [24].

9.1.2 TARGETING AMPK

At present, a number of approaches to the AMPK activation have been developed; they can be divided into direct and indirect ones. Natural compounds activate AMPK mainly in indirect way, but among them, there are direct activators too. The mechanism of action for direct activators consists in: (1) binding to the AMP-binding sites, which leads to the allosteric activation; (2) interaction with the allosteric sites different from the AMP-binding sites; and (3) blocking of the autoinhibitory domain function. Indirect activators can provide the activating effect on AMPK through: (1) stimulation of kinases, which phosphorylate the main activation site of AMPK; (2) inhibition of protein phosphatases, which dephosphorylate the main activation site of AMPK; and (3) affecting various pathways and enzymes that leads to an increase in the AMP/ATP ratio.

To date, quite a large number of experimental AMPK activators have been developed, and studies on the activators are carried out by both research institutes and leading pharmaceutical companies [25]. Despite the fact that direct AMPK activators are in the spotlight, indirect activators, which include the vast majority of AMPK-active natural compounds, do not represent insignificant objects. This is due to the ability to fold increase in the AMPK activity under synergetic effect of direct and indirect activators [26, 27].

9.1.3 AMPK ACTIVATORS OF PLANT ORIGIN

Goat's rue (*Galega officinalis*) is known for its hypoglycemic properties since the middle ages; it is used to treat the symptoms of diabetes in traditional medicine. Both positive and negative effects of Goat's rue are due to the presence of guanidine and its derivatives. In 1914, galegine (Fig. 9.1), which possesses a hypoglycemic effect, was isolated and then identified from Goat's rue; moreover, it turned out to be less toxic than guanidine [28, 29]. Preclinical trials of galegine were unsuccessful, but its synthetic derivatives, biguanides (buformin, phenformin, and metformin), developed later and have contributed to the clinical practice. Metformin (Fig. 9.1) is of a special interest; it has been used as a drug more than half a century, and currently, it is the first choice drug for treatment of diabetes type 2 [30]. For a long time, its mechanism of action remained unknown, until it was shown that the inhibition of gluconeogenesis in the liver by metformin is mediated by the AMPK activation [31]. Later, it was suggested that metformin causes a change in the AMP/ATP ratio and activates the LKB1-AMPK signaling pathway [32, 33]. Although the mechanism of metformin still has not been clearly established, it is the only activator of AMPK used in the clinical practice nowadays.

Galegine   Metformin

FIGURE 9.1   Natural and synthetic guanidine derivatives.

The ability to activate AMPK was found for salicylate (Fig. 9.2). It is an active component of willow bark and produced by many plants to fight the infection [34]. Its synthetic derivatives, aspirin and salsalate, are used in the clinical practice for the treatment of headaches and diabetes. It is assumed that salicylate directly interacts with AMPK by binding to an alternative binding site, causing the allosteric activation of AMPK [35].

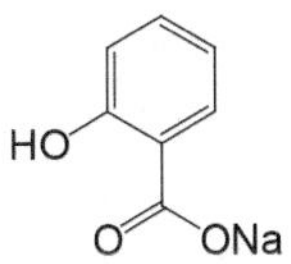

FIGURE 9.2   Sodium salicylate.

A large number of compounds that activate AMPK have been found among plant polyphenols and lignans. Curcumin (Fig. 9.3) is the major curcuminoid of turmeric rhizomes (*Curcuma longa*). Despite a poor solubility and low bioavailability, it is in phase II clinical trials for pancreatic cancer [36]. Curcumin induces the AMPK activation in a dose-dependent manner, probably, via LKB1 [37]. Antidiabetic effect of curcumin is manifested in suppression of gluconeogenesis gene expression and improvement of glucose uptake in peripheral tissues, which is realized by stimulation of the AMPK signaling cascade [38, 39]. Other curcuminoid, demethoxycurcumin (see Fig. 9.3), also found in turmeric rhizomes, showed greater AMPK activity compared with curcumin and 5-aminoimidazole-4-carboxamide ribonucleotide (AICAR), a well-known direct activator, and turned out to be more stable in cellular conditions [40].

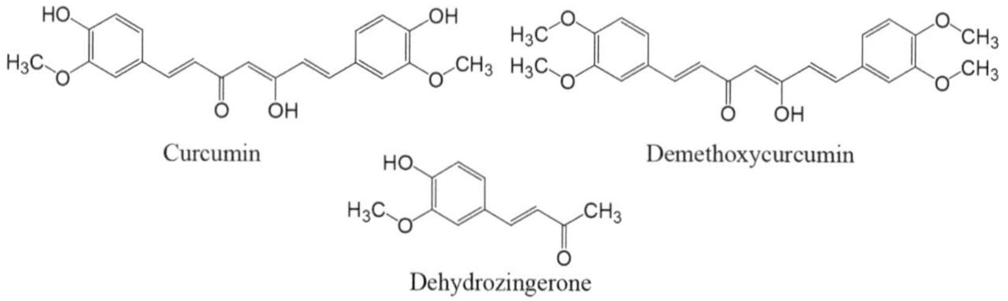

Curcumin   Demethoxycurcumin

Dehydrozingerone

FIGURE 9.3   Curcuminoids and related compounds.

Dehydrozingerone, isolated from rhizomes of ginger (*Zingiber officinale*), is a structural fragment of curcumin, representing a half of the molecule of the latter. Ten years back, a cytotoxic effect on cancer cells, along with antiinflammatory, antioxidant, and antimicrobial properties, was shown for dehydrozingerone and related compounds [41]. Recently, researchers noticed its beneficial metabolic effects. It was shown that dehydrozingerone prevents weight gain, lipid accumulation, and hyperglycemia and stimulates glucose uptake. The mechanism of the metabolic effects is determined by the activation of AMPK: dehydrozingerone stimulates the phosphorylation of the main activation site of AMPK. There is a speculation that dehydrozingerone is a fragment that determines the effect of curcuminoids on glucose metabolism [42].

Chlorogenic acid (Fig. 9.4), the ester of caffeic and quinic acids, is widely distributed in nature and is present in large quantities in green coffee beans. It is the main phenolic compound identified in peach. It was demonstrated that chlorogenic acid stimulates glucose transport in a dose- and time-dependent manner in skeletal muscle. This effect is determined by the AMPK activation, which manifests in increasing phosphorylation of AMPK and ACC and explains beneficial effects of coffee in diabetes type 2 [43]. It was found that chronic administration of chlorogenic acid allows to improve glucose and lipid metabolism, in particular, to normalize lipid profile and increase glucose uptake in skeletal muscle [44]. Chlorogenic acid is shown to be effective in late diabetes; the long-term treatment with chlorogenic acid in mouse models showed no adverse side effects [45]. Despite the data available in the literature, there is evidence that it is caffeic acid (Fig. 9.4), but not chlorogenic acid, which determines a stimulating effect on AMPK and glucose transport. Moreover, the specificity to the α2-subunit of AMPK was shown for caffeic acid [46]. Interestingly, cinnamic acid (Fig. 9.4) is also able to stimulate the phosphorylation of AMPK [47].

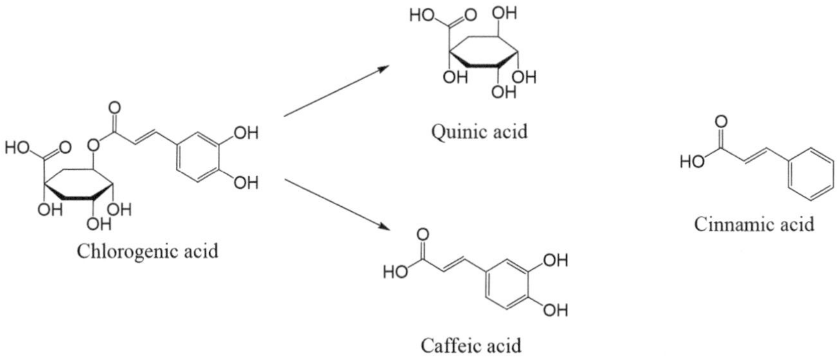

FIGURE 9.4   Cinnamic acid derivatives.

The use of capsaicin (Fig. 9.5), which is structurally similar to curcuminoids and is the active compound of chili pepper, allows to attenuate the symptoms of metabolic syndrome, in particular, reduces blood glucose, insulin, and triglyceride levels in blood plasma and liver. It was shown that these beneficial metabolic properties are accompanied by the activation of AMPK [48]. It is believed that capsaicin stimulates the intracellular reactive oxygen species release, which leads to the AMPK activation [49].

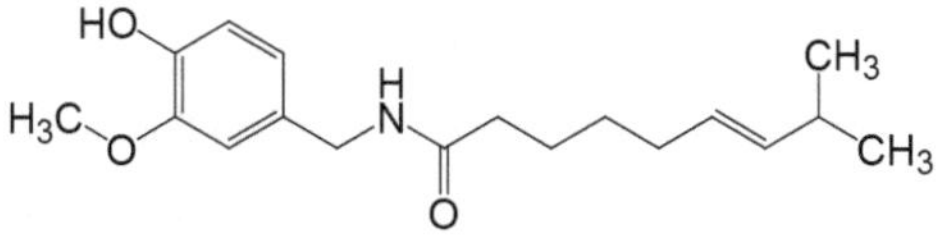

FIGURE 9.5   Capsaicin.

Honokiol (Fig. 9.6), the main active compound of the bark extract of the plants, belonging to the genus *Magnolia*, also was found as a natural activator of AMPK. It was shown that honokiol attenuates intracellular fat overloading and triglyceride accumulation. The activation of the AMPK-dependent signaling pathway by honokiol is mediated by the phosphorylation of LKB1 and, therefore, AMPK [50].

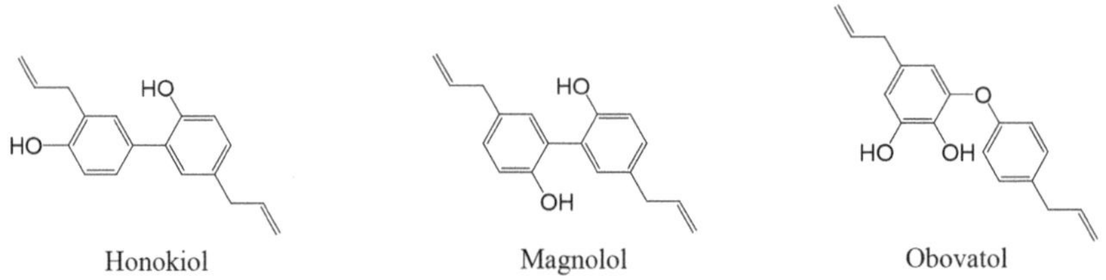

FIGURE 9.6 Active compounds of Magnolia.

In turn, antiproliferative properties mediated by the activation of AMPK were shown for magnolol (Fig. 9.6), which is an isomer of honokiol. It was found that magnonol induces the AMPK activation in a dose- and time-dependent manner, whereas an AMPK inhibitor, compound C, abrogates the effect of magnolol on the AMPK activation and, thus, suppression of proliferation [51]. The long-term combined administration of honokiol and magnolol reduces steatosis and liver dysfunction due to the inhibition of the expression of lipogenesis genes, which is realized through the AMPK-dependent signaling cascade [52]. An AMPK activator, obovatol (Fig. 9.6), structurally related to magnolol and honokiol, was also isolated from the bark of *Magnolia obovate*. It was proposed for the treatment of diseases associated with metabolic syndrome [53].

Resveratrol (Fig. 9.7) is a natural polyphenol, which has been intensively studied over the past two decades. Resveratrol is produced by some plants against bacterial and fungal infection. The highest content of resveratrol is found in grape skins, it is also found in cocoa, nuts, and other fruits. The ability of resveratrol to activate AMPK was detected along with its diverse biological activity. Initially, it was assumed that the activation of AMPK by resveratrol is associated with changes in the AMP/ATP ratio. It was shown that resveratrol causes inhibition of ATP synthesis in mitochondria by inhibition of F-ATPase [54]. This assumption was subsequently disproved as it was found that resveratrol does not affect the ATP level. Studies showed that resveratrol activates AMPK also in LKB1-deficient cells, while increasing the level of Ca^{2+} in the cytoplasm. The experiments on the inhibition of CaMKKβ identified critical decrease in the AMPK phosphorylation under the treatment with resveratrol and also confirmed the direct involvement of CaMKKβ in the activation mechanism. Interestingly, the study on the mechanism of AMPK activation by resveratrol was carried out within the drug development against Alzheimer's disease [55].

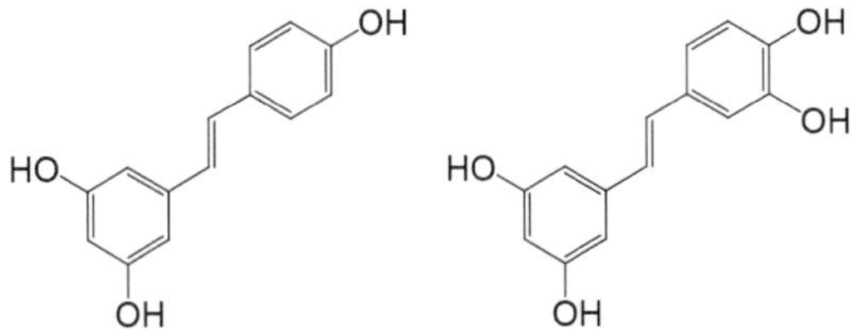

Resveratrol   Piceatannol

FIGURE 9.7   Plant *trans*-stilbenes.

Anticancer activity is typical for many plant stilbenes [56]. Piceatannol (Fig. 9.7), found in mycorrhizal and nonmycorrhizal roots of Norway spruces (*Picea abies*) and also in red wine, can be noted among them. It was shown that its antiproliferative activity is mediated by cell cycle arrest at low concentrations and apoptosis induction at high concentrations [57]. Recently, a hypoglycemic effect of piceatannol was revealed. It was found that piceatannol improves glucose uptake in myocytes through the phosphorylation of AMPK and translocation of GLUT4 (glucose transporter 4) to the plasma membrane, as well as prevents an increase in blood glucose levels at early stages and improves the impaired glucose tolerance at late stages of diabetes in diabetic mouse models. The ability of piceatannol to stimulate the activation of AMPK, GLUT4 translocation, and glucose uptake in muscle cells in the absence of insulin determines the possibility for this compound to overcome the insulin resistance, which makes it a promising candidate for the prevention and treatment of diabetes [58].

Salidroside (Fig. 9.8) is an active glycoside of *Rhodiola rosea*, which provides antidepressant and anxiolytic properties for the extracts of this plant. It was shown that salidroside dose-dependently stimulates glucose uptake via the activation of AMPK [59]. Salidroside reduces blood glucose and serum insulin levels and alleviates insulin resistance. It is believed that salidroside inhibits Complex I of the electron transport chain, and thereby increases the AMP/ATP ratio, which is an activating factor for AMPK [60].

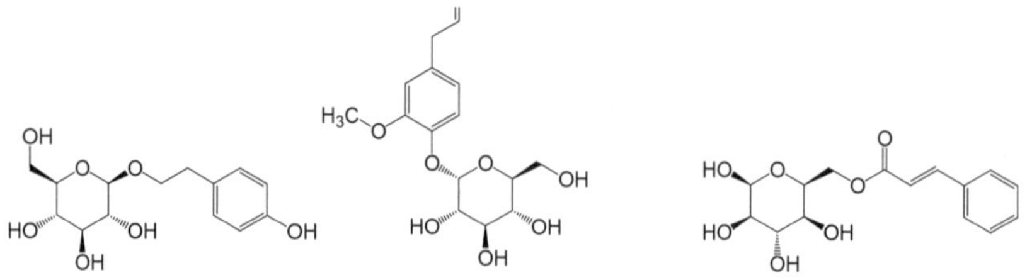

| Salidroside | Eugenyl β-D-glucopyranoside | 6-O-cinnamoyl-D-glucopyranose |

FIGURE 9.8 Glycosidic activators of AMPK.

An unusual approach for the search of AMPK activators among the drugs of traditional Chinese medicine was applied in the paper [61]. The scientists used the most comprehensive database of traditional Chinese medicine compounds (TCM Database@Taiwan, http://tcm.cmu.edu.tw) for virtual screening of AMPK activators, which interact with the AMP-binding sites. The primary selection criterion was calculated binding energy, and then the selection was carried out according to bioavailability and maximum structural similarity to AMP. As a result, two compounds of natural origin, eugenyl β-D-glucopyranoside and 6-*O*-cinnamoyl-D-gluco-pyranose (Fig. 9.8), were proposed as activators of AMPK. Their ability to bind AMPK in a AMP-like manner was demonstrated using molecular dynamic simulations. Despite the fact that there is no information on the activity of the identified compounds, such approach to the analysis of natural compound libraries is very promising.

Neuroprotective properties were found in oleuropein aglycone (Fig. 9.9), the main polyphenol of extra virgin olive oil. It was shown in the model of amyloid beta accumulation that the use of this compound significantly alleviates the symptoms associated with Alzheimer's disease due to induction of autophagy [62]. When trying to understand the mechanism of induction, it was found that oleuropein aglycone activates AMPK-dependent signaling cascade through the bicyclic release of Ca^{2+} ions, which activates CaMKKβ and causes the AMPK phosphorylation [63].

Panduratin A (Fig. 9.9), isolated from rhizomes of *Boesenbergia pandurata*, was identified as a natural activator of AMPK. Studies showed its therapeutic potential for the treatment of obesity; it was determined that the mechanism of the AMPK activation is caused by the LKB1 stimulation. Oral administration of panduratin A reduced weight gain, fat mass, fatty liver and improved serum lipid profiles in obese mice [64].

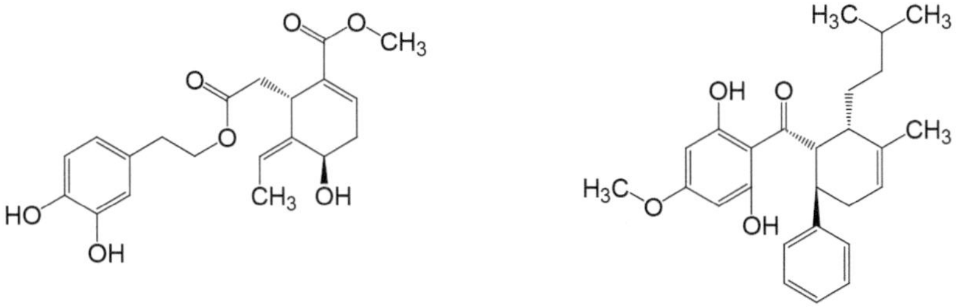

Oleuropein aglycone

Panduratin A

FIGURE 9.9   Chalcones.

Arctiin and its aglycone arctigenin (Fig. 9.10) are the main lignans isolated from burdock seeds (*Arctium lappa*). Arctiin was shown to decrease body weight in diet-induced obese mice due to the activation of the AMPK-dependent cascade [65]. At the same time, it was demonstrated that arctigenin dose-dependently activates AMPK in vivo, but fails to activate recombinant AMPK in vitro, suggesting an indirect activation

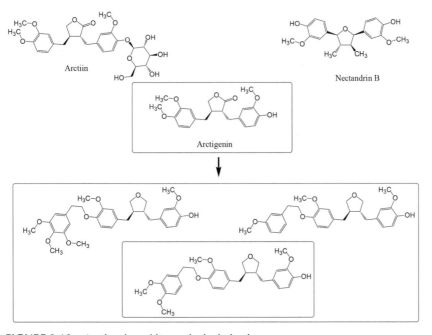

Arctiin

Nectandrin B

Arctigenin

FIGURE 9.10   Arctigenin and its synthetic derivatives.

mechanism. It is believed that arctigenin activates AMPK by stimulating LKB1- and CaMKKβ-dependent pathways [66]. A series of derivatives (Fig. 9.10) was designed and synthesized based on arctigenin. It was shown that the synthetic derivatives of arctigenin activate AMPK in a greater extent compared with the parent compound. It was found that 2-(3,4-dimethoxyphenyl)ethyl ether moiety is the potential functional group to increase the AMPK phosphorylation [67].

Nectandrin B (Fig. 9.10), a lignan structurally related to arctigenin, was isolated from the total extract of *Myristica fragrans*, which was found to possess antiobesity effect mediated by the AMPK activation. Nectandrin B is considered as a potential candidate for the main active compound of the extract affecting AMPK [68]. The effect of nectandrin B on the AMPK activation was confirmed by abolishment of its effect in a mouse model when inhibiting AMPK by compound C. It was assumed that nectandrin B activates AMPK via the activation of CaMKKβ [69].

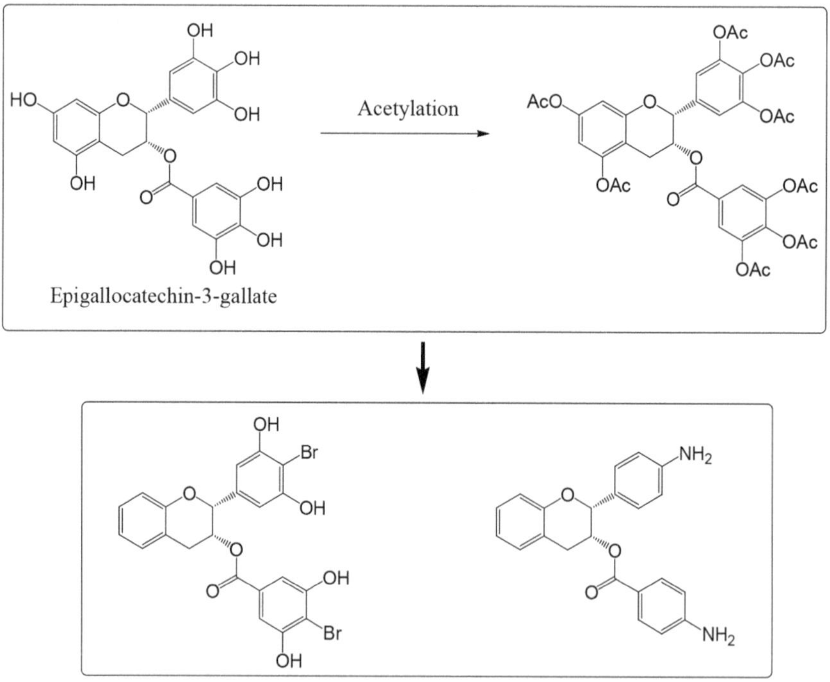

Epigallocatechin-3-gallate

FIGURE 9.11 Epigallocatechin-3-gallate (EGCG) and its synthetic derivatives.

Epigallocatechin-3-gallate (EGCG, Fig. 9.11) is a natural catechin, found in large quantities in green tea. EGCG is widely used as a food supplement due to its beneficial effects on the human organism. It was found that EGCG activates liver AMPK through the CaMKKβ stimulation [70]. A low stability of EGCG under physiological conditions and poor bioavailability resulted in further modification to improve the pharmacological characteristics. More active prodrug with improved bioavailability, which showed anti-tumor properties, was obtained by acetylation of free hydroxyl groups (see Fig. 9.11) [71]. In addition, a focused library of analogs was synthesized based on EGCG. Novel compounds (see Fig. 9.11) were identified as more effective activators of AMPK compared with metformin and acetylated EGCG among the analogs by screening [72].

It is known that phlorotannins such as phlorofucofuroeckol A, dieckol, and dioxinodehydroeckol are potent antioxidants and also possess antiinflammatory properties [73]. In particular, it was shown that dieckol (Fig. 9.12), isolated from a species of edible brown algae of the genus *Ecklonia* (*Ecklonia cava*), also possesses antidiabetic and adipogenic activity. It lowers lipid accumulation, suppresses the expression of key transcription factors of adipogenesis and reduces weight, blood glucose, and insulin levels. It was confirmed that these beneficial effects are mediated by the activation of AMPK [74, 75]. It was shown that another phlorotannin, dioxinodehydroeckol (Fig. 9.12), isolated from *Ecklonia stolonifera*, inhibits adipogenesis via the activation of AMPK and AMPK-dependent pathways [76].

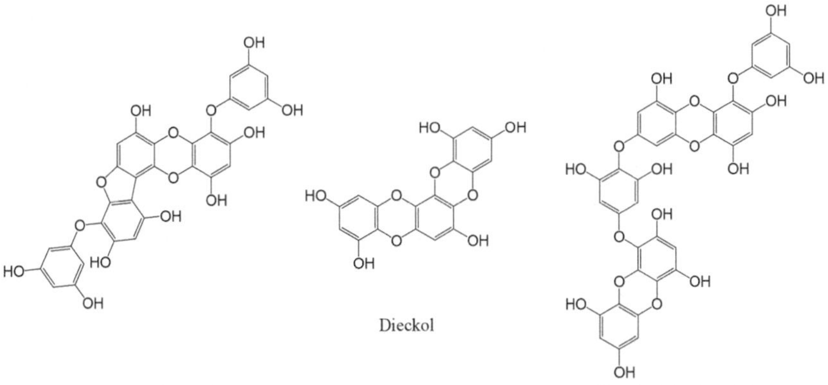

Phlorofucofuroeckol A

Dieckol

Dioxinodehydroeckol

FIGURE 9.12 Phlorotannins.

Among the natural flavonoids, the ability to activate AMPK was observed for quercetin and hispidulin (Fig. 9.13) [77]. It was shown that quercetin inhibits adipogenesis in obesity and induces apoptosis in cancer cells due to the stimulation of the AMPK-signaling cascade [78, 79]. It was found that quercetin activates AMPK through the inhibition of phosphatase 2C, the primary enzyme, which dephosphorylates and inactivates AMPK [80]. Hispidulin is able to activate AMPK in cancer cells, as well as sensitize cancer cells to apoptosis due to the activation of AMPK [81, 82]. Both compounds are used in traditional Chinese medicine and widely used as food supplements.

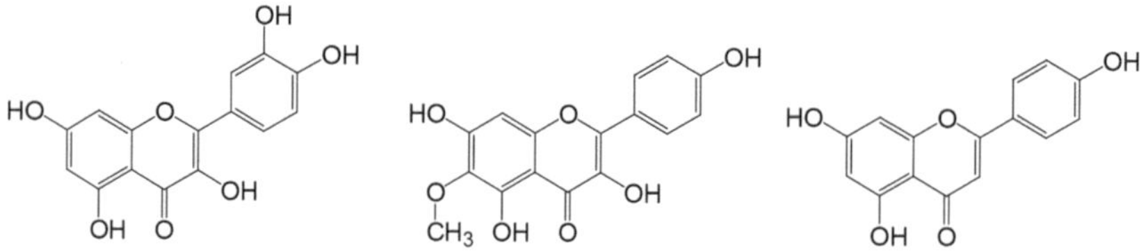

                    Quercetin   Hispidulin   Naringenin

FIGURE 9.13   Flavonoids.

Naringenin (Fig. 9.13) is a flavonoid found in citrus fruit, especially in grapefruit, and also tomatoes. It has been reported to possess antioxidant, anticancer, and antiatherogenic properties, as well as effects on lipid metabolism and plasma glucose levels. It showed an insulin-like effect in studies in vitro. Naringenin does not have a significant effect on glucose transport in myoblasts suggesting that naringenin affects GLUT4 translocation or activity. It was shown that the mechanism of naringenin action involves the activation of AMPK and is similar to that of resveratrol [83]. It was hypothesized that antidiabetic and antiobesity effects of naringenin in vivo reported recently may be mediated by AMPK [84].

Saponarin (Fig. 9.14) is a flavone glucoside, found in *Saponaria officinalis* and *Strongylodon macrobotrys*, giving the characteristic jade color to the flowers of the latter. It was shown that saponarin does not directly interact with AMPK, but activates AMPK in a Ca^{2+}-dependent way. In experiments saponarin demonstrated the ability to reduce the rate of gluconeogenesis and increase glucose uptake through the regulation of the AMPK activity [85].

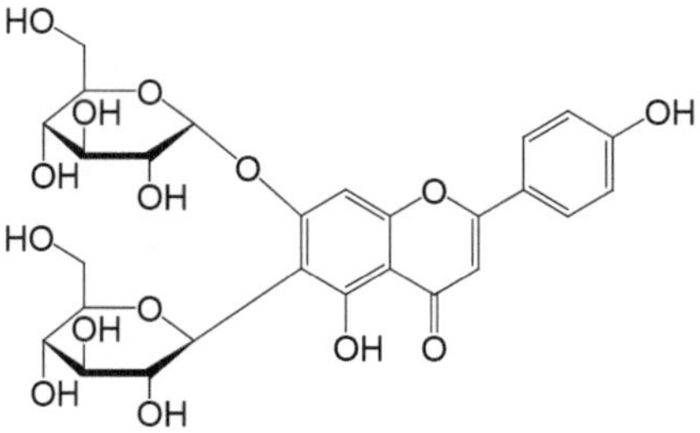

FIGURE 9.14 Saponarin.

Baicalin, a flavonoid of Baikal skullcap (*Scutellaria baicalensis*), and its aglycone baicalein (Fig. 9.15) also show the activity against AMPK. It was shown that the long-term administration of baicalin stimulates the phosphorylation of liver AMPK that enables to ameliorate hepatic steatosis and metabolic disorders [86]. It was found that at least partially the activation of AMPK by baicalin is mediated by the stimulation of the CaMKKβ-dependent pathway [87]. The activation of AMPK in vivo was also shown for baicalein; it facilitated to minimize the manifestations of metabolic disorders [88, 89]. Also, new compounds based on baicalein,

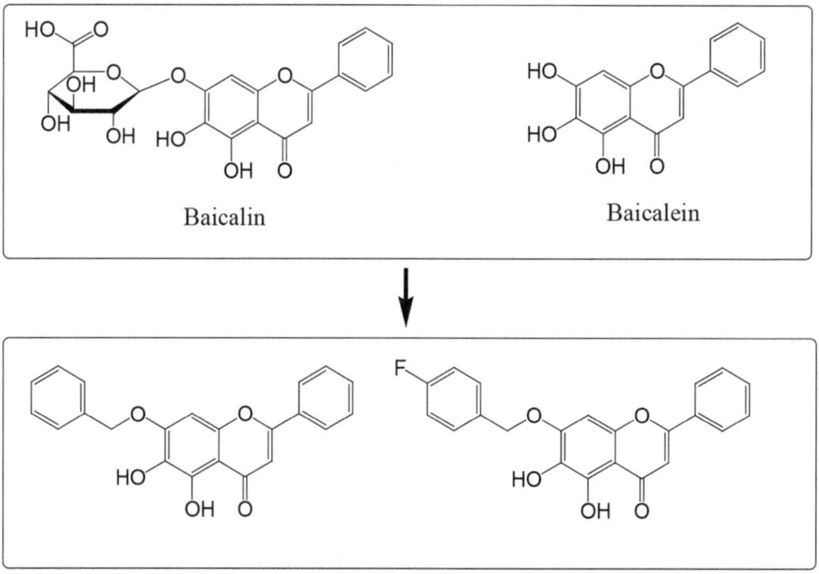

FIGURE 9.15 Baicalein and its derivatives.

capable of activating AMPK, were designed and synthesized (Fig. 9.15). The compounds showed a higher AMPK activity compared with baicalein and AICAR when studying their antiproliferative properties [90].

Isoflavone genistein (Fig. 9.16) was first isolated from *Genista tinctoria*, and thus it is so named. Despite a wide spectrum of biological activity, antitumor properties of this compound are most known [91]. It was shown that genistein causes the activation of AMPK in adipocytes; it is assumed that the activation is determined by the stimulation of the intracellular release of reactive oxygen species, as in the case of capsaicin [49]. It is believed that its antioxidant effect is mediated by the activation of AMPK [92].

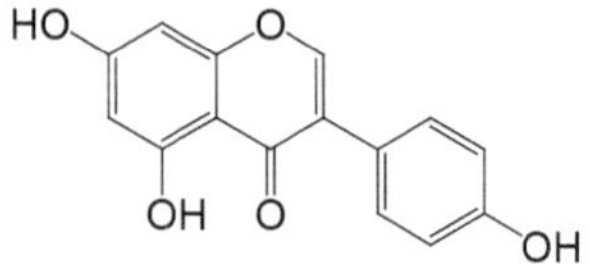

FIGURE 9.16   Genistein.

A number of compounds belonging to dihydrochalcones and homoisoflavonoids (Fig. 9.17) was isolated from the root of angular Solomon's seal (*Polygonatum odoratum*). The compounds showed significant activation of AMPK when studying their activating effect [93].

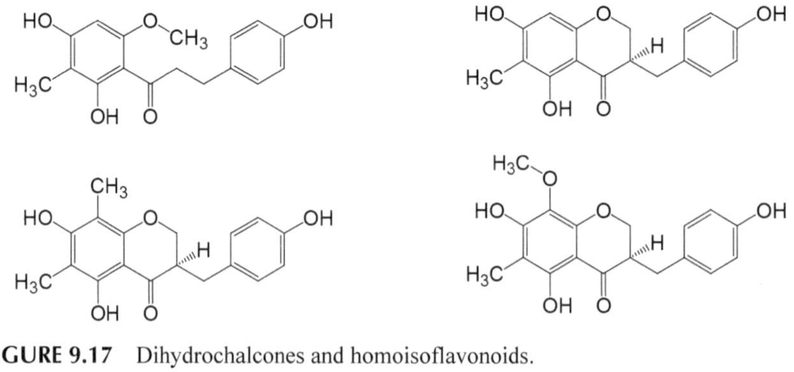

FIGURE 9.17   Dihydrochalcones and homoisoflavonoids.

Screening of chemical compound libraries using an in vitro fluorescence resonance energy transfer (FRET)-based assay enabled to identify

benzophenanthridine alkaloid sanguinarine (Fig. 9.18) as a direct activator of AMPK [94]. Sanguinarine can be found in many plants, including
Sanguinaria canadensis. Despite the toxicity of the alkaloid, its ability
to induce apoptosis in cancer cells, in particular, through the generation
of reactive oxygen species makes it a drug candidate for the treatment of
various cancers [95, 96]. As a result of the molecular docking, the most
probable binding site of AMPK, distinct from the binding site of salicylate, was determined; it indicates the possibility that there are a number of
additional allosteric sites, in addition to the AMP-binding sites. Considering the direct mechanism of the interaction with AMPK, the activity of
sanguinarine, however, is 5–10 times lower than that of AMP [95].

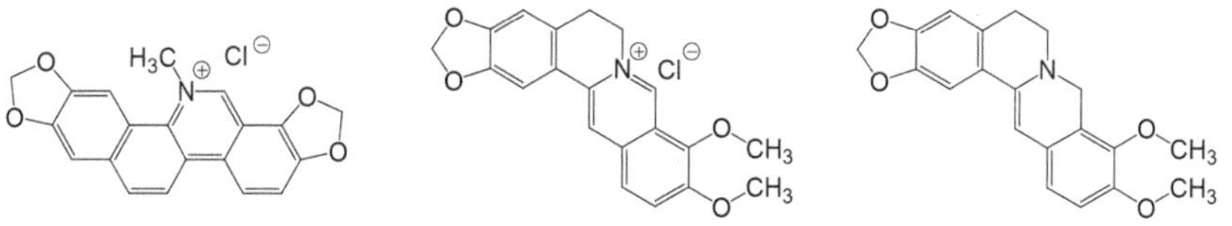

| Sanguinarine | Berberine | Dihyidroberberine |

FIGURE 9.18   Plant alkaloids.

Berberine (Fig. 9.18) is a plant alkaloid, found in many plants,
including *Berberis vulgaris*. In the form of bisulfate, berberine is used as
a cholagogue. It was shown that berberine increases the AMPK activity to
a greater extent compared with AICAR. In addition, it was found that its
therapeutic effect in the treatment of diabetes is at least partially mediated
by the activation of AMPK [97]. It is believed that the mechanism of the
AMPK activation by berberine and more bioavailable dihyidroberberine
(Fig. 9.18) does not involve the LKB1- or CaMKKβ-signaling pathway
and is associated with changes in the AMP/ATP ratio as a result of the
inhibition of Complex I of the electron transport chain [98]. It should be
noted that berberine and its derivative have a significant structural similarity with sanguinarine, which possesses a direct activating effect.

Structures capable of activating AMPK are also found among the terpenoids. Nootkatone (Fig. 9.19), a sesquiterpene ketone, is interesting for its
strong smell of grapefruit and varied biological activity; it is extensively
used as a repellent and insecticide. It was found that nootkatone increases
the AMPK activity along with an increase of the AMP/ATP ratio and the
phosphorylation of AMPK and its downstream targets. It is assumed that

the effect of nootkatone is also mediated by CaMKKβ, as the inhibition of this kinase facilitates to reduce the phosphorylation degree of both AMPK and ACC. It was shown that nootkatone stimulates fatty acid oxidation in muscle and liver, as well as energy metabolism at the whole-body level. The use of nootkatone as a dietary supplement enables to inhibit fat accumulation, suggesting its application in obesity [99].

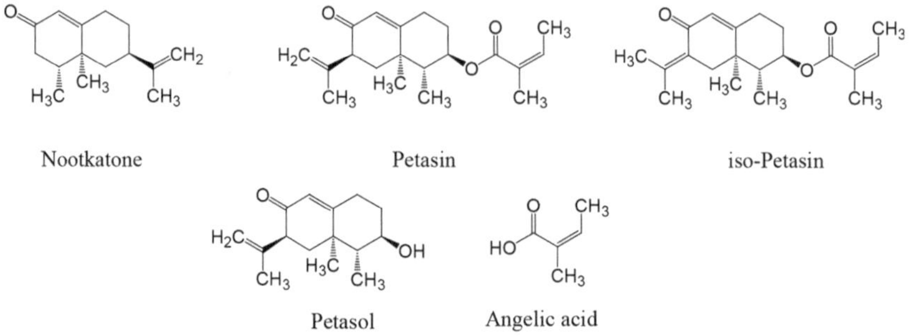

FIGURE 9.19   Sesquiterpene ketones.

When studying an extract of *Petasites japonicus*, it was found that this extract possesses the ability to increase the degree of the AMPK phosphorylation in vitro. A sesquiterpene ketone, petasin (Fig. 9.19), is an ester of petasol and angelic acid; it was identified as the main active compound, which activates AMPK. The activating effect was achieved due to the change in the AMP/ATP ratio as a result of the inhibition of Complex I of the electron transport chain in mitochondria. In mice, AMPK activation was observed in liver, skeletal muscle, and adipose tissue when petasin was orally administered. Furthermore, it was shown that petasin is able to enhance glucose tolerance in experimental animals. However, structurally related isopetasin (Fig. 9.19) is three times less potent in the AMPK activation [100].

Daphne genkwa is an evergreen shrub, included in the list of 50 fundamental herbs used in traditional Chinese medicine. It was shown that diterpenoid yuanhuacine (Fig. 9.20), isolated from the flowers of the plant, is also able to activate AMPK. It was found that it significantly activates the AMPK-dependent signaling pathway, and in particular, increases the expression of the phosphorylated AMPK. The inhibition of downstream targets of AMPK, involved in the tumor growth, enables to consider yuanhuacine as a potential natural chemotherapeutic agent [101].

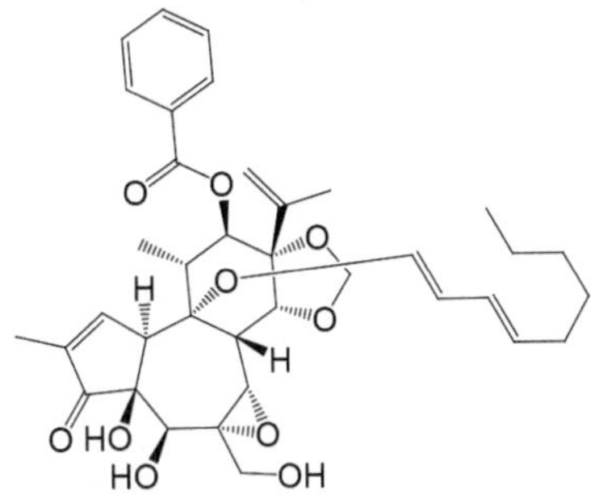

FIGURE 9.20   Yuanhuacine (YC).

Bitter melon (*Momordica charantia*), rich in iron and vitamins, is widely used to treat various diseases in tropical countries and is also used as a folk remedy for diabetes due to its hypoglycemic effect. In particular, it was shown that triterpenoids and their glycosides (Fig. 9.21), isolated from bitter melon, are able to activate AMPK in myotubes and adipocytes and stimulate glucose uptake by peripheral tissues due to the translocation of GLUT4 to the plasma membrane. The mechanism of these triterpenoids is probably mediated by LKB1, because the activation of AMPK was not observed in LKB1-deficient cells [102].

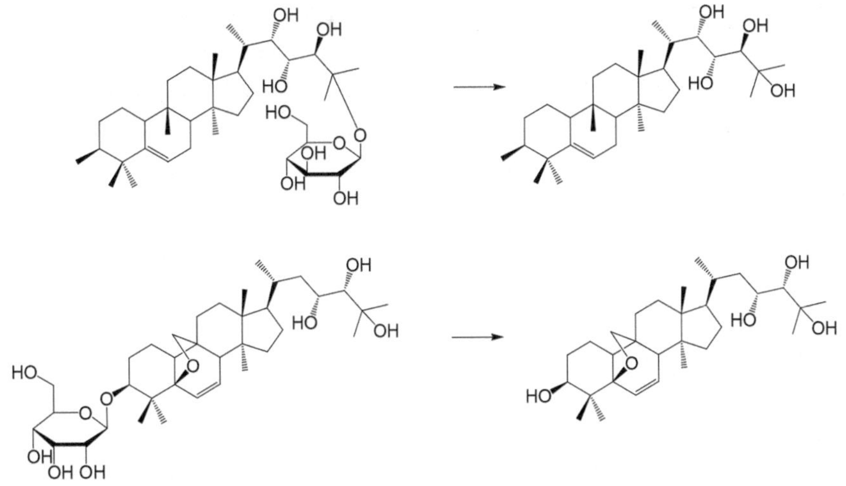

FIGURE 9.21   Triterpenoids and glycosides isolated from bitter melon.

Other representatives of triterpenoids, possessing a hypoglycemic effect, were isolated from the extract of *Siraitia grosvenorii*. This plant is widely used in traditional Chinese medicine, whereas the extract obtained from its fruits is used as a natural sweetener over thousands of years. Mogrol and its structural analogs (Fig. 9.22) were found as potent activators of liver AMPK among the known and first-identified mogrosides, isolated from the extract of *Siraitia grosvenorii*. It was shown that hypoglycemic and hypolipidemic properties of *S. grosvenorii* are at least partially mediated by the activation of AMPK [103].

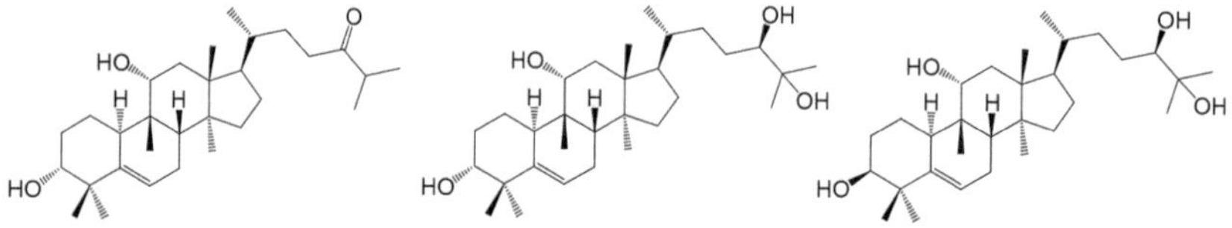

Mogrol

FIGURE 9.22   Mogrosides of *Siraitia grosvenorii*.

Oleanolic acid (Fig. 9.23) is a natural triterpenoid, which can be found in olive oil. In particular, it is used in traditional Chinese medicine for the treatment of diabetes and ischemic heart disease. It is assumed that cardioprotective effects of oleanolic acid are mediated by the activation of the AMPK-dependent signaling cascade. It was shown that this compound stimulates the phosphorylation of AMPK and protects cardiomyocytes from contractile dysfunction [104]. Recently, the combination of oleanolic acid and metformin was proposed for the synergistic treatment of diabetes [105].

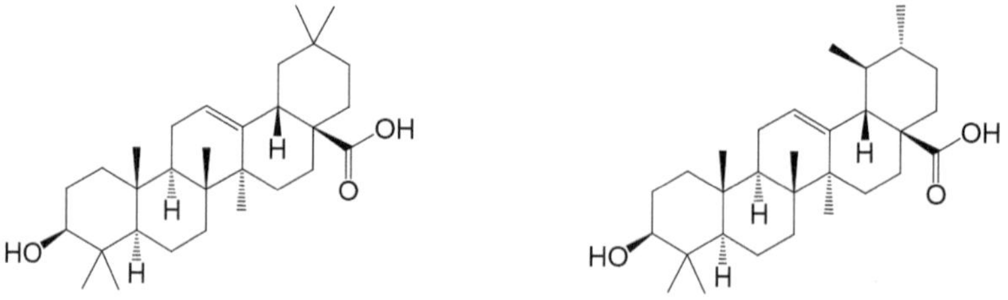

Oleanolic acid   Ursolic acid

FIGURE 9.23   Sapogenins.

Ursolic acid (Fig. 9.23) is present in many plants, as well as fruits and berries, such as basil, apple, cranberry, and others. It was shown that it inhibits growth and induces apoptosis in hepatic carcinoma cells through the activation and phosphorylation of AMPK with subsequent phosphorylation of its downstream targets [106, 107]. A dose-dependent activation of AMPK, abolished by silence of LKB1, was demonstrated for ursolic acid, indicating the direct involvement of LKB1 in the activation mechanism [108].

Finally, lycopene (Fig. 9.24) was found as an AMPK activator among tetraterpenoid. It is a linear tetraterpenoid consisting of eight isoprenoid units; it determines the color of orange-red parts of a number of plants including tomato. It was shown that lycopene induces the phosphorylation of AMPK via activation of LKB1 and CaMKKβ, which causes increased accumulation of the phosphorylated form of AMPK in the nucleus and possesses antiinflammatory effect in microglia cells. It is expected that lycopene may be useful for the treatment of neuroinflammation-associated disorders [109].

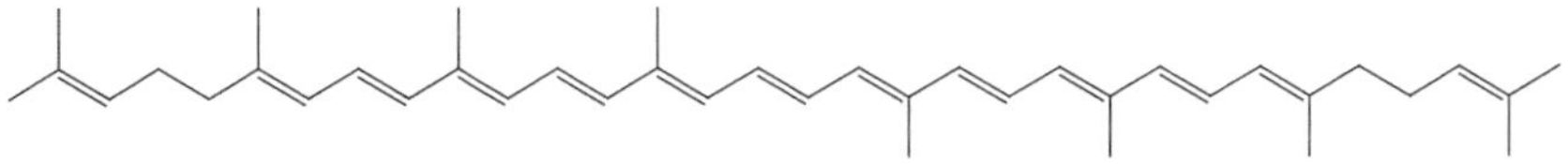

FIGURE 9.24   All-*trans* lycopene.

9.2 CONCLUSIONS

Plant activators of AMPK reflect objectively the overall picture of biologically active natural compounds. The vast majority of them are substances with multiple activities. The most outstanding examples include resveratrol and quercetin; dozens of papers are devoted to the activity of each of them. In most cases, plant AMPK activators are indirect activators, which as a rule trigger numerous and complex signaling cascades. If it is determined that the compound affects AMPK through LKB1, this merely means that the signal for the activation is transmitted by LKB1, but does not define LKB1 as a direct target. However, the absence of a systematic search for small molecules that activate LKB1 and CaMKKβ makes natural compounds, possessing such activity, particularly valuable.

Identified activators, despite all the drawbacks, are able to serve as a basis for the development of selective indirect activators of AMPK.

KEYWORDS

- **AMPK**
- **polyphenols**
- **flavonoids**
- **alkaloids**
- **terpenoids**
- **natural compounds**

REFERENCES

1. Hardie, D. G.; Ashford, M. L. AMPK: Regulating Energy Balance at the Cellular and Whole Body Levels. *Physiology (Bethesda)*. **2014,** *29*(2), 99–107.
2. Ghillebert, R.; Swinnen, E.; Wen, J.; Vandesteene, L.; et al. The AMPK/SNF1/ SnRK1 Fuel Gauge and Energy Regulator: Structure, Function and Regulation. *FEBS J.* **2011,** *278*(21), 3978–3990.
3. Steinberg, G. R.; Kemp, B. E. AMPK in Health and Disease. *Physiol. Rev.* **2009,** *89*(3), 1025–1078.
4. Kurokawa, H.; Sugiyama, S.; Nozaki, T.; Sugamura, K.; et al. Telmisartan Enhances Mitochondrial Activity and Alters Cellular Functions in Human Coronary Artery Endothelial Cells via AMP-Activated Protein Kinase Pathway. *Atherosclerosis*. **2015,** *239*(2), 375–385.
5. Ulgherait, M.; Rana, A.; Rera, M.; Graniel, J.; Walker, D. W. AMPK Modulates Tissue and Organismal Aging in a Non-Cell-Autonomous Manner. *Cell Rep.* **2014,** *8*(6), 1767–1780.
6. Gureev, M. A.; Davidovich, P. B.; Tribulovich, V. G.; Garabadzhiu, A. V. Natural Compounds as a Basis for the Design of Modulators of p53 Activity. *Russ. Chem. Bull.* **2014,** *63*(9), 1963–1975.
7. Crute, B. E.; Seefeld, K.; Gamble, J.; Kemp, B. E.; Witters, L. A. Functional Domains of the Alpha1 Catalytic Subunit of the AMP-Activated Protein Kinase. **1998,** *273*(52), 35347–35354.
8. McBride, A.; Ghilagaber, S.; Nikolaev, A.; Hardie, D. G. The Glycogen-Binding Domain on the AMPK Beta Subunit Allows the Kinase to Act as a Glycogen Sensor. *Cell Metab.* **2009,** *9*(1), 23–34.

9. Cheung, P. C.; Salt, I. P.; Davies, S. P.; Hardie, D. G.; Carling, D. Characterization of AMP-Activated Protein Kinase Gammasubunit Isoforms and their Role in AMP Binding. *Biochem. J.* **2000,** *346,* 659–669.

10. Suter, M.; Riek, U.; Tuerk, R.; Schlattner, U.; et al. Dissecting the Role of 5'-AMP for Allosteric Stimulation, Activation, and Deactivation of AMP-Activated Protein Kinase. *J. Biol. Chem.* **2006,** *281*(43), 32207–32216.

11. Hawley, S. A.; Davison, M.; Woods, A.; Davies, S. P.; et al. Characterization of the AMP-Activated Protein Kinase Kinase from Rat Liver and Identification of Threonine 172 as the Major Site at which it Phosphorylates AMP-Activated Protein Kinase. *J. Biol. Chem.* **1996,** *271*(44), 27879–27887.

12. Woods, A.; Johnstone, S. R.; Dickerson, K.; Leiper, F. C.; et al. LKB1 is the Upstream Kinase in the AMP-Activated Protein Kinase Cascade. *Curr. Biol.* **2003,** *13*(22), 2004–2008.

13. Hawley, S. A.; Selbert, M. A.; Goldstein, E. G.; Edelman, A. M. et al. 5'-AMP Activates the AMP-Activated Protein Kinase Cascade, and Ca2+/Calmodulin Activates the Calmodulin-Dependent Protein Kinase I Cascade, via three Independent Mechanisms. *J. Biol. Chem.* **1995,** *270*(45), 27186–27191.

14. Xie, M.; Zhang, D.; Dyck, J. R.; Li, Y.; et al. A Pivotal Role for Endogenous TGF-beta-Activated Kinase-1 in the LKB1/AMP-Activated Protein Kinase Energy-Sensor Pathway. *Proc. Natl. Acad. Sci. U S A.* **2006,** *103*(46), 17378–17383.

15. Pang, T.; Zhang, Z. S.; Gu, M.; Qiu, B. Y.; et al. Small Molecule Antagonizes Autoinhibition and Activates AMP-Activated Protein Kinase in Cells, *J. Biol. Chem.* **2008,** *283*(23), 16051–16060.

16. Xiao, B.; Sanders, M. J.; Carmena, D.; Bright, N. J.; et al. Structural Basis of AMPK Regulation by Small Molecule Activators. *Nat. Commun.* **2013,** *4,* 3017.

17. Viollet, B.; Horman, S.; Leclerc, J.; Lantier, L. et al. AMPK Inhibition in Health and Disease. *Crit. Rev. Biochem. Mol. Biol.* **2010,** *45*(4), 276–295.

18. Marsin, A. S.; Bertrand, L.; Rider, M. H.; Deprez, J.; et al. Phosphorylation and Activation of Heart PFK-2 by AMPK has a Role in the Stimulation of Glycolysis During Ischaemia. *Curr. Biol.* **2000,** *10*(20), 1247–1255.

19. McGee, S. L.; van Denderen, B. J.; Howlett, K. F.; Mollica, J.; et al. AMP-Activated Protein Kinase Regulates GLUT4 Transcription by Phosphorylating Histone Deacetylase 5. *Diabetes.* **2008,** *57*(4), 860–867.

20. Lochhead, P. A.; Salt, I. P.; Walker, K. S.; Hardie, D. G.; Sutherland, C. 5-Aminoimidazole-4-carboxamide Riboside Mimics the Effects of Insulin on the Expression of the 2 Key Gluconeogenic Genes PEPCK and Glucose-6-phosphatase. Diabetes. **2000,** *49*(6), 896–903.

21. Davies, S. P.; Carling, D.; Munday, M. R.; Hardie, D. G. Diurnal Rhythm of Phosphorylation of Rat Liver Acetyl-CoA Carboxylase by the AMP-Activated Protein Kinase, Demonstrated Using Freeze-Clamping. Effects of High Fat Diets. *Eur. J. Biochem.* **1992,** *203*(3), 615–623.

22. Merrill, G. F.; Kurth, E. J.; Hardie, D. G.; Winder, W. W. AICA Riboside Increases AMP-Activated Protein Kinase, Fatty Acid Oxidation, and Glucose Uptake in Rat Muscle. *Am. J. Physiol.* **1997,** *273*(6), E1107–1112.

23. Clarke, P. R.; Hardie, D. G. Regulation of HMG-CoA Reductase: Identification of the Site Phosphorylated by the AMP-Activated Protein Kinase In Vitro and in Intact Rat Liver. *EMBO J.* **1990,** *9*(8), 2439–2446.

24. Novikova, D. S.; Garabadzhiu, A. V.; Melino, G.; Barlev, N. A.; Tribulovich, V. G. AMP-Activated Protein Kinase: Structure, Function, and Role in Pathological Processes. *Biochemistry (Moscow).* **2015,** *80*(2), 127–144.

25. Novikova, D. S.; Garabadzhiu, A. V.; Melino, G.; Barlev, N. A.; Tribulovich, V. G. Small-Molecule Activators of AMP-Activated Protein Kinase as Modulators of Energy Metabolism. *Russ. Chem. Bull.* **2015,** *64*(7), 1497–1517.

26. O'Brien, A. J.; Villani, L. A.; Broadfield, L. A.; Houde, V. P.; et al. Salicylate Activates AMPK and Synergizes with Metformin to Reduce the Survival of Prostate and Lung Cancer Cells Ex Vivo through Inhibition of De Novo Lipogenesis. *Biochem. J.* **2015,** *469*(2), 177–187.

27. Mounier, R.; Theret, M.; Lantier, L.; Foretz, M.; Viollet, B. Expanding Roles for AMPK in Skeletal Muscle Plasticity. *Trends Endocrinol. Metab.* **2015,** *26*(6), 275–286.

28. Barger, G.; White, F. D. The Constitution of Galegine. *Biochem. J.* **1923,** *17*(6), 827–835.

29. Witters, L. A. The Blooming of the French Lilac. *J. Clin. Investig.* **2001,** *108*(8), 1105–1107.

30. Bailey, C. J.; Turner, R. C. Metformin. *New Engl. J. Med.* **1996,** *334*(9), 574–579.

31. Zhou, G.; Myers, R.; Li.Y.; Chen, Y.; et al. Role of AMP-Activated Protein Kinase in Mechanism of Metformin Action. *J. Clin. Investig.* **2001,** *108*(8), 1167–1174.

32. Owen, M. R.; Doran, E.; Halestrap, A. P. Evidence that Metformin Exerts Its Anti-Diabetic Effects through Inhibition of Complex 1 of the Mitochondrial Respiratory Chain. *Biochem. J.* **2000,** *348*, 607–614.

33. Shaw, R. J.; Lamia, K. A.; Vasquez, D.; Koo, S. H.; et al. The Kinase LKB1 Mediates Glucose Homeostasis in Liver and Therapeutic Effects of Metformin. *Science.* **2005,** *310*(5754), 1642–1646.

34. Reymond, P.; Farmer, E. E. Jasmonate and Salicylate as Global Signals for Defense Gene Expression. *Curr. Opin. Plant Biol.* **1998,** *1*(5), 404–411.

35. Hawley, S. A, Fullerton, M. D.; Ross, F. A.; Schertzer, J. D.; et al. The Ancient Drug Salicylate Directly Activates AMP-Activated Protein Kinase. *Science.* **2012,** *336*(6083), 918–922.

36. Dhillon, N.; Aggarwal, B. B.; Newman, R. A.; Wolff, R. A. et al. Phase II Trial of Curcumin in Patients with Advanced Pancreatic Cancer. *Clin. Cancer Res.* **2008,** *14*(14), 4491–4499.

37. Pan, W.; Yang, H.; Cao, C.; Song, X.; et al. AMPK Mediates Curcumin-Induced Cell Death in CaOV3 Ovarian Cancer Cells. *Oncol. Rep.* **2008,** *20*(6), 1553–1559.

38. Kim, J. H.; Park, J. M.; Kim, E. K.; Lee, J. O.; et al. Curcumin Stimulates Glucose Uptake through AMPK-p38 MAPK Pathways in L6 Myotube Cells. *J. Cell. Physiol.* **2010,** *223*(3), 771–778.

39. Kim, T.; Davis, J.; Zhang, A. J.; He, X.; Mathews, S. T. Curcumin Activates AMPK and Suppresses Gluconeogenic Gene Expression in Hepatoma Cells. *Biochem. Biophys. Res. Commun.* **2009,** *388*(2), 377–382.

40. Hung, C. M.; Su, Y. H.; Lin, H. Y.; Lin, J. N.; et al. Demethoxycurcumin Modulates Prostate Cancer Cell Proliferation via AMPK-Induced Down-Regulation of HSP70 and EGFR. *J. Agric. Food Chem.* **2012**, *60*(34), 8427–8434.

41. Tatsuzaki, J.; Bastow, K. F.; Nakagawa-Goto, K.; Nakamura, S.; et al. Dehydrozingerone, Chalcone, and Isoeugenol Analogues as In Vitro Anticancer Agents. *J. Nat. Prod.* **2006**, *69*(10), 1445–1449.

42. Kim, S. J.; Kim, H. M.; Lee, E. S.; Kim, N.; et al. Dehydrozingerone Exerts Beneficial Metabolic Effects in High-Fat Diet-Induced Obese Mice via AMPK Activation in Skeletal Muscle. *J. Cell. Mol. Med.* **2015**, *19*(3), 620–629.

43. Ong, K. W.; Hsu, A.; Tan, B. K. Chlorogenic Acid Stimulates Glucose Transport in Skeletal Muscle via AMPK Activation: a Contributor to the Beneficial Effects of Coffee on Diabetes. *PLoS ONE.* **2012**, *7*(3), e32718.

44. Ong, K. W.; Hsu, A.; Tan, B. K. Anti-Diabetic and Anti-Lipidemic Effects of Chlorogenic Acid are Mediated by AMPK Activation. *Biochem. Pharmacol.* **2013**, *85*(9), 1341–1351.

45. Jin, S.; Chang, C.; Zhang, L.; Liu, Y. et al. Chlorogenic Acid Improves Late Diabetes through Adiponectin Receptor Signaling Pathways in db/db Mice. *PLoS ONE.* **2015**, *10*(4), e0120842.

46. Tsuda, S.; Egawa, T.; Ma, X.; Oshima, R.; et al. Coffee Polyphenol Caffeic Acid But not Chlorogenic Acid Increases 5′AMP-Activated Protein Kinase and Insulin-Independent Glucose Transport in Rat Skeletal Muscle. *J. Nutr. Biochem.* **2012**, *23*(11), 1403–1409.

47. Kopp, C.; Singh, S. P.; Regenhard, P.; Müller, U. et al. Trans-Cinnamic Acid Increases Adiponectin and the Phosphorylation of AMP-Activated Protein Kinase through G-Protein-Coupled Receptor Signaling in 3T3-L1 Adipocytes. *Int. J. Mol. Sci.* **2014**, *15*(2), 2906–2915.

48. Kang, J. H.; Tsuyoshi, G.; Le Ngoc, H.; Kim, H. M.; et al. Dietary Capsaicin Attenuates Metabolic Dysregulation in Genetically Obese Diabetic Mice. *J. Med. Food.* **2011**, *14*(3), 310–315.

49. Hwang, J. T.; Park, I. J.; Shin, J. I.; Lee, Y. K.; et al. Genistein, EGCG, and Capsaicin Inhibit Adipocyte Differentiation Process via Activating AMP-Activated Protein Kinase. *Biochem. Biophys. Res. Commun.* **2005**, *338*(2), 694–699.

50. Seo, M. S.; Kim, J. H.; Kim, H. J.; Chang, K. C.; Park, S. W. Honokiol Activates the LKB1-AMPK Signaling Pathway and Attenuates the Lipid Accumulation in Hepatocytes. *Toxicol. Appl. Pharmacol.* **2015**, *284*(2), 113–124.

51. Park, J. B.; Lee, M. S.; Cha, E. Y.; Lee, J. S.; et al. Magnolol-Induced Apoptosisin HCT-116 Colon Cancer Cells Is Associated with the AMP-Activated Protein Kinase Signaling Pathway. *Biol. Pharm. Bull.* **2012**, *35*(9),1614–1612.

52. Lee, J. H.; Jung, J. Y.; Jang, E. J.; Jegal, K. H.; et al. Combination of Honokiol and Magnolol Inhibits Hepatic Steatosis through AMPK-SREBP-1c Pathway. *Exp. Biol. Med. (Maywood, N. J.)* **2015**, *240*(4), 508–518.

53. Huh, T. L.; Song, H.; Kim, J. E.; Kwon, B. M.; et al. Composition for the Treatment of Diabetes and Metabolic Syndrome Containing Obovatol and Its Synthesized Derivatives, US Patent US20100125103, 2010.

54. Gledhill, J. R.; Montgomery, M. G.; Leslie, A. G.; Walker, J. E. Mechanism of Inhibition of Bovine F1-ATPase by Resveratrol and Related Polyphenols. *Proc. Natl. Acad. Sci. U S A.* **2007**, *104*(34), 13632–13627.

55. Vingtdeux, V.; Giliberto, L.; Zhao, H.; Chandakkar, P.; et al. AMP-activated Protein Kinase Signaling Activation by Resveratrol Modulates Amyloid-Beta Peptide Metabolism. *J. Biol. Chem.* **2010,** *285*(12), 9100–9113.

56. Papandreou, I.; Verras, M.; McNeil, B.; Koong, A. C.; Denko, N. C. Plant Stilbenes Induce Endoplasmic Reticulum Stress and Their Anti-Cancer Activity Can Be Enhanced by Inhibitors of Autophagy. *Exp. Cell Res.* **2015,** *339*(1), 147–153.

57. Kita, Y.; Miura, Y.; Yagasaki, K. Antiproliferative and Anti-Invasive Effect of Piceatannol, a Polyphenol Present in Grapes and Wine, Against Hepatoma AH109A Cells. *J. Biomed. Biotechnol.* **2012,** *2012*, 672416.

58. Minakawa, M.; Miura, Y.; Yagasaki, K. Piceatannol, a Resveratrol Derivative, Promotes Glucose Uptake through Glucose Transporter 4 Translocation to Plasma Membrane in L6 Myocytes and Suppresses Blood Glucose Levels in Type 2 Diabetic Model db/db Mice. *Biochem. Biophys. Res. Commun.* **2012,** *422*(3), 469–475.

59. Li, H. B, Ge, Y. K.; Zheng, X. X.; Zhang, L. Salidroside Stimulated Glucose Uptake in Skeletal Muscle Cells by Activating AMP-Activated Protein Kinase. *Eur. J. Pharmacol.* **2008,** *588*(2–3), 165–169.

60. Zheng, T.; Yang, X.; Wu, D.; Xing, S.; et al. Salidroside Ameliorates Insulin Resistance through Activation of a Mitochondria-Associated AMPK/PI3K/Akt/GSK3β Pathway. *Br. J. Pharmacol.* **2015,** *172*(13), 3284–3301.

61. Tang, H. C.; Chen, C. Y. In Silico Design for Adenosine Monophosphate-Activated Protein Kinase Agonist from Traditional Chinese Medicine for Treatment of Metabolic Syndromes. *Evid. Based Complement. Altern. Med.* **2014,** *2014*, 928589.

62. Casamenti, F.; Grossi, C.; Rigacci, S.; Pantano, D.; et al. Oleuropein Aglycone: A Possible Drug against Degenerative Conditions. In Vivo Evidence of its Effectiveness against Alzheimer's Disease. *J. Alzheimer's Dis.* **2015,** *45*(3), 679–688.

63. Rigacci, S.; Miceli, C.; Nediani, C.; Berti, A.; et al. Oleuropein Aglycone Induces Autophagy via the AMPK/mTOR Signalling Pathway: a Mechanistic Insight. *Oncotarget.* **2015,** *6*(34), 35344–35357.

64. Kim, D.; Lee, M. S.; Jo, K.; Lee, K. E.; Hwang, J. K. Therapeutic Potential of Panduratin A, LKB1-Dependent AMP-Activated Protein Kinase Stimulator, with Activation of PPARα/δ for the Treatment of Obesity. *Diab. Obes. Metab.* **2011,** *13*(7), 584–593.

65. Min, B.; Lee, H.; Song, J. H.; Han, M. J.; Chung, J. Arctiin Inhibits Adipogenesis in 3T3-L1 Cells and Decreases Adiposity and Body Weight in Mice Fed a High-Fat Diet. *Nutr. Res. Pract.* **2014,** *8*(6), 655–661.

66. Tang, X.; Zhuang, J.; Chen, J.; Yu, L.; et al. Arctigenin Efficiently Enhanced Sedentary Mice Treadmill Endurance. *PLoS ONE.* **2011,** *6*(8), e24224.

67. Shen, S.; Zhuang, J.; Chen, Y.; Lei, M.; et al. Synthesis and Biological Evaluation of Arctigenin Ester and Ether Derivatives as Activators of AMPK. *Bioorg. Med. Chem.* **2013,** *21*(13), 3882–3893.

68. Nguyen, P. H.; Le, T. V.; Kang, H. W.; Chae, J.; et al. AMP-Activated Protein Kinase (AMPK) Activators from *Myristica fragrans* (Nutmeg) and their Anti-Obesity Effect. *Bioorg. Med. Chem. Lett.* **2010,** *20*(14), 4128–4131.

69. Hien, T. T.; Oh, W. K.; Nguyen, P. H.; Oh, S. J.; et al. Nectandrin B Activates Endothelial Nitric-Oxide Synthase Phosphorylation in Endothelial Cells: Role of the AMP-Activated Protein Kinase/Estrogen Receptor α/Phosphatidylinositol 3-Kinase/Akt Pathway. *Mol. Pharmacol.* **2011,** *80*(6), 1166–1178.

70. Collins, Q. F.; Liu, H. Y.; Pi, J.; Liu, Z.; et al. Epigallocatechin-3-gallate (EGCG), a Green Tea Polyphenol, Suppresses Hepatic Gluconeogenesis through 5'-AMP-Activated Protein Kinase, *J. Biol. Chem.* **2007,** *282*(41), 30143–30149.

71. Landis-Piwowar, K. R.; Huo, C.; Chen, D.; Milacic, V.; et al. A Novel Prodrug of the Green Tea Polyphenol (-)-Epigallocatechin-3-gallate as a Potential Anticancer Agent. *Cancer Res.* **2007,** *67*(9), 4303–4310.

72. Chen, D.; Pamu, S.; Cui, Q.; Chan, T. H.; Dou, Q. P. Novel Epigallocatechin Gallate (EGCG) Analogs Activate AMP-Activated Protein Kinase Pathway and Target Cancer Stem Cells. *Bioorg. Med. Chem.* **2012,** 20(9), 3031–3037.

73. Kim, A. R.; Shin, T. S.; Lee, M. S.; Park, J. Y.; et al. Isolation and Identification of Phlorotannins from *Ecklonia stolonifera* with Antioxidant and Anti-Inflammatory Properties. *J. Agric. Food Chem.* **2009,** *57*(9), 3483–3489.

74. Ko, S. C.; Lee, M.; Lee, J. H.; Lee, S. H.; et al. Dieckol, a Phlorotannin Isolated from a Brown Seaweed, *Ecklonia cava*, Inhibits Adipogenesis through AMP-Activated Protein Kinase (AMPK) Activation in 3T3-L1 Preadipocytes. *Environ. Toxicol. Pharmacol.* **2013,** *36*(3), 1253–1260.

75. Kang, M. C.; Wijesinghe, W. A.; Lee, S. H.; Kang, S. M.; et al. Dieckol Isolated from Brown Seaweed *Ecklonia cava* Attenuates Type II Diabetes in db/db Mouse Model. *Food Chem. Toxicol.* **2013,** *53*, 294–298.

76. Kim, S. K.; Kong, C. S. Anti-Adipogenic Effect of Dioxinodehydroeckol via AMPK Activation in 3T3-L1 Adipocytes. *Chem. Biol. Interact.* **2010,** *186*(1), 24–29.

77. Hwang, J. T.; Kwon, D. Y.; Yoon, S. H. AMP-Activated Protein Kinase: a Potential Target for the Diseases Prevention by Natural Occurring Polyphenols. *New Biotechnol.* **2009,** *26*(1–2), 17–22.

78. Ahn, J.; Lee, H.; Kim, S.; Park, J.; Ha, T. The Anti-Obesity Effect of Quercetin is Mediated by the AMPK and MAPK Signaling Pathways. *Biochem. Biophys. Res. Commun.* **2008,** *373*(4), 545–549.

79. Lee, Y. K.; Hwang, J. T.; Kwon, D. Y.; Surh, Y. J.; Park, O. J. Induction of Apoptosis by Quercetin is Mediated through AMPKalpha1/ASK1/p38 Pathway. *Cancer Lett.* **2010,** *292*(2), 228–236.

80. Lu, J.; Wu, D. M.; Zheng, Y. L.; Hu, B.; et al. Quercetin Activates AMP-Activated Protein Kinase by Reducing PP2C Expression Protecting Old Mouse Brain against High Cholesterol-Induced Neurotoxicity. *J. Pathol.* **2010,** *222*(2), 199–212.

81. Yang, J. M.; Hung, C. M.; Fu, C. N.; Lee, J. C.; et al. Hispidulin Sensitizes Human Ovarian Cancer Cells to TRAIL-Induced Apoptosis by AMPK Activation Leading to Mcl-1 Block in Translation. *J. Agric. Food Chem.* **2010,** *58*(18), 10020–10026.

82. Wang, Y.; Liu, W.; He, X.; Fei, Z. Hispidulin Enhances the Anti-Tumor Effects of Temozolomide in Glioblastoma by Activating AMPK. *Cell Biochem. Biophys.* **2015,** *71*(2), 701–706.

83. Zygmunt, K.; Faubert, B.; MacNeil, J.; Tsiani, E. Naringenin, a Citrus Flavonoid, Increases Muscle cell Glucose Uptake via AMPK. *Biochem. Biophys. Res. Commun.* **2010,** *398*(2), 178–183.

84. Ke, J. Y.; Cole, R. M.; Hamad, E. M.; Hsiao, Y. H.; et al. Citrus Flavonoid, Naringenin, Increases Locomotor Activity and Reduces Diacylglycerol Accumulation in Skeletal Muscle of Obese Ovariectomized Mice. *Mol. Nutr. Food Res.* **2016,** *60*(2), 313–324.

85. Seo, W. D.; Lee, J. H.; Jia, Y.; Wu, C.; Lee, S. J. Saponarin Activates AMPK in a Calcium-Dependent Manner and Suppresses Gluconeogenesis and Increases Glucose Uptake via Phosphorylation of CRTC2 and HDAC5. *Bioorg. Med. Chem. Lett.* **2015,** *25*(22), 5237–5242.

86. Guo, H. X.; Liu, D. H.; Ma, Y.; Liu, J. F.; et al. Long-Term Baicalin Administration Ameliorates Metabolic Disorders and Hepatic Steatosis in Rats Given a High-Fat Diet. *Acta Pharmacol. Sin.* **2009,** *30*(11), 1505–1512.

87. Ma, Y.; Yang, F.; Wang, Y.; Du, Z.; et al. CaMKKβ is Involved in AMP-Activated Protein Kinase Activation by Baicalin in LKB1 Deficient Cell Lines. *PLoS ONE.* **2012,** *7*(10), e47900.

88. Pu, P.; Wang, X. A.; Salim, M.; Zhu, L. H.; et al. Baicalein, a Natural Product, Selectively Activating AMPKα(2) and Ameliorates Metabolic Disorder in Diet-Induced Mice. *Mol. Cell. Endocrinol.* **2012,** *362*(1–2), 128–38.

89. Aryal, P.; Kim, K.; Park, P. H.; Ham, S.; et al. Baicalein Induces Autophagic Cell Death through AMPK/ULK1 Activation and Downregulation of mTORC1 Complex Components in Human Cancer Cells. *FEBS J.* **2014,** *281*(20), 4644–4658.

90. Ding, D.; Zhang, B.; Meng, T.; Ma, Y.; et al. Novel Synthetic Baicalein Derivatives Caused Apoptosis and Activated AMP-Activated Protein Kinase in Human Tumor Cells. *Org. Biomol. Chem.* **2011,** *9*(21), 7287–7291.

91. Ganai, A. A.; Farooqi, H. Bioactivity of Genistein: A Review of In Vitro and In Vivo Studies. *Biomed. Pharmacother.* **2015,** *76*, 30–38.

92. Park, C. E.; Yun, H.; Lee, E. B.; Min, B. I.; et al. The Antioxidant Effects of Genistein are Associated with AMP-Activated Protein Kinase Activation and PTEN Induction in Prostate Cancer Cells. *J. Med. Food.* **2005,** *13*(4), 815–820.

93. Guo, H.; Zhao, H.; Kanno, Y.; Li, W.; et al. A Dihydrochalcone and Several Homoisoflavonoids from Polygonatum Odoratum are Activators of Adenosine Monophosphate-Activated Protein Kinase. *Bioorg. Med. Chem. Lett.* **2013,** *23*(11), 3137–3139.

94. Choi, J.; He, N.; Sung, M. K.; Yang, Y.; Yoon, S. Sanguinarine is an Allosteric Activator of AMP-Activated Protein Kinase. *Biochem. Biophys. Res. Commun.* **2011,** *413*(2), 259–263.

95. Xu, J. Y.; Meng, Q. H.; Chong, Y.; Jiao, Y.; et al. Sanguinarine Inhibits Growth of Human Cervical Cancer Cells through the Induction of Apoptosis. *Oncol. Rep.* **2012,** *28*(6), 2264–2270.

96. Han, M. H.; Kim, G. Y.; Yoo, Y. H.; Choi, Y. H. Sanguinarine Induces Apoptosis in Human Colorectal Cancer HCT-116 Cells through ROS-Mediated Egr-1 Activation and Mitochondrial Dysfunction. *Toxicol. Lett.* **2013,** *220*(2), 157–166.

97. Lee, Y. S.; Kim, W. S.; Kim, K. H.; Yoon, M. J.; et al. Berberine, a Natural Plant Product, Activates AMP-Activated Protein Kinase with Beneficial Metabolic Effects in Diabetic and Insulin-Resistant States. *Diabetes.* **2006,** *55*(8), 2256–2264.

98. Turner, N.; Li, J. Y.; Gosby, A.; To, S. W.; et al. Berberine and Its More Biologically Available Derivative, Dihydroberberine, Inhibit Mitochondrial Respiratory Complex I: A Mechanism for the Action of Berberine to Activate AMP-Activated Protein Kinase and Improve Insulin Action. *Diabetes.* **2008,** *57*(5), 1414–1418.

99. Murase, T.; Misawa, K.; Haramizu, S.; Minegishi, Y.; Hase, T. Nootkatone, a Characteristic Constituent of Grapefruit, Stimulates Energy Metabolism and Prevents

Diet-Induced Obesity by Activating AMPK, American Journal of Physiology. *Endocrinol. Metab.* **2010**, *299*(2), E266–E275.

100. Adachi, Y.; Kanbayashi, Y.; Harata, I.; Ubagai, R.; et al. Petasin Activates AMP-Activated Protein Kinase and Modulates Glucose Metabolism. *J. Nat. Prod.* **2014**, *77*(6), 1262–1269.

101. Yang, Y.; Zhao, Z.; Liu, Y.; Kang, X.; et al. Suppression of Oxidative Stress and Improvement of Liver Functions in Mice by Ursolic Acid via LKB1-AMP-Activated Protein Kinase Signaling. *J. Gastroenterol. Hepatol.* **2015**, *30*(3), 609–618.

102. Tan, M. J.; Ye, J. M.; Turner, N.; Hohnen-Behrens, C.; et al. Antidiabetic Activities of Triterpenoids Isolated from Bitter Melon Associated with Activation of the AMPK Pathway. *Chem. Biol.* **2008**, *15*(3), 263–273.

103. Chen, X. B.; Zhuang, J. J.; Liu, J. H.; Lei, M.; et al. Potential AMPK Activators of Cucurbitane Triterpenoids from Siraitia Grosvenorii Swingle. *Bioorg. Med. Chem.* **2011**, *19*(19), 5776–5781.

104. Wang, J.; Ma, H.; Zhang, X.; He, L.; et al. A Novel AMPK Activator from Chinese Herb Medicine and Ischemia Phosphorylate the Cardiac Transcription Factor FOXO3. *Int. J. Physiol. Pathophysiol. Pharmacol.* **2009**, *1*(2), 116–126.

105. Wang, X.; Chen, Y.; Abdelkader, D.; Hassan, W.; et al. Combination Therapy with Oleanolic Acid and Metformin as a Synergistic Treatment for Diabetes. *J. Diabetes Res.* **2015**, *2015*, 973287.

106. Yie, Y.; Zhao, S.; Tang, Q.; Zheng, F.; et al. Ursolic Acid Inhibited Growth of Hepatocellular Carcinoma HepG2 Cells through AMPKα-Mediated Reduction of DNA Methyltransferase. *Mol. Cell. Biochem.* **2014**, *402*(1–2), 63–74.

107. Son, H. S.; Kwon, H. Y.; Sohn, E. J.; Lee, J. H.; et al. Activation of AMP-Activated Protein Kinase and Phosphorylation of Glycogen Synthase Kinase3 β Mediate Ursolic Acid Induced Apoptosis in HepG2 Liver Cancer Cells. *Phytother. Res.* **2013**, *27*(11), 1714–1722.

108. Kang, J. I.; Hong, J. Y.; Lee, H. J.; Bae, S. Y, et al. Anti-Tumor Activity of Yuanhuacine by Regulating AMPK/mTOR Signaling Pathway and Actin Cytoskeleton Organization in Non-Small Cell Lung Cancer Cells. *PLoS ONE.* **2015**, *10*(12), e0144368.

109. Lin, H. Y.; Huang, B. R.; Yeh, W. L.; Lee, C. H.; et al. Antineuroinflammatory Effects of Lycopene via Activation of Adenosine Monophosphate-Activated Protein Kinase-α1/Heme Oxygenase-1 Pathways. *Neurobiol. Aging.* **2014**, *35*(1), 191–202.

CHAPTER 10

EFFECTS OF LOW DOSES OF SAVORY ESSENTIAL OIL DIETARY SUPPLEMENTATION ON LIFETIME AND THE FATTY ACID COMPOSITION OF THE AGEING MICE TISSUES

TAMARA A. MISHARINA[1,2*], VALERY N. YEROKHIN[1], and LUJDMILA D. FATKULLINA[1]

[1]Emanuel Institute of Biochemical Physics of Russian Academy of Sciences, 4, Kosygin St., Moscow 119334, Russia

[2]Plekhanov Russian University of Economics, 36, Stremyanny lane, Moscow 117997, Russia

*Corresponding author. E-mail: tmish@rambler.ru

CONTENTS

ABSTRACT

The effect of long-term administration of essential oil from savory *Satureja hortensis* L. in low doses with drinking water (150 ng/ml) on the lifetime and on the fatty acids composition in liver and brain of mice of the high-cancer strain AKR was studied. It was found for the first time that long-term administration of an essential oil in low doses increased the average lifetime of mice by 26%. The age and the development of leukemia were accompanied by significant changes in the total proportion of saturated, mono-, and polyunsaturated acids in the tissues of liver and brain of mice. In response to administration of savory essential oil at low doses, the synthesis of polyunsaturated fatty acid in the liver increased, the deviations in these characteristics from the normal values were reduced, that is, the savory essential oil performed a preventive role and stabilized the biochemical parameters. The results of this study suggest that the antioxidant-containing volatile fractions of plants (in particular, savory essential oil) at low doses are promising for therapeutic and prophylactic purposes.

10.1 INTRODUCTION

The search for natural bioactive substances protecting the organism from unfavorable action of the environment, particularly various carcinogenic factors, has become very important nowadays. Antioxidants are of particular interest as these are by biologically active substances [1]. It has been shown that a synthetic antioxidant from the class of hindered phenols [β-(4-hydroxy-3,5-ditretbutylphenyl) propionic acid (phenozan)] shows significant antitumor activity both in low and very low doses when administered into the organism of leukemic mice [2].

It is known that some plant products, such as herbs, spices, and their essential oils and extracts, possess wide biological activity, including antioxidant and pharmacological ones [3–5]. These compounds are applied in small doses; they have low toxicity and are recommended for decreasing the risk of disease caused by increased oxidation of cell components. The addition of some flavonoids, and also products containing these compounds (dried apples and onions) in the feed of mice, has increased the content of reduced glutathione and decreased the content of oxidized glutathione and mixed disulphide protein-glutathione in the animal liver. The antioxidant activity of plant flavonoids is caused by their ability to inhibit prooxidant enzymes, which gives complexes with the cations of iron and copper and catches radicals of oxygen and nitrogen being the donor of hydrogen [6].

Among natural antioxidants of plant origin, an important place is taken by essential oils which are a mixture of volatile compounds isolated from aromatic plants. The presence of antioxidant properties in many essential oils has been proved in model experiments [5, 7, 8]. Thymol, carvacrol, and eugenol from some essential oils have shown a dose dependent decrease of mitochondrial activity of cancer cells [9]; these are able to reduce the consequences of oxidative stresses [10].

Carvacrol and thymol are the major components in the summer savory essential oil (*Satureja hortensis* L.). This herb has been used to enhance the flavor of food for over 2000 years. In England, savory became popular as a medicine and also as a cooking herb with pungent, spicy taste. The Germans called savory the herb bean because it complemented green beans, dried beans, and lentils so well. William Shakespeare, in his "The Winter's Tale," mentions savory along with lavender and marjoram [3]. The Italians are probably the first to grow savory as a garden herb. Summer savory is now cultivated in France, Spain, Germany, other parts of Europe, Canada, and USA. The Yugoslavian savory is considered premium grade. This herb may be cultivated too in Russia. Summer savory is an annual herbaceous plant up to 30 cm high, the leaves are elliptical, leathery, petiolate, and dark green. The flowers are fragrant, white, pink, or lilac and appear in small spikes in the leaf axils. It has well-developed taproot. The other savory is the winter savory (*Satureja montana* L.), and it grows wild in southern Europe [3].

The parts used include the fresh or dried leaves and tender stems. The bright green leaves are used as spice. The flowering tops are used for oil extraction. Savory leaves are used whole or crushed. Leaves have a strongly aromatic, sweet, spicy, and herbaceous aroma. The taste is strongly aromatic, sweet, and peppery thyme. The herb has a thyme-like flavor. Summer savory is popular in teas, herb butters, flavored vinegars, and with shell beans, lentils, chicken soups, creamy soup, beef soup, eggs, beans, peas, eggplant, asparagus, onions, cabbage, Brussels sprouts, squash, garlic, liver, fish, and chutneys.

The savory leaves contain about 1% essential oil with carvacrol as the major component [3]. Carvacrol and the essential oil of summer savory have wide spectrum of activities, including antimicrobial, antimutagenic, antigenotoxic, analgesic, antispasmodic, antiinflammatory, angiogenic, and antiparasitic [3, 11, 12]. The ground material, essential oil, and extracts of summer and winter savory possess strong antioxidant activity, which could be compared with the synthetic antioxidant agent–butylated hydroxytoluene [13–17]. It was shown that long-term administration of essential oil from savory *S. hortensis* L. in low doses with drinking water (150 ng/ml) decreased the hemolysis level and the content of lipid peroxidation products in erythrocytes of mice of the high-cancer strain AKR by 20% [14].

The goal of the present work was to study the biological activity of the essential oil of summer savory (*S. hortensis* L.) in vivo. The effect of low dose of savory oil consumed by mice of the high-cancer strain AKR with drinking water on the fatty acid composition in the liver and brain of these mice and on their lifetime was studied.

10.2 MATERIALS AND METHODOLOGY

Mice of the AKR line at the age of 3 months were separated into two equal groups. The mice were kept at the room temperature at the level of 20–22°C with natural light. Control group of mice were given usual drinking water and standard laboratory feed (pellets), which contained wheat, corn, barley, soy oil meal, sunflower cake, fish flour, dry milk, feed yeast, limestone flour, vitamin complex, malt sprouts, and a mineral mixture (PK120 receipt, Laboratorkorm LLC, Moscow) without limitation. Mice of the experimental group got drinking water, in which the essential oil of summer savory *S. atureja hortensis* L. (Lionel Hitchen Ltd., Great Britain) was added, and standard laboratory feed was also added without limitation. The content of the essential oil in drinking water was 150 ng/ml. The drinking water was placed into waterers in a sufficient amount.

The presence of leukemia in dead animals was determined by the increased size of the thymus (more than 30 mg) and spleen (more than 150 mg). The experiment was performed for 17 months until the natural death of the last animal. On the basis of data on lifetime, the mortality curves, which represented the dependence of the portion of surviving animals on age, were plotted (Fig. 10.1). Antileukemic activity of the savory essential oil under study was estimated by the mortality curves.

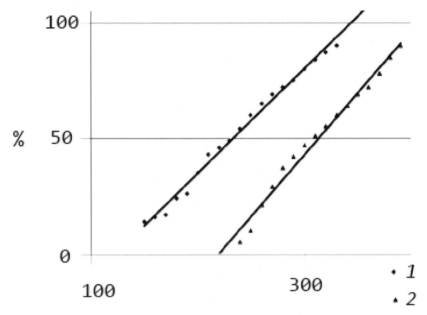

FIGURE 10.1   The mortality curves of AKR mice with spontaneous leucosis in control (1) and experimental (2) groups. Axes: X—age, days; Y—mortality, %.

The study of the fatty acids content in the brain and liver cells was performed by gas–liquid chromatography and combined gas chromatography mass-spectrometry methods. Liver was perfused by the separation medium and fragmented by scissors. The tissues of liver and brain were homogenized manually in a glass homogenizator with a teflon pestle. A volume of 5 ml methanol was added to 200 mg of tissue placed into a glass tube with a hermetic screw top, left in the refrigerator for 1 h; then, 200 µl of acetylchloride was added with cooling and intensive mixing and boiled in a water bath for 1 h. Then, 5 ml of 6% water solution of K_2CO_3 were added to the sample, shaken, 2 ml of hexane was added, and fatty acid methyl esters were extracted by shaking for 5 min and centrifuged for 5 min at 3000 rpm.

Hexane phase, containing fatty acid methyl esters, was analyzed using the Kristall 2000 M chromatograph (Russia) with a flame ionization detector and a DB-1 fused silica column (50 m × 0.32 mm, 0.25 µm phase layer; Supelco, United States). The analysis was performed with the programing of the column temperature from 120 to 270°C with a speed of

4°C/min with the temperature of the injector and the detector of 270°C. The speed of the helium carrier gas through the column was 1.5 ml/min. The quantitative content of fatty acid methyl esters in samples was calculated as the ratio of the peak area of a corresponding fatty acid methyl ester to the sum of peak areas of all fatty acid methyl esters and was expressed in percent. Component identification in samples was performed on the basis of the retention indexes and mass spectra obtained after the separation of fatty acid methyl esters in conditions analogous to those for GC-analysis on an HP 5890/5980 instrument (Hewlett Packard, United States). Mass spectra were obtained in the regime of the electron hit with an ionizing voltage of 70 eV. The data were processed and averaged using the ANOVA software at $p < 0.05$. Plots were constructed and processed using the Origin 7.5 program.

10.3   RESULTS AND DISCUSSION

The component composition of savory essential oil used in this work is given in Table 10.1. For comparison, the composition of thyme essential oil is also provided in the table.

Savory essential oil contained 0.5–1.7% of each of the following monoterpene hydrocarbons: α-thujene, α-pinene, camphene, β-pinene, β-myrcene, sabinene, α-phellandrene, α-terpinene, and 4.2% sesquiterpene hydrocarbon—β-caryophyllene. The content of γ-terpinene was 10.5%. This essential oil contained two isomeric phenols—thymol (17.5%) and carvacrol (35.2%). A high content of thymol, carvacrol, and γ-terpinene was respondent for the antioxidant properties of the savory oil [16, 17]. It was revealed earlier that the addition of thyme oil (1200 mg per 1 kg of mass) into rat feed increased the general antioxidant status of the animals and kept a high level of polyunsaturated fatty acids in cell membranes during the process of their ageing [18]. Thyme and savory essential oils have a close content of the main components, both essential oils showed the same antiradical properties. As can be seen in Table 10.1, the values of antiradical efficiencies AE are close for two oils [16]. That is why we hoped that the oil of savory, which is successfully grown in central Russia, would also possess biological activity. It should be noted that savory oil doses in our work were by a factor of 100 lower than in the study by Youdim and Deans [18].

TABLE 10.1 Composition of Components in the Essential Oils from Savory and Thyme.

Retention index	Compound	Content in savory essential oil (%)	Content in thyme essential oil (%)
925	α-Thujene	1.00	1.02
933	α-Pinene	0.70	0.53
946	Camphene	0.54	1.15
961	Sabinene	0.62	0.42
974	β-Pinene	0.43	1.94
983	β-Myrcene	1.70	–
1000	α-Phellandrene	0.28	0.24
1012	α-Terpinene	1.91	0.81
1015	p-Cymene	10.73	20.75
1023	1,8-Cineole	0.86	1.50
1026	Limonene	1.02	–
1038	Ocimene	0.20	–
1052	γ-Terpinene	10.49	11.48
1085	Linalool	0.54	5.21
1152	Isoborneol	1.86	1.94
1165	Terpinene-4-ol	0.88	1.20
1175	α-Terpineol	0.18	0.50
1240	Bornyl acetate	5.93	0.62
1271	Thymol	17.48	45.11
1283	Carvacrol	35.23	2.36
1420	β-Caryophyllene	4.19	1.72
1431	α-Bergamotene	0.52	–
1496	Bicyclogermacrene	1.82	–
Values of antiradical efficiency, (AE, l/(g s)) [16]		2.77×10^{-3}	3.09×10^{-3}
The total content of phenols (%)		52.71	47.47

It was found that savory essential oil affected general antioxidant status of the animals, their lifetime, and the level of polyunsaturated fatty acids in cell membranes during the process of their ageing. Figure 10.1 represents the survival curves of the AKR line of mice in the control and in the case of administration of savory essential oil with drinking water.

It is seen that the essential oil shows remarkable antileukemic action: the mortality curve of the experimental group of mice is significantly shifted to the right in comparison with the control. These curves have been described by linear question $Y = A + BX$, where Y—part of dead animals,%; X—time of life. Coefficients of questions are given in Table 10.2. Using these questions, the parameters of mice life for two groups were calculated; these values are given in Table 10.2. It was found that savory essential oil administration reduced the incidence of leukemia by 35% in comparison with control group. The obtained data give evidence that the constant consumption of savory essential oil significantly increases the latent period—up to 82 days (see Table 10.2). The maximum lifetime of mice from the experimental group was 41 days (12%) longer than in the control. The average lifetime of mice increased by 62 days (26%) in the case of the consumption of savory essential oil with drinking water in comparison with the control. It is possible that the savory essential oil has prophylactic action putting off the terms of contraction of leukemia and mass animal mortality.

TABLE 10.2   The Parameters of the Course of Spontaneous Leukosis in the AKR Line of Mice in the Control and in the Case of Consumption of Savory Essential Oil with Drinking Water.

Parameters of question and lifetime	Control group	Experiment group	Experiment control
Coefficient A	−74.4	−150.8	−
Coefficient B	0.52	0.67	−
Leucosis extend, %	98	63	−35
Latent time, days	143	225	+82
Maximum lifetime, days	315	376	+41
Average lifetime, days	239	301	+62

Spontaneous leukosis appears in mice of the high cancer strain AKR at the age of 6–11 months in 65–90% of cases. It should be noted that precisely spontaneous leukoses of mice is the closest to human leukoses according to the origin and clinical aspects and the similarity of pathological and morphological special features [19]. Detailed study of the development kinetics of this process has been performed [20]. Mice of the high-cancer AKR strain suffered from leukemia upon ageing. This disease develops spontaneously with different rate in different animals. It is determined by the content of leukocytes and the weight of the thymus and spleen. Judging by these parameters, mice in our study became ill with leukemia after 8 months of age; this disease was pronounced at an age of 10 months [20]. For this reason, in this group of mice, we took into account not only the age-related changes but also the changes that occurred due to cancer.

The ratio of saturated, monounsaturated, and polyunsaturated fatty acids in cell membrane lipids determines their common structural characteristics and functions. Changes in the composition of fatty acids affect the activity of receptors, transport of metabolites in and out of cells, as well as hormonal and other signal transduction processes. Upon ageing, the content of polyunsaturated fatty acids in lipids (especially in the brain) decreases, which affects its function [18, 21, 22]. Tables 10.3 and 10.4 show the content of saturated, monounsaturated, and polyunsaturated fatty acids in the brain and liver of mice upon ageing from 4 to 10 months.

TABLE 10.3 Composition of Fatty Acids in the Mice Brain at the Different Age.

Trivial name of fatty acid, shorthand name, ω	Acid content (%) at the age			
	4 months	6 months	8 months	10 months
Saturated fatty acids				
Palmitic acid, 16:0	25.0	23.5	21.7	24.2
Heptadecanic acid, 17:0	0.2	0.3	0.1	0.2
Stearic acid, 18:0	21.8	20.6	19.0	19.9
Arachidic acid, 20:0	0.4	0.5	0.6	0.4
Behenic acid, 22:0	0.6	0.7	0.8	0.7
Lignoceric acid, 24:0	0.5	0.9	1.1	2.3
Total amount saturated acids, %	48.5	46.5	43.3	47.7
Monounsaturated fatty acids				
Palmitoleic acid, 16:1ω9	0.5	0.5	0.7	0.3
Oleic acid, 18:1ω9	13.6	15.5	21.6	16.2
Vaccenic acid, 18:1ω7	3.5	3.5	4.3	3.4
Eicosaenoic acid, 20:1ω9	2.4	2.3	3.2	2.8
Erucic acid, 22:1ω9	0.4	0.3	0.5	0.9
Nervonic acid, 24:1ω9	3.4	2.7	2.3	2.0
Total amount monounsaturated acids, %	23.8	24.8	32.6	25.6

TABLE 10.3 *(Continued)*

Trivial name of fatty acid, shorthand name, ω	Acid content (%) at the age			
	4 months	6 months	8 months	10 months
Polyunsaturated fatty acids				
Linoleic, 18:2ω6	0.4	0.3	0.2	0.5
Arachidonic, 20:4ω6	8.4	9.1	7.8	9.8
Eicosatrienoic, 20:3ω6	0.3	0.2	0.3	0.3
Docosahexaenoic, 22:6ω3	16.6	16.1	13.1	14.8
Docosatetraenoic, 22:4ω6	2.0	2.8	2.7	1.6
Total amount polyunsaturated acids, %	27.7	28.5	24.1	27.0

Note: ω—position of double bond at the end of molecule of fatty acid.

TABLE 10.4 Composition of Fatty Acids in the Mice Liver at the Different Age.

Trivial name of fatty acid, shorthand name, ω	Fatty acid content (%) at the age			
	4 months	6 months	8 months	10 months
Saturated fatty acids				
Palmitic acid, 16:0	20.6	21.5	23.2	23.1
Heptadecanic acid, 17:0	0.3	0.2	0.2	0.3
Stearic acid, 18:0	10.4	12.7	11.0	11.3
Arachidic acid, 20:0	0.1	0.1	0.3	0.2
Behenic acid, 22:0	0.2	0.2	0.3	0.4
Lignoceric acid, 24:0	0.3	0.2	0.3	0.3
Total amount saturated acids, %	31.9	34.9	35.3	35.6
Monounsaturated fatty acids				
Palmitoleic acid, 16:1ω9	1.8	2.6	1.9	4.1
Oleic acid, 18:1ω9	22.2	17.4	16.1	14.8
Vaccenic acid, 18:1ω7	2.5	1.8	1.9	2.8
Eicosaenoic acid, 20:1ω9	0.3	0.5	0.4	0.5
Erucic acid, 22:1ω9	0.3	0.7	0.8	0.9
Total amount monounsaturated acids, %	27.3	23.0	21.1	23.1
Polyunsaturated fatty acids				
Linolenic acid, 18:3ω6	0.2	0.2	0.2	0.2
Linoleic acid, 18:2ω6	15.2	15.3	16.2	13.9

TABLE 10.4   *(Continued)*

Trivial name of fatty acid, shorthand name, ω	Fatty acid content (%) at the age			
	4 months	6 months	8 months	10 months
Arachidonic acid, 20:4ω6	15.3	15.5	15.1	17.8
Eicosatrienoic acid, 20:3ω6	1.0	0.9	0.9	0.8
Eicosadienoic acid, 20:2ω6	0.2	0.2	0.4	0.3
Docosapentaenoic acid, 22:5ω6	0.6	0.7	0.4	0.9
Docosahexaenoic acid, 22:6ω3	7.6	7.6	8.2	6.5
Docosatetraenoic acid, 22:4ω6	0.4	0.4	0.4	0.7
Total amount polyunsaturated acids, %	40.5	40.8	41.8	41.1

Note: ω—position of double bond at the end of molecule of fatty acid.

The proportion of saturated acids in brain lipids of 4-month-old mice was 48.5%; monounsaturated acids, 23.8%; and polyunsaturated acids, 27.7%. In the group of saturated acids, palmitic (C16:0) and stearic (C18:0) acids prevailed and accounted for 25.0 and 21.8%, respectively.

As seen from Table 10.3, the total content of saturated acids decreased from 100% at an age of 4 months to 91.4% at an age of 8 months and then increased again. In the group of monounsaturated acids, the proportion of oleic acid (C18:1ω9) accounted for 13.6%, whereas the content of all other acids was less than 4%. The content of monounsaturated acids increased with age. For example, the total content of these acids in mice at an age of 8 months increased and accounted for 32.6% than decreased by 10 months, but accounted for 107.6% compared to the content in 4-month-old animals. The group of polyunsaturated fatty acids of the brain was represented primarily by arachidonic (C20:4ω6) and docosahexaenoic (C22:6ω3) acids: the content of the former and the latter was 8.4 and 16.6%, respectively. By the age of 8 months, the total polyunsaturated fatty acids content decreased by 13% but increased again by 10 months and reached 97% of the initial level. Similar changes were observed with ageing in the content of C20:4ω6 and C22:6ω3 acids (Table 10.3). Thus, the proportion of saturated and polyunsaturated acids in the brain of AKR mice decreased with age, conversely, the proportion of monounsaturated acids increased. The development of leukemia was accompanied by an increase in the total proportion of saturated and polyunsaturated acids and a decrease in the content of monounsaturated acids. Similar changes were also observed for individual fatty acids.

In the liver, the proportion of saturated acids increased with the age and upon leukemia development in mice (see Table 10.4). The proportion of monounsaturated acids decreased by the age of 8 months by 21.1% and increased again at 10 months, reaching 84% of the content in the liver of 4-month-old mice (see Table 10.4). The total content of polyunsaturated fatty acids almost did not change with age and with leukemia progression. However, significant changes were detected in the content of individual fatty acids. For example, by the age of 10 months, the content of oleic (C18:1ω9) acid in mouse liver monotonically decreased by 34%, whereas the content of arachidonic (C20:4ω6) acid increased at the age of 10 months. The content of docosahexaenoic (C22:6ω3) acid increased at the age of 8 months but decreased again to the initial level with leukemia development (see Table 10.4).

Thus, changes in the content of major fatty acids in the liver and brain of ageing mice and mice with leukemia differ. It is known that, in mammals, the synthesis of the main polyunsaturated fatty acids takes place in the liver. It is also known that ageing and leukemia development are accompanied by enhancement of oxidative processes. Possibly, to maintain the level of polyunsaturated fatty acids in the brain of mice and to compensate for polyunsaturated fatty acids destroyed as a result of oxidation, the synthesis of these acids in the liver of mice increases. It was shown earlier that the content of C18:2ω6, C20:1ω9, C22:4ω6, and C22:5ω3 acids in the rat brain increased with age, whereas the proportion of C22:6ω3 acid decreased [18]. The brain is an organ where the concentration of polyunsaturated fatty acids (especially docosahexaenoic acid C22:6ω3) is highly important. Normally, ageing causes a decrease in the concentration of C22:6ω3 acid and an increase in the concentration of C20:4ω6 acid [18, 22]. The maintenance of the content of the C22:6ω3 acid at a constant level is extremely important because it is involved in the electrophysiological functions as well as in learning, memory, and behavior. Maintaining the content of this acid at the optimal level harmonizes the activities of the higher nervous system and cerebral development [21, 22].

Earlier, it was shown that the addition of thyme essential oil to food of rats at an age of 7–28 months increased the antioxidant status of the brain and maintained the content of polyunsaturated fatty acids at a high level [18]. It was assumed that this effect of essential oil may be due to its antioxidant properties. In our earlier study [14], we found that daily consumption of 300 ng of savory essential oil with drinking water by AKR mice increased the lifespan of mice by 20% and reduced the incidence of leukemia by 35%. To study the influence of savory essential oil on the biochemical processes in mice in greater detail, we determined the fatty acids composition of liver and brain cells in the control and experimental mice at the age of 4 and 6 months (Table 10.5).

TABLE 10.5 Changes in the Content of Some Fatty Acids in the Brain and Liver of AKR Mice after the Consumption of Savory Essential Oil for 1 and 3 Months Compared with the age Matched Control Animals (%).

Fatty acid	Content of fatty acid in the brain and liver of mice compared with control (%) after consumption of savory oil during 1 and 3 months			
	Brain		Liver	
	1 month	3 months	1 month	3 months
Palmitic acid, 16:0	95.3 ± 0.6	97.7 ± 1.4	102.6 ± 1.8	102.4 ± 1.5
Stearic acid, 18:0	91.5 ± 2.0	105.4 ± 1.9	107.0 ± 2.4	106.1 ± 1.7
Saturated acids, average	93.8 ± 1.9	103.6 ± 1.7	105.3 ± 2.1	103.2 ± 1.2
Oleic acid, 18:1ω9	117.2 ± 1.2	101.2 ± 1.9	70.0 ± 1.2	65.8 ± 1.5
Eicosaenoic acid, 20:1ω9	154.0 ± 3.6	92.3 ± 2.5	78.0 ± 1.3	63.5 ± 2.1
Monounsaturated acids, average	132.0 ± 2.2	94.0 ± 1.8	70.4 ± 1.9	64.6 ± 1.8
Linoleic acid, 18:3ω6	85.2 ± 1.3	92.8 ± 1.9	110.8 ± 2.3	112.8 ± 1.0
Arachidonic acid, 20:4ω6	73.4 ± 1.5	104.4 ± 1.4	103.1 ± 1.1	114.5 ± 1.7
Docosahexaenoic acid, 22:6ω3	77.1 ± 1.8	89.3 ± 1.5	105.9 ± 1.9	111.6 ± 1.9
Polyunsaturated acids, average	83.5 ± 0.7	102.0 ± 1.4	112.5 ± 2.0	113.0± 1.8

It was found that the administration of savory essential oil to AKR mice for 1 month starting from the age of 3 months was accompanied by a decrease in the level of saturated fatty acids and polyunsaturated fatty acids and an increase in the content of monounsaturated fatty acids by 32% in the brain of mice (see Table 10.5). In addition, during this time, the content of oleic acid increased, whereas the content of arachidonic and docosahexaenoic acids decreased. Long-term (for 3 months) administration of savory essential oil to mice slightly increased the level of saturated fatty acids and polyunsaturated fatty acids in the brain and reduced the total content of monounsaturated fatty acids by 6% and the content of docosahexaenoic acid by 11%. The content of polyunsaturated fatty acids (including the arachidonic acid) slightly increased; as a result, the fatty acid composition in brain lipids was stabilized. The administration of savory essential oil also influenced the fatty acid composition in the liver of mice (Table 10.5). For example, after 1 and 3 months of administration of this essential oil, the proportion of saturated and polyunsaturated acids increased by 3–5% and 13%, respectively, whereas the content of mono-unsaturated fatty acids significantly decreased by 30–35%. The content of oleic acid in the liver of the experimental animals was 30–35% smaller than that in the liver of the control mice. At the same time, the content of arachidonic and docosahexaenoic acids slightly increased after 1 month of administration of the essential oil and more significantly (by 12–15%) increased after 3 months of its administration compared to the control animals.

10.5 CONCLUSIONS

Thus, changes in the content of major fatty acids in the liver and brain of ageing mice and mice with leukemia differ. The proportion of saturated and polyunsaturated acids in the brain of AKR mice decreased with age, conversely, the proportion of monounsaturated acids increased. The development of leukemia was accompanied by an increase in the total proportion of saturated and polyunsaturated acids and a decrease in the content of monounsaturated acids. Similar changes were also observed for individual fatty acids. In response to administration of savory essential oil at low doses, the synthesis of polyunsaturated fatty acids in the liver increased. The differences in the content of polyunsaturated fatty acids in the liver, detected in this study, indicate that the administration of savory essential oil stabilized the level of polyunsaturated fatty acids or reduced the deviations in these characteristics from the normal values, that is, performed a preventive role and stabilized the biochemical parameters. The results of this study suggest that the antioxidant-containing volatile fractions of plants (in particular, savory essential oil) at low doses are promising for therapeutic and prophylactic purposes.

KEYWORDS

- ageing
- savory essential oil
- leukemia
- leukosis
- phenols
- unsaturated acids

REFERENCES

1. Burlakova, E. B.; Alesenko, A. V.; Molochkina, E. M.; Palmina, N. P.; Chrapova, N. G. *Bioantioxidants in Radiation Injury and Malignant Growth*. Nauka: Moscow, 1975; p 214 (In Russian).
2. Erokhin, V. N.; Krementsova, A. V.; Semenov, V. A.; Burlakova, E. B. Effect of Antioxidant β-(4-Hydroxy-3,5-Ditertbutylphenyl) Propionic Acid (Phenosan) on the Development of Malignant Neoplasms. *Biol. Bull.* **2007,** *34,* 485–491 (In Russian).
3. Charles, D. J. *Antioxidant Properties of Spices, Herbs and Others Sources*. Springer; New York, 2013; p 610.
4. Dragland, S.; Senoo, H.; Wake, K. Several Culinary and Medicinal Herbs are Important Sources of Dietary Antioxidants. *J. Nutr.* **2003,** *133,* 1286–1290.
5. Edris, A. E. Pharmaceutical and Therapeutic Potentials of Essential Oils and Their Individual Volatile Constituents: A Reviews. *Phytother. Res.* **2007,** *21,* 308–323.
6. Meyers, K. J.; Rudolf, J. L.; Mitchell, A. E. Influence of Dietary Quercetin on Glutatione Redox Status in Mice. *J. Agric. Food Chem.* **2008,** *56,* 830–836.
7. Miguel, G. Antioxidant Activity of Medicinal and Aromatic Plants. *Flavour Fragr. J.* **2010,** *25,* 91–312.
8. Misharina, T. A.; Terenina, M. B.; Krikunova, N. I. Antioxidant Properties of Essential Oils. *Appl. Biochem. Microbiol.* **2009,** *45,* 642–647.
9. Mastelic, J.; Jercovic, I.; Blazevic, I. Comparative Study on the Antioxidant and Biological Activities of Carvacrol, Thymol, and Eugenol Derivatives. *J. Agric. Food Chem.* **2008,** *56,* 3989–3996.
10. Danesi, F.; Elementi, S.; Neki, R. Effect of Cultivaron the Protection of Cardiomyocytes from Oxidative Stress by Essential Oils and Aqueous Extracts of Basil (*Ocimumbasilicum* L.). *J. Agric. Food Chem.* **2008,** *56,* 9911–9917.
11. Baser, K. H. Biological and Pharmacological Activities of Carvacrol and Carvacrol Bearing Essential Oils. *Curr. Pharm. Des.* **2008,** *14,* 3106–3119.
12. Gursoy, U. K.; Gursoy, M.; Gursoy, O. V.; Cakmakci, L.; Kononen, E.; Uitto, V.J.; Aristatile, B.; Al-Numair, K. S.; Veeramani, C.; Pugalendi, K. V. Antihyperlipidemic

Effect of Carvacrol on D-Galactosamine-Induced Hepatotoxic Rats. *J. Basic Clin. Physiol. Pharmacol.* **2009**, *20,* 25–27.

13. Kim, I. S.; Yang, M. R.; Lee, O. H.; Kang, S. N. Antioxidant Activities of Hot Water Extracts from Various Spices. *Int. J. Mol. Sci.* **2011**, *12,* 4120–4131.

14. Burlakova, E. B.; Erokhin, V. N.; Misharina, T. A.; Fatkullina, L. D.; Krementsova, A. V.; Semenov, V. A.; Terenina, M. B.; Vorob'eva, A. K.; Goloshchapov, A. N. The Effect of Volatile Antioxidants of Plant Origin on Leukemogenesis in Mice. *Izv. Akad. Nauk. Ser. Biol.* **2010**, *6,* 711–718 (In Russian).

15. Serrano, C.; Matos, O.; Teixeira, B.; Ramos, C.; Neng, N.; Nogueira, J.; Nunes, M. L.; Marques, A. Antioxidant and Antimicrobial Activity of *Satureja montana* L. Extracts. *J. Sci. Food Agric.* **2011**, *91,* 1554–1560.

16. Alinkina, E. S.; Misharina, T. A.; Fatkullina, L. D. Antiradical Properties of Oregano, Thyme, and Savory Essential Oils. *Appl. Biochem. Microbiol.* **2013**, *49,* 73–78.

17. Ruberto, G.; Baratta, M. Antioxidant Activity of Selected Essential Oil Components in Two Lipid Model Systems. *Food Chem.* **2002**, *69,* 167–174.

18. Youdim, K. A.; Deans, S. G. Effect of Thyme Oil and Thymol Dietary Supplementation on the Antioxidant Status and Fatty Acid Composition of the Ageing Rat Brain. *Br. J. Nutr.* **2000**, *83,* 87–93.

19. Bergol'ts, V. M.; Rumyantsev, N. V. *Comparative Pathology and Etiology of Leukosis of Human and Animals.* Meditsina: Moscow, 1996; p 278 (In Russian).

20. Erokhin, V. N.; Burlakova, E. B. Spontaneous Leukemia—A Model for Studying the Effects of Low and Ultralow Doses of Physical and Physicochemical Factors on Tumorigenesis. *Radiats. Biol. Radioekol.* **2003**, *43,* 237–241. (In Russian).

21. Uauy, R.; Hoffman, D. R.; Peirano, P.; Birch, D. G.; Birch, E. E. (2001). Essential Fatty Acids in Visual and Brain Development. *Lipids.* **2001**, *36,* 885–895.

22. Nakamura, M. T.; Cho, H. P.; Xu, J.; Tang, Z.; Clarke, S. D. (2001). Metabolism and Functions of Highly Unsaturated Fatty Acids: An Uptake. *Lipids.* **2001**, *36,* 961–964.

CHAPTER 11

TECHNOLOGY FOR OBTAINING OF BIOPREPARATIONS AND INVESTIGATION OF THEIR EFFECTIVENESS

TATYANA V. KHURSHKAINEN* and ALEXANDER V. KUTCHIN

Institute of Chemistry of Komi Scientific Center of Ural Branch of Russian Academy of Science, 48, Pervomaiskaya St., Syktyvkar 167000, Russia

Corresponding author. E-mail: hurshkainen@chemi.komisc.ru

CONTENTS

ABSTRACT

This chapter focuses on investigation of emulsion method of extraction low-molecular compounds from *Abies* and *Picea* wood greenery. Comparison of efficiency various extraction equipment was lead, and the chemical composition of the extracted compounds was studied. It has been shown that the developed technology allows us to obtain various classes of bioactive substances with a high yield. Efficiency of the fodder additive Verva obtained by emulsion method was tested on young cattle, pigs, and quails.

11.1 INTRODUCTION

Renewable plant raw materials are the source of bioactive substances, which are used to obtain specimen for medicine, pharmacology, veterinary medicine, agriculture, etc.

Extraction methods based on water, organic solvents, and condensed gases are used for extraction substances from plant raw materials. The advantage of raw extraction is environment-friendly process and products. However, to achieve a high degree of extraction, one should have a long-time extraction at high temperature when bioactive components are destructed. If water is used as an extractant, one can get only hydrophilic compounds.

Organic solvents, on the contrary, can extract compounds of different polarity. Petrol, ethanol, and isopropanol are mainly used for extraction. The disadvantage of this method is that it consists of residual amounts of solvents. Moreover, the process of extraction is fire hazardous.

Condensed gases are selective and environmental-safety extractants. However, the production cost is rather high because of complicated high-priced equipment [1].

The method of emulsion extraction developed in the Institute of Chemistry, Komi Research Center of the Ural Division RAS, is environmentally safe, and the production cost is rather low.

One of the branches devoted to the use of plant extractive substances is obtaining fodder additives. It is chlorophyll–carotene paste which is sum of substances extracted from coniferous wood greenery with the help of benzine [2]. Chlorophyll–carotene paste is used as a fodder additive for pigs, calves, and chickens to encourage the weight gain and better assimilation of nutrients. The paste is also used in veterinary medicine to fight cow infertility and to treat gastrointestinal diseases of young cattle.

Fodder-additive Abisib obtained from *Abies* wood greenery by the water extraction possesses antiinflammatory, immunostimulatory, and adaptogenic properties [3]. Veterinary preparation Florabis, which is the sum of complexes of acids (from *Abies* wood greenery) with cobalt ions, is used to protect animals from viruses and bacterial infections [4]. Carbon dioxide extract—Pichtovit—is an antioxidant and hemostimulant. Its usage increases zootechnical indices of the broilers and requires less fodder [5].

Extractive compounds of *Abies* wood greenery are an active substance of fodder additive Verva (state certification No. PVR-2-5.0/0260) developed in Institute of Chemistry of Komi SC of the Ural Division RAS. The preparation is obtained with the help of environment-friendly emulsion extraction method from *Abies* wood greenery.

11.2 MATERIALS AND METHODOLOGY

11.2.1 CHEMICAL EXPERIMENTAL PART

The wood greenery of *Abies* and *Picea*, which meet the requirements of State Standard 21679-84, gathered in the suburbs of Syktyvkar were used for the experiment. The humidity of the raw material was determined by Dean–Stark method [6].

The raw material was milled on a disk breaker until the fraction is not exceeding 80–100 mm, then on a spiral breaker, until the fraction is not exceeding 1.0–5.0 mm.

Emulsion extraction of the raw material was carried out at a temperature of 45–50°C using pulsing, gravity, and modernized extraction and filtration equipment.

Neutral and acid compounds were extracted from emulsion solutions using the method described in this chapter [7]. The chemical composition of neutral compounds was studies like in this chapter [8]. Essential oils were extracted by hydrodistillation. The amount of caratenoids was analyzed by ultraviolet spectrometry [9]. The amount of flavonoids in terms of rutin (company Alfa Aesar) was determined using this method [10].

The isolation of compounds was carried out by the method of column chromatography on silica gel (company Alfa Aesar) 70–230 with the help of solvent system petroleum ether–diethyl ether with the increasing amount of the latter. Components were identified according to infrared- and magnetic nuclear resonance spectroscopy in comparison with the spectra of the standard samples and with the published data.

11.2.2 BIOLOGICAL EXPERIMENTAL PART

For the tests, proof samples of the fodder additive Verva (*Abies* wood greenery) with concentration 5 g/L (in terms of active substances amount) were made. The water-diluted (in 1:10 ratio) fodder additive given to the test animal was added to their diet. The fodder-additive Verva was enriched by water-soluble cobaltic salt. To feed quails, the preparation was added to the drinking water in the concentration of 0.50, 0.33, and 0.25%. The quails' growth and development was measured according to the changings in their body weight in every two weeks.

11.3 RESULTS AND DISCUSSION

The rational of the research is to develop a complex technology of processing plant raw material by eco-friendly emulsion method for obtaining natural biopreparations. Coniferous wood greenery is waste products of lumbering, which are not used. Meanwhile, extractive compounds of wood greenery possess wide spectrum of biological activity. The reason which hardens the industrial processing of wood greenery is insufficient development of technological schemes of complex processing for highly effective production. The improvement in techniques of extraction will allow the isolated valuable ready-made compounds more useful. These compounds frequently cannot be obtained synthetically, or their synthesis is expensive and difficult.

The chemical composition of coniferous wood greenery is complicated and various. Extractive substances consist of different classes of organic compounds, which are divided into neutral and acid groups.

The group of neutral compounds includes carotene, the predecessor of vitamin A, which is an antioxidant. Tocopherol (vitamin E) is also important antioxidant. Essential oils are widely used in perfumes and cosmetic industry and in medicines. The main components of essential oils are α- and β-pinene, camphene, Δ^3-carene, bornyl acetate, and borneol. *Abies* wood greenery contains five times more volatile oils than a *Pinus* and *Picea* ones combined. Essential oils possess antibacterial activity.

Polyprenols are a group of low-molecular bioregulators. Being immunopotentiating and having an extremely low toxity, poliprenols can be used for treating different diseases of immune system. Preparations based on polyprenols have a wide action spectrum. They can reduce blood pressure and can be used as burntreating, antiulcer, and vulnerary agents.

A special place in acid group is occupied by polyunsaturated fatty acids. These acids are called irreplaceable. *Picea* wood greenery has most of them. Mammals cannot synthase linoleic and linolenoic acid. Like vitamins, they must come as ready-made products. The most active among them are arachidonic acid (it is synthesized from linoleic acid) and lilolenoic acid. Fatty acids are known as being cardioprotective. They influence a condition of coagulation system of blood and inflammatory processes; in addition, they also have antiarrhythmic effect. Linolenoic acid–possessing antioxidant action is one of the most significant nutrients; it plays an important role in maintaining vital activity of cells.

The *Abies* wood greenery contains triterpenic acids, which the other coniferous species do not have. Biological tests of the sum of triterpenic acids of *Abies* wood greenery have shown high efficiency of their influence for agriculture crops. Phenol compounds possess antioxidant and antifungal effects. The method of obtaining of biologically active compounds from plants include extraction of raw material, filtration, and concentration extract and using them as a ready-made product or a raw material for isolation of compounds with valuable features.

11.3.1 INVESTIGATION EMULSION METHOD OF EXTRACTION

The method of emulsion extraction is based on raw material processing with water solution of alkali. Water-soluble salts of resin acids and higher fatty acids contained in the raw material are formed in the processing. Salts of resin acid are surface-active agents; salts of higher fatty acids belong to micelle-forming lipids. Surface-active agents and micelles in the reaction media form a so-called "oil-in-water" emulsion, that is, isolation of extractive substances from the raw material is a result of emulsion extraction [7].

The studies of emulsion method of extracting biologically active substances from a plant raw material are the bases for devising a complex technology of processing the coniferous wood greenery [11]. The emulsion method of extracting substances from *Abies* wood greenery was carried out with pulsing and filtering extractor, 500 L (AIS 2.952.020, manufacturer "IREA-Penzmash"). Optimal conditions of extraction are presented in Table 11.1.

TABLE 11.1 Conditions for Extracting the *Abies* Wood Greenery in a Pulsing and Filtering Machine.

Technological index	Value index
Degree of milling of the raw material, mm	1–5
Process temperature, °C	50 ± 5
Hydromodule	1/10
Concentration of extraction solution, %	5
Time of extraction, h	4

The yield of extractive substances of *Abies* wood greenery in these conditions is 6–7% on a dry basis. The disadvantage of this equipment discovered during the experiment is formation of dead zones in the equipment.

The research went on to improve the extraction technology for coniferous wood greenery. It showed that the yield of extractive substances to a considerably degree depends on the equipment used in the extraction process. There are various types of equipment for extracting biologically active substances from plant raw materials [12]. We used delta-rotor (rotor and pulsing equipment, manufactured at limited liability corporation, scientific production company "Aerotechnics ", Kazan), gravity, and modernized extraction and filtering equipment devised at a small-business enterprise of Chemistry Institute (Fig. 11.1).

Delta-rotor is a rotor and pulsing equipment which has a hydromechanics influence on chemical and technological processes. The equipment helps to intensify extracting plant raw materials at the expense of multiple processing by crushing, attrition, and impact loads [13]. The disadvantages which prevent the use of this equipment were discovered in the experiments. First of all, hard-to-degrade suspension is formed during the extraction. Moreover, cracks in a rotor and stator are blocked which results in necessity to stop and clean the machine. As a result, a lot of technological difficulties appear while filtering a heterogeneous solution.

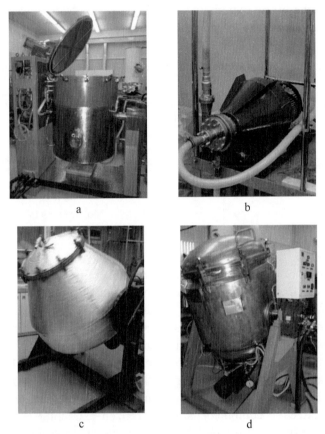

FIGURE 11.1   Extractors: a—pulsing and filtering, b—delta-rotor, c—gravity, and d—modernized extracting and pulsing.

We devised a special mobile gravity equipment to carry out emulsion extraction at the lumbering place. Its carcass is a pear-shaped cylinder rotating at some angle. While rotating, the raw material goes up and then down because of gravitation, which intensifies mass exchanging processes because of improving hydrodynamic conditions of raw material extraction.

A modernized extracting and filtering machine was devised to avoid disadvantages of pulsing and filtering machine that is dead zones formed while processing raw materials. A modernized machine is similar to pulsing and filtering one: it has a rotating cylindrical carcass with a flap lid which contains a filter. A pulse bell is replaced by a mechanic mixing machine.

The results of the experiments which study emulsion extraction of a coniferous raw material are presented in Table 11.2.

TABLE 11.2   The Yield of Extractive Substances from Oil-in-Water Emulsion of *Abies* and *Picea* by Using Different Equipments.

Extractor	Yield extractive substances, % on a dry basis		Technological peculiarities
	Abies wood greenery	*Picea* wood greenery	
Pulsing and filtering	6–7	3–4	The forming of dead zones
Delta-rotor	10–11	7–8	Filtering problems
Gravity	6–7	4–5	Mobile extractor
Modernized	8–9	5–6	Mechanic mixing

The yields of extractive substances are sums of neutral and acid ones which were extracted from emulsion solutions.

At consecutive extraction by organic solvents (benzene, ethanol, and acetone), the total yield of extractive substances from *Abies* wood greenery is 8%, from *Picea*—6% [14]. In our experiments, the maximum yield of extractive substances was reached by using delta-rotor. However, the drawbacks of a rotor and pulsing machine mentioned above prevent us from using it for emulsion method of coniferous raw material processing. The most effective equipment is a modernized extracting and filtering machine. It extracts substances not worse that in case where organic solvents are used.

To develop the technology of emulsion extraction, experiments were carried out in a modernized extractor to define effective factors (Figs. 11.2–11.4).

The concentration of extractive solution was varied from 3% to 5%, hydromodule—from 1/8 to 1/12, the time of extraction—from 2 to 4 h The results show that the optimal conditions are 3–3.5 h, hydromodule—1/8–1/10, and concentration of alkaline solution—3.5–5%. The yield of extractive substance from *Abies* wood greenery is 9%, from *Picea*—6% on a raw material mass basis.

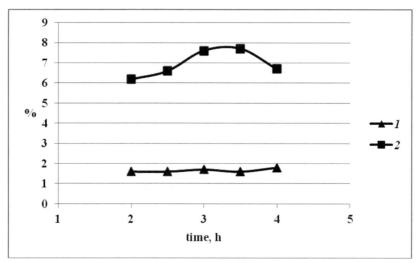

FIGURE 11.2 The dependence of *Abies* wood greenery extractive substances yield from the time of emulsion extraction, 1—neutral compounds and 2—acid compounds.

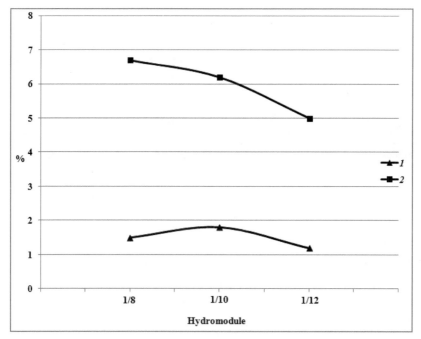

FIGURE 11.3 The dependence of *Abies* wood greenery extractive substances yield from the hydromodule of emulsion extraction, 1—neutral compounds and 2—acid compounds.

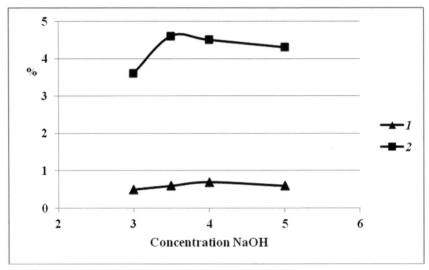

FIGURE 11.4 The dependence of *Picea* wood greenery extractive substances yield from the concentration of NaOH solution, 1—neutral compounds and 2—acid compounds.

Thus, experiments show that the most effective eco-friendly method of emulsion extraction is a modernized extracting and filtering machine. Biologically, active components are extracted from a raw material with high yields without organic solvents (Table 11.3).

TABLE 11.3 The Composition of Bio-Active Components of Wood Greenery, % on a Raw-Material Mass Basis.

Components	*Abies* wood greenery	*Picea* wood greenery
Hydrophobic components		
Essential oils	1.0–1.5	0.5–0.7
Polyprenols	0.1–0.2	0.1–0.2
Sitosterol	0.8–0.9	0.7–0.8
Saturated alcohols	0.07–0.08	0.02–0.03
Carotenoids	1.60 mg%	1.65 mg%
Hydrophilic components		
Flavonoids	1.1–1.2	0.7–0.8
Acid components	5.5–7.0	3.5–4.5

11.3.2 THE STUDIES OF FODDER ADDITIVE VERVA AND ITS EFFECTIVENESS

The fodder additive Verva is obtained from *Abies* wood greenery using the method of emulsion extraction. Extractive compounds are active substances of a specimen. Biologically, active compounds of the fodder additive are immunomodulating and adaptogenious. Minerals essential for animals and poultry such as Fe, Mn, Cu, Zn, Ca, and P are available in the fodder additive.

The effectiveness of fodder additive Verva in plant-growing and cattle-breeding was presented above [15]. The further studies show that including fodder, additive Verva into the cow diet increases milk yield and its fatness. The fodder additive was given to high-productive milk cows of Kholmogor breed in the farm "Prigoroniy" of Komi Republic where the milk yield was increased by 4%.

This chapter presents the effect fodder additive Verva has on calves, pigs, and quails.

The experiment was carried out on calves in the farm Palevetskaya of Komi Republic in summer and autumn 2012. A day amount of the fodder additive was 5 mg (in terms of active substance amount) during the first month, 10 mg—during the second month, and 20 mg—during the third month. The calves' weight was measured every month. The average weight the calves put on was 6.4%.

The animals had fodder additive Verva (5 mg in terms of active substance amount) enriched with cobalt (daily—2.5 mg on head). Lack of cobalt in the diet of different ruminant animals in the Komi Republic is 25–75%. The fodder additive was given to the animals for 90 days. The weight of the calves which were given the fodder additive enriched with cobalt was 5.5% higher than the average.

The biochemical tests of blood were carried out. They showed that fodder additive Verva did not have any negative effects for the health of the animals. All biochemical factors were standard.

The effectiveness of industrial pig-breeding is measured by their reproductive and productive abilities. One of the main reasons which cause the decrease in pig productivity is lack of minerals and biologically active substances in animals' diet. At the present stage, in practice of livestock, breeding with the purpose of strengthening of physiological processes in an organism, improvements of a metabolism, and increase of nonspecific

natural resistance of an organism use various biological products. Therefore, a new natural fodder additive Verva is interesting for scientific theory
and practice in the sphere of pig-breeding.

The influence of the fodder additive on the meat productivity of pigs of
breeds Landras and Big White was studied in the farm "Zarechie" (Kirov
region) [16]. A total of 15 pigs each were divided into three test groups
and one control group. They had fodder additive Verva for 30 days. The
day amount was 5 mg (in terms of active substance) in the first group, 15
mg in the second, and 25 mg in the third group. After that period, the mass
productivity and the weight of internal organs were measured.

According to the results, fodder additive Verva increases the weight of
the animals. The best results had the pigs which took the dose of 1 mL of
the supplement a day. The prelethal alive weight of pigs of the first group
has made 107 kg (that on 14–17% of more weight of animal other groups).
According to the experiment, we can suggest that fodder additive Verva
is accumulated in the organism which can be the bases for the following
studies [16].

The experiments were also conducted in the agricultural firm
"Doronichi" (Kirov region). The fodder additive was given to reproducing
pigs Big White breed at the age of 1.5–2 years old for 30 days. A day
amount was 25 mg (in term of active substance) for an animal [17]. The
use of the fodder additive improved the animals' sexual instinct, biological
quality of sperm, and its quantity. According to sperm count of pigs, the
volume of ejaculate was increased, and the concentration of spermium
was 6% more than the one of the control group.

The influence of the fodder additive on the reproduction and prevention after-birth diseases was studied on 40 sows of Big White breed with
a pregnancy period of 80 days [18]. The animals were divided into three
test groups and one control group. The day's amount was 5 mg (in term
of active substance) in the first group, 15 mg in the second, and 25 mg
in the third group. The gestation course, farrowing period, and health
state in and after birth-giving periods were observed. Sows reproduction
functions and newborn calves' survival, growth, and development were
studied.

The experiments show that the fodder additive included in the diet
prevents some after-birth complications and intensifies calves' survival,
growth, and development. The fodder additive did not have any negative effects on the animals' pregnancy. More than that, the first group

demonstrated the decrease in diseases by 10%, the second, and the third groups by 20%.

The amount of newborn pigs increased in the second and third test groups by 15.5% in comparison with control. The survival in the third test group was 94.5% which is 12.4% more that in the control group. The piglets from the second group grew faster than their peers. Milking ability of sows of the second and third groups was higher in 10.4–11% in comparison with control.

Thus, the use of fodder additive Verva in small doses in pig farming positively effects productive qualities, reproduction, and health of the animals.

Nowadays, the amount of non-traditional kinds of birds such as turkeys, ducks, quails, and guinea fowls is 4% in the total amount of fowl production in Russia. According to the concept of development of poultry farming till 2020, growth of a share of these kinds of birds in total amount up to 10% is supposed. More than that, compared to chicken and duck eggs quail eggs are low allergic. They contain a small amount of cholesterol and have rich vitamins A, PP, B1, B2, and minerals. The parity of proteins, fats, carbohydrates, and mineral substances is unique, which has practically ideal nutrition for a person.

The goal of the experiment conducted in vivarium of Vyatka State agricultural academy was to study how fodder additive Verva effects the growth and development of young Japanese quails [19].

The test groups had the fodder additive with the water for 30 days. The first test group had the fodder additive in concentration of 0.50%, the second group—0.33%, and the third group—0.25%. The growth and development of quails were diagnosed by the weight they put on. The beginning of egg producing ability and the quality of the eggs were also taken into consideration. Male and female species starting from 16-week's period were analyzed because of their sexual dimorphism.

Thus, taking the water solution with fodder additive Verva positively effects the growth and development of quails and the earlier beginning of egg producing ability [19]. The male quails from the third test group grew much faster. Their weight was 10% more than that of the control group. The weight of the female quails from the second test group was 13% more than that of the control group. The earliest ability to produce eggs was observed in the second test group at the age of 54 days, that is, 7–11 days before the other groups. The quality of the eggs did not change.

11.4   CONCLUSIONS

The experiments have shown that the most effective eco-friendly method of extraction of coniferous wood greenery is a modernized extracting and filtering machine. Biologically, active components are extracted from the raw material with a high yield without taking organic solvents.

Fodder additive Verva is a practical issue because it increases the productivity and reproduction of the animals and birds and decreases their diseases. The fodder additive is an alternative to fodder synthetic antibiotics.

KEYWORDS

- **emulsion extraction**
- *Abies*
- *Picea*
- **extractor**
- **fodder additive**
- **Verva**
- **calves**
- **pigs**
- **quails**

REFERENCES

1. Rubchevskaya, L. P.; Ushanova, L. N. Biologically Active Substances of Carbon-Dioxide and Propan-Butan Extractants of Wood Greenery. *Russ. Chem. J.* **2004,** *3*, 80–83 (In Russian).
2. Levin, E. D.; Repyakh, S. M. *The Processing of Wood Greenery.* Forest Industry: Moscow, 1984; p 120 (In Russian).
3. Kostesha, N. Ya.; Lukjanenok, P. I.; Chardynzeva, N. V.; Matveewa, L. A.; Strelis, A. K. Extract Siberian Fir ABISIB and Its Application in Medicine and Veterinary. *Adv. Curr. Nat. Sci.* **2010,** *12*, 11–13 (In Russian).
4. Lantseva, N. N.; Martyshenko. A. E.; Shvydkov. A. N.; Ryabukha, L. A.; Smirnov, P. N.; Kotlyarova, O. V.; Chebakov, V. P. The Influence of the Functional Properties of

Probiota Vitabiotics on Productivity and Functional Status of Broiler Chicken. *Fund. Res.* **2015,** 2, 1417–1423 (In Russian).

5. Kulikova, A. V.; Khokhlova, A. V. The Influence of Pichtovit on Priductivity and Antioxidant Status of Broilers. *Vet. Med.* **2007,** 2, 12–15 (In Russian).

6. Kolesnikov, A. L. *Technical Analysis of Organic Synthesis Products.* Higher School: Moscow, 1966; 232p (In Russian).

7. Kutchin, A. V.; Karmanova, L. P.; Koroleva. A. A.; Khurshkaynen, T. V.; Sychev, R. L. Emulsion Method of Extracting Lipids, Patent of the Russian Federation No 2117487, Published 20.08.1998, Bul. # 23 (In Russian).

8. Koroleva A. A.; Karmanova L. P.; Kutchin A. V. Extraction Poliprenols from a Raw Material, News of Higher Schools, Series. *Chem. Chem. Technol.* **2005,** *48*(3), 97 (In Russian).

9. Briton, G. *Biochemistry of Natural Pigments.* Mir: Moscow, 1986; p 422 (In Russian).

10. Muzychkina, P. A.; Korulkin. D. U.; Abilov, G. A. Quality and Quantity Analysis of the Main Groups of BAS in a Medical Raw Material and Herbal Formulations. Al-Farabi Kazakh National University: Almaty, 2004; p 285 (In Russian).

11. Khurshkaynen, T. V.; Skripova, N. N.; Kutchin. A. V. Highly Productive Technology of Plant Raw Material Processing and Obtaining of Specimens for Agriculture. *Theor. Appl. Ecol.* **2007,** *1*, 74–77 (In Russian).

12. Ponomarev, V. D. *Extraction of Medical Raw Material.* Medicine: Moscow, 1976; p 202 (In Russian).

13. Balabudkin, M. A. *Rotor and Pulsing Machines in Chemical and Pharmacy Industry.* Medicine: Moscow, 1983; p 160 (In Russian).

14. Roshchin, V. I.; Vasiliev, S. N.; Pavlutskaya, I. S.; Baranova, R. A.; Skachkova, N. M. The Processing Method of Coniferous wood greenery, Patent of the Russian Federation No 2015150, Published 30.06.1994, Bul. No. 10 (In Russian).

15. Khurshkaynen, T. V.; Kutchin A. V. *Woodchemistry for Innovation in Agriculture.* In Proceedings of the Komi Science Centre of the Ural Division of the Russian Academy of Sciences, 2011, *1*, 17–23 (In Russian).

16. Philatov, A. V.; Shemuranova, N. A.; Khurshkaynen, T. V.; Kutchin, A. V. Pigs Productivity Indices by Using VERVA. *Vest. Veterinar.* **2014,** *2*, 81–84 (In Russian).

17. Philatov, A. V.; Shemuranova, N. A.; Ponomarev. I. N.; Khurshkaynen, T. V. *Spermoproduction Boars Producers in Applying the Drug VERVA.* Material of International Conference Important Problems of Veterinary Obstetrics and Animal Reproduction: Belarus, Gorki, 2013; pp 114–116 (In Russian).

18. Philatov A. V.; Kubasov O. S.; Khurshkaynen T. V.; Kutchin A. V. Of Sow Postnatal Pathologies and Increase of Piglets Viability, *Questions of Normative-Legal Regulation in Veterinary Medicine.* **2014,** *3*, 171–174 (In Russian).

19. Philatov, A. V.; Sapozhnikov A. F.; Khurshkaynen T. V.; Kutchin A. V. Use of Liquid Fodder Additive VERVA for Growing of Japanese Quails. *Eur. Sci. Union.* **2014,** *4*, 36–39 (In Russian).

CHAPTER 12

PLANT GROWTH AND DEVELOPMENT REGULATORS AND THEIR EFFECT ON THE FUNCTIONAL STATE OF MITOCHONDRIA

IRINA V. ZHIGACHEVA[*] and ELENA B. BURLAKOVA

Emanuel Institute of Biochemical Physics of Russian Academy of Sciences, 4, Kosygin St., Moscow 119334, Russia

[*]*Corresponding author. E-mail: zhigacheva@mail.ru*

CONTENTS

ABSTRACT

This chapter discusses the problem of the protection of plant cells from oxidative stress and possible mechanisms to reduce the generation of reactive oxygen species by plant mitochondria in these conditions. Protective properties of some plant growth and development regulators are considered in terms of the impact of these drugs on the functional state of mitochondria.

12.1 ABIOTIC STRESS AND SYSTEM OF ENERGY DISSIPATION IN PLANT MITOCHONDRIA

Throughout the process of vegetation, the plants exposed to action not one but multiple stressful environmental factors. The plants, unlike animals, are not able to escape or hide themselves from the action of a stress factor. They are forced to adapt to changing environmental conditions. The reaction of plants to changing environmental conditions is composite, which include changes in biochemical and physiological processes. These changes can be nonspecific and specific. It is known that the implementation of antistress programs require large energy expenditure [1]. Therefore, the energy exchange plays an important role in adaptive reactions of the organism. In this review, we focus mainly on the mitochondria, as these organelles in plants and animals play a major role in the body's response to the action of stress factors. They play a key role in energy, redox, and metabolic processes of cells [2]. Mitochondria, as an energy metabolism regulator, play one of the basic roles in the organism response to the action of stressors. About 1–3% of oxygen consumed by mitochondria form, as the result of 1–2-electron reduction, reactive oxygen species (ROS), which participate in the cellular redox-signaling. Normally, the bound level of ROS in organs and tissues is rather low (on the order of 10^{-10}–10^{-11} M) due to the enzymatic and nonenzymatic systems of regulation of the accumulation and elimination of ROS. A shift in the antioxidant–prooxidant relationship toward increasing the ROS generation is a result of the action of stressors and leads to the disturbance of physiological functions of plant organisms (reduction of growth processes, crop yield, etc.). ROS can inhibit or reduce the activity of enzymes of mitochondria containing Fe–S clusters, such as NADH-dehydrogenase

(Nicotinamide adenine dinucleotide reduced form) (complex I), ATP (adenosine triphosphate)-synthetase (complex V), and aconitase [3]. The accumulation of H_2O_2 in these organelles may induce the interaction of hydrogen peroxide with Fe^{2+} of mitochondria, which will evidently promote the formation of OH• via the Fenton's reaction [4]. The interaction of OH˙ with polyunsaturated fatty acids of membrane lipids, that is., linoleic acid, linolenic acid, and arachidonic acid, leads to activation of lipid peroxidation (LPO). Appearance of hydrophilic products oxidation as a result of LPO alters the structure of the lipid bilayer membranes in the hydrophobic areas. This creates conditions for passive transport of ions and metabolites and, thus, to a certain extent (depending on the intensity of the oxidative process), disturbed coordination and specificity of membrane processes [5], which leads to mitochondrial dysfunction. In addition, as a result of LPO, toxic for plant cells the aldehydes and (4-hydroxy-2, 3-nonenals) are formed. These toxicants inhibit enzymes involved in the key metabolic pathways, one of them in photorespiration while others-in the citric acid cycle, which affects the electron transport in the mitochondrial respiratory chain due to the depletion of NADH pool in the mitochondrial matrix [2]. The main sources of ROS under stress conditions are the mitochondria and chloroplasts [2]. It can be assumed that the basic mechanism of action of preparations-adaptogens is the reduction of generation of ROS by these organelles. The main candidates for this role are antioxidants. However, reducing the generation of ROS by mitochondria can be achieved in other ways. ROS generation is increased in condition of a high degree the reduced intermediates of the respiratory chain. "Soft"uncoupling, ie.that is, an increase in proton conductivity, which does not disrupt the synthesis of ATP, but leads to a decrease in the redox potential of the inner mitochondrial membrane by 13–15% by 80% decreases ROS generation in mitochondria [6]. To understand this mechanism, let us consider the scheme of the structure of the respiratory chain of mitochondria. The composition of the main respiratory chain, that is, common to animals, plants, and fungi consist of five protein complexes: I—NADH-dehydrogenase, II—succinate dehydrogenase, III—bc1 complex, IV—cytochromoxidase, and V—ATP-synthase. The connection between protein complexes performs two-electron carriers— ubiquinone and cytochrome C. Complexes I, III, and IV act as proton pumps, they transfer protons from the matrix to the intermembrane space, which are bound with transport of electrons (Fig. 12.1).

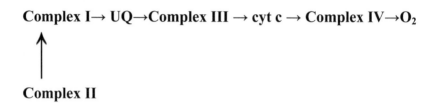

Complex I→ UQ→Complex III → cyt c → Complex IV→O$_2$

Complex II

FIGURE 12.1   The main respiratory chain.

This creates a transmembrane potential difference, which used ATP-synthetase for ATP synthesis. Naturally, the increase in proton conductivity will lead to a decrease in membrane potential and, consequently, to the reducing the generation of ROS by mitochondria.

12.2   UNCOUPLING PROTEINS AND THE MITOCHONDRIAL ATP-SENSITIVE POTASSIUM (MITOK$^+_{ATP}$) CHANNEL

Cell protection of plants from excessive production of ROS often occurs owing to a change in the energetics of mitochondria: synthesis of cold shock proteins and HSPs, uncoupling of oxidative phosphorylation with fatty acids. They are mediators of thermoregulatory uncoupling, which gives a large contribution to the increase of heat during the short-term cooling. The transfer of fatty acid anions from the matrix into the intermembrane space of mitochondria in some cases releases protein, which has been found in brown adipose tissue of mammals and has received the name thermogenin (UCP) (uncoupling protein)[7]. The presence of UCP in plant mitochondria was first found in potato, and it was given the name pUCP (plant uncoupling protein) [8]. Later, pUCP has been found in various tissues of higher plants, including monocotyledonous and dicotyledonous plants. The mechanism of uncoupling is reduced to the following: fatty acid anions are transferred to the external part of the inner mitochondrial membrane with the help of pUCP1. Here, they become protonated and returned to the neutral form;

in this case, they transfer the hydrogen ions into the matrix. This stage is accompanied by uncoupling of oxidative phosphorylation and the release of heat. Thus, according to the hypothesis "the cyclic turnover of fatty acids," fatty acids act as cyclic protonophores. Note that purine nucleotides inhibit the transfer H^+ by fatty acids with the participation of UCP. Mokhova *et al.* shows participation of another carrier—ADP/ATP antiporter in the "cyclic turnover of fatty acids (FFA) (free fatty acids)." "Proof of this is to inhibit the potent inhibitor of ADP / ATP antiporter-carboxyatractylate (Catr) reduction $\Delta\psi$, caused by free fatty acids. In this case, uncoupler (FCCP-Carbonyl cyanide-4-(trifluoromethoxy) phenylhydrazone) has not rendered such effect [9]. The participation of ADP/ATP antiporter in the function of the adaptive uncoupling of respiration and phosphorylation by fatty acids bind to the lack of selectivity of this carrier for hydrophobic anions, in contrast to hydrophilic anions. That is why, ADP/ATP antiporter is capable to transport anions of fatty acids [10]. FFA and ADP/ATP antiporter indeed participate in the protection of plant cells from oxidative stress. So due to the cooling of the plants, the cells significantly increases the activity of phospholipases [11], the functioning of which provides a constant flow of FFA, especially linoleic acid, from the cytoplasm to the mitochondria. It is shown that the mitochondria winter wheat by a high content of fatty acids may use them, particularly linoleic acid as the substrate oxidation [12], which leads to lower generation of ROS. These same results lead to oxidative stress caused by hydrogen peroxide or antimycin A in cell culture of tomato in vitro [13]. In "mild uncoupling," fatty acids are involved in ATP/ADP antiporter and pUCP-like proteins. Contribution pUCP and ATP/ADP antiporter in uncoupling of oxidation and phosphorylation caused by the unsaturated and the saturated fatty acids differ. The uncoupling of oxidation and phosphorylation caused by unsaturated fatty acids, realizes pUCPs, and the uncoupling caused by saturated fatty acids, realizes through ATP/ADP-antiporter [11]. On the basis of these data, Skulachev [6] suggested that the mechanism of participation of UCP and ADP/ATP antiporter in thermogenesis and in protection against oxidative stress is the same and is reduced to "cyclic turnover of fatty acid."

According to some authors, it contributes to the protection of cells from oxidative stress and mitochondrial ATP making-sensitive potassium (mitoK$^+_{ATP}$) channel. The mitochondrial ATP-sensitive potassium (mitoK$^+_{ATP}$) channel carry out the entrance of potassium in mitochondria that was detected by patch clamp method in the inner membrane of mitochondria in 1991 [14]. This channel was first discovered in the mitochondria-etiolated durum wheat and named by analogy with mitoK$^+_{ATP}$ mammals PmitoK$^+_{ATP}$ [15]. Later, PmitoK$^+_{ATP}$ was found in the mitochondria of etiolated seedlings peas, soybeans, and other crops [10, 15]. It has been shown that opening mitoK$^+_{ATP}$ channel, observed in the presence of cyclosporine, is regulated by redox status and is inhibited by ATP and ADP [10]. Superoxide anion radicals, as well as FFA (FFA) and acetyl-CoA [16] activate themselves. Possible functions of this channel may be a reduction of ROS and the regulation of mitochondrial volume [15, 17]. The ability of PmitoK$^+_{ATP}$ to reduce ROS production by mitochondria has been demonstrated in the pea mitochondria. ATP stimulates while cyclosporin A inhibits succinate-dependent formation of H$_2$O$_2$ by mitochondria that serves as an indirect confirmation of the participation PmitoK$^+_{ATP}$ channel in regulating the generation of ROS by these organelles [18]. As a result of the opening of the channel decrease, the generation of ROS is almost in 35 folds [16]. In the mitochondria of higher plants, the operation of PmitoK$^+_{ATP}$ is closely linked with the K+/H+ exchanger (Fig. 12.2).

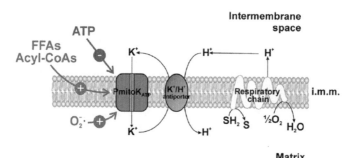

FIGURE 12.2   ATP-sensitive potassium (mitoK+ATP)-channel. (Reprinted from Trono, D.; Laus, M. N.; Soccio, M; Alfarano, M.; Pastore, D. Modulation of Potassium Channel Activity in the Balance of ROS and ATP Production by Durum Wheat Mitochondria-An Amazing Defense Tool Against Hyperosmotic Stress. *Front Plant Sci.* **2015**, http://www.ncbi.nlm.nih.gov/pmc/articles/PMC4664611 https://creativecommons.org/licenses/by/3.0)

This allows realize a "potassium cycle" which provide the return of H+ to the mitochondrial matrix, that is, decrease of Δψ. The decrease Δψ during activation mitoK$^+_{ATP}$ has no effect on the synthesis of ATP by oxidative phosphorylation, indicating a possible involvement of the channel in adaptation of plants to stress factors [16]. Note that in mammalian cells, potassium cycle "does not significantly contribute to uncoupling of mitochondrial respiration, as the maximum activity of this cycle is 20% of the total pool of protons generated by the respiratory chain [17]." In plant cells, the activity of this cycle is comparable to the activity of proton pumps of the respiratory chain of mitochondria. In plant cells, the activity of this cycle is comparable to the activity of proton pumps mitochondrial respiratory chain. It is noteworthy that ATP inhibits this channel in the mitochondria of plants with an efficiency of 10 times less than in mammals, and Mg^{2+}, which inhibits this channel in a mammal, is not effective for durum wheat. Possibly, this channel can act as an antioxidant system in response to stress factors, preventing damage to plants by reducing the formation of the mitochondrial ROS. Indeed, experiments on mitochondria isolated from seedlings of durum wheat, which subjected to salt stress, show the possibility of participation $PmitoK^+_{ATP}$ channel for reducing the generation of ROS by the respiratory chain [15, 19]. In plant organisms, protection from oxidative stress is carried out by antioxidant system cells. Reducing of the generation of ROS by mitochondria is achieved thanks to the activation of the alternative pathway oxidation involving alternative oxidase (AOX), which branches off from the main respiratory chain at the level of ubiquinone and transfers electrons directly to oxygen to form water (Fig. 12.3).

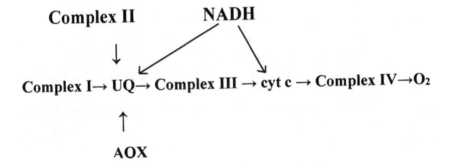

FIGURE 12.3 Respiratory chain of plant mitochondria. Numbers (I–V) identifies the large respiratory complexes located on the inner mitochondrial membrane. AOX; UQ—ubiquinone/ubiquinole pool. (Original drawing courtesy Nicolas Taylor & Harvey Millar, Plants in Action, Chapter 2)

Electrons transfer via path with AOX does not involve the synthesis of ATP. The importance of this pathway is in competition with auto-oxidation of the carriers with the formation of ROS. Note that the mitochondria of plants, unlike animal mitochondria, apart from the rotenone-sensitive complex I contains a number of rotenone-insensitive NADH and NAD(P) H (Nicotinamide adenine dinucleotide phosphate) dehydrogenases. They are localized on the outer ("external" NAD(P)H-dehydrogenase) and the internal ("inner" NAD(P)H-dehydrogenase) mitochondrial membrane. Rotenone-insensitive NADH and NAD(P)H-dehydrogenase transmit electrons to ubiquinone. In contrast to the rotenone-sensitive complex I, the rotenone-insensitive NADH and NAD(P)H-dehydrogenase do are not constitute the proton pump, therefore, their work is not created by the transmembrane gradient of protons. In resting plant cells at low concentrations of Ca^{2+}, the "external" NAD(P)H-dehydrogenase is practically inactive. Under stress conditions, when the concentration of Ca^{2+} increases and simultaneously the synthesis of polyamines increases, the activity of these enzymes increases dramatically [20]. In the method of immunoblotting, the presence in the protein spectrum of mitochondria of winter wheat seedling proteins, NDA (internal rotenone-insensitive NAD(P)H-dehydrogenase), and NDB (external rotenone-insensitive NAD(P)H-dehydrogenase) during hardening to cold were shown [21]. There is evidence that activation of NAD(P)H dehydrogenase, which is located on the outer mitochondrial membrane, may affect the subsequent sections of the respiratory chain, for example, cause an increase in expression of AOX [22]. It is assumed that functional status of NAD(P)H-dehydrogenase may have an impact on the level of restored NADP in the matrix and thus to influence a number of biochemical processes in which NADP is a coenzyme, particularly in the folate biosynthesis, turnover of glutathione and thioredoxin and activation AOX [23]. On the light in the green tissues of plants between chloroplasts and mitochondria, there is a constant exchange of metabolites. At the same time, light induces a spatial dependence of the location of mitochondria in the immediate area of the chloroplasts, which indicates the presence of exchange of metabolites between these organelles [24]. It is shown that under intensive irradiation with light, the activity of the external rotenone-insensitive NADH dehydrogenase dramatically increases [25]. In the conditions of proceeding rapidly of photosynthesis, the main functions of mitochondria are production of ATP for sucrose synthesis, supply of carbon skeletons for biosynthetic processes (especially for nitrogen assimilation), and the transformation of glycine into serine in the process of photorespiration. Thus, the role of the external rotenone-insensitive NADH dehydrogenase is reoxidation of the synthesized NADH and maintenance of the photosynthetic activity of chloroplasts [24]. In addition, the experiments Grabelnych [21] show activation of external rotenone-insensitive NADH dehydrogenase in the hardening of plants (wheat shoots). According to the author in the period of action tempering temperatures, the high activity of the external NADH dehydrogenase may compensate for the reduced rates of oxidation of malate and succinate, as well as allows you to maintain the required level of NAD+ in the cytoplasm. In the protection against oxidative stress in plant cells, essential value plays an AOX (alternative oxidase). Direct proof of such a function is the data on the accumulation of ROS in transgenic tobacco plants with reduced levels of AOX in the presence of inhibitors of the electron transport chain of mitochondria. Moreover, on mitochondria isolated from cotyledons of pea and soybean, it was shown that in tissues, which have high-level expression of AOX, under the inhibition, which realized by salicylhydroxamic acid (SHAM) and by propylgallate, highly stimulated the generation of hydrogen peroxide by mitochondria[26, 27]. At the same time, 2 μM antimycin A increases SHAM-dependent generation of H_2O_2 by mitochondria. Application of SHAM and propylgallate on the rat liver mitochondria does not increase the generation of H_2O_2, which confirms the conclusion of the significance of AOX, which is absent in animal tissues, in reduction of the rate of generation of H_2O_2. Transgenic tobacco plants (*Nicotiana tabacum*) which have been knocking out regarding AOX, or suspensions of cells deficient in the AOX, have enhanced expression of numerous genes encoding antioxidant enzymes. This observation is consistent with the hypothesis that AOX reduces the generation of ROS by the electron transport chains of mitochondria, and the absence of this transporter increases the generation of ROS, the mitochondria, but triggers the synthesis of antioxidant enzymes [28]. This carrier is activated in response to a large number of different types of external influences on plants and probably participates in the response of plants to different types of stresses. Induction of AOX is observed at low temperatures [21, 29], it is the combined action of water deficiency and lowering the temperature to 14–15°C [30], in oxidative stress [31, 32] and the limitation of food. It is activated in yeast cells after the exhaustion of glucose in the medium [33] in response to heat stress and ethanol (yeast

Yarrowia lipolytica) [34], and at a variety of infections [34]. Under conditions of low temperature stress, when attacking pathogens or treatment by salicylic acid (SA) in plant cells, an increase in hydrogen peroxide and a gene expression *Aox1* is observed, which is responsible for the synthesis of the AOX—an enzyme that prevents the excessive production of superoxide radicals by mitochondria. In tobacco cell culture, inhibition of respiration of mitochondria by antimycin A leads to increased generation of $O_2^{\cdot}$ and the accumulation of H_2O_2. In the culture of tobacco cells inhibition mitochondrial respiration by antimycin A leads to increased O_2 generation. and accumulation of H_2O_2. Under these conditions, after 60 minutes, was increasing content of mRNA Aox1 and a few hours and was increasing of synthesis and content AOX, thus confirming the hypothesis of the occurrence of the signal for de novo synthesis of AOX in mitochondria and its transfer into the nucleus In tobacco, cell culture inhibition of respiration of mitochondria by antimycin A leads to increased generation of O_2^{-}. and the accumulation of H_2O_2. In these conditions, after 60 minutes, the content of Aox1 mRNA gets increased, and after a few hours increases the content and synthesis of AOX increase, thus confirming the hypothesis of the occurrence of the signal for de -novo synthesis of AOX in mitochondria and its transfer into the nucleus [35]. Moreover, the inhibition of complex III by antimycin and as a result of increased generation of ROS, an induction of genes encoding AOX from tobacco (gene *NtAox1*), soybean (gene *GmAox*), corn (genes *ZmAox1*, *ZmAox2*, *ZmAox3*), and Arabidopsis (gene *AtAoh1a*) takes place [36, 37]. Tobacco and Arabidopsis genes, which encode this carrier, have also been induced after inhibition of aconitase in the TCA cycle(The tricarboxylic acid cycle or Citric acid cycle) by monofluoroacetate [38, 39]. Exogenously added organic acids of TCA (citrate, malate, and 2-oxoglutarate) at concentrations close to physiological (0.1 mm) significantly increase the level of mRNA AOX, and this increase is observed regardless of the generation of ROS. On this basis, the expression of AOX involves two paths of signal transmission in the core, one involving ROS and the other involving the organic acids [40]. The ability of the AOX to influence the redox status, energy metabolism, and the interrelationship of the organelles gave the possibility to assume that AOX is involved in the reorganization of cell metabolism in response to changing environmental factors. So, the change of expression *aox1α* in *Arabidopsis* leads to a 20% reduction in growth when it is cultivated at low temperatures [41]. In this over-expression (over-expression) of AOX, tobacco increases its sensitivity with the treatment of ozone, at the same time on line of tobacco with cytoplasmic male sterility AOX increases resistance to this factor [42]. Note the activity of energy-dissipating system mitochondria regulated with the content of FFA. They inhibit the activity of AOX and activate pUCP and ATP/ADP antiporter [43], the opening of $PmitoK^+_{ATP}$ channel [44] and PTP (Mitochondrial Permeability Transition pore) [45, 46]. However, when the activation-listed systems of energy dissipation mitochondria are not able to cope with increasing generation of ROS, opening of the PTP (Mitochondrial Permeability Transition pore) occurs. At the same time, drop in membrane potential and increasing the rate of respiration up to the maximum quantities which are limited only by the activity of respiratory enzymes are observed [6]. When the oxygen concentration falls, and the rate of ROS accumulation is reduced, it is time locked [6]. Thus, the opening and closing of PTP, apparently, fluctuate, and may be, this is a physiological mechanism which protects cells from oxidative stress.

12.3 PLANT HORMONES AND THEIR SYNTHETIC ANALOGS AS REGULATORS OF PLANT GROWTH AND DEVELOPMENT

The plant growth regulator (PGR) is an extensive group of natural and synthetic organic compounds that in small doses affect actively the metabolism of higher plants. They stimulate the growth and development of plants by accelerating cell division or stretch in length. Uses of PGRs are one of the most effective ways of increasing the yield, quality of crops, and increase their resistance to stress and pathogens. The natural stimulants of plant growth include the phytohormones—auxins, gibberellins, cytokines, and their synthetic analogs. Application of phytohormones in agriculture meets a number of difficulties. For this purpose, synthesis and selection of analogs of natural phytohormones with desired properties to ensure the prevention of lodging of crops, acceleration maturation, and improve setting fruit, for facilitating mechanical harvesting, increasing productivity and quality of agricultural products are conducted [47, 48]. The tread properties of the RRR are likely are associated with a reduction in the generation of ROS by mitochondria. This reduction can be achieved through the activation of antioxidant systems of cells or other mechanisms described previously.

12.3.1   CYTOKININ 6-BENZYLAMINOPURINE

One promising a synthetic analog of phytohormones is a synthetic analog of cytokinins–cytokinin 6-benzylaminopurine (BAP). Its gross formula is $C_{12}H_{11}N_5$. As an analog of the natural cytokinins, BAP in low concentrations (10^{-5}–10^{-9} M) stimulate cell division, growth, and differentiation of plant cells. This PGR stimulates the synthesis of chlorophyll and carotene from the *Dunaliella salina teodoresco.* Processing of BAP already after 10 days increases the content of carotene on 16–77% depending on the concentration of the drug. The content of chlorophyll *a* was increased by 34–80% and of chlorophyll *b*—237–275% compared with controls [49]. The drug increases mitotic activity root meristem of maize and barley by 29.6 and 12.1%, respectively [50], increasing the resistance of plants to salinity [51]. It is obvious that the activation of synthetic processes requires significant energy, thus is closely related to energy metabolism. As mitochondria are the main energy suppliers, it is likely that the BAP must affect their functional state. Indeed, the introduction of 6-BAP to the suspension of mitochondria from six species of plants (beans, peas, corn, soy, wheat) leads to suppression of the alternative pathway of oxidation [52]. Furthermore, these PGRs can inhibit the activity of the complex I of the respiratory chain of mitochondria. However, by using different concentrations of 6-benzyladenine, the obtained results revealed different mechanisms of the influence of 6-BAP on the transport of electrons in respiratory chain. It turned out that after the 12-h processing the cotyledon and seedlings of yellow lupine *Lupinus luteus* (cv. "Friendly") and *Glycine max* (L.) *Merrill* 22 μM BAP increases the maximum rate of electron transport on the main respiratory chain at 128% (via cytochrome oxidase).It has been found that the changes in electron transport in the mitochondrial respiratory chain of lupine were carried out in two different ways: at a low concentration (1–22 μM), BAP stimulated electron transport through main path (through the cytochrome oxidase complex). At concentrations of 50–100 μM, PGR reduced speed of electron transport through the AOX. Along with the activation of electron transport, the drug in a concentration of 22 μM reduces the intensity of transcription of *atp9* gene, encoding a subunit of ATP synthase. Furthermore, 6-BAP reduces more than two times the intensity of transcription of the genes *cox3* and *cox1* (complex IV); *ccmB* (associated with the biosynthesis of cytochrome *C*); *nad6* (complex I); *rps13, rrn26*, and *trnI* (protein-synthesizing system of mitochondria); *atpE*; and *rrn16* (plastid genes). The suppression of the intensity of transcription against the background activation of the main respiratory path under the action of BAP, according to Belozerova occurs due to the increased time half-life of mRNA of the corresponding genes. The author believes that the accumulation of matrix, probably, may lead to suppression of the synthesis mRNA data [53]. In addition, the increase in the rate of respiration by main cytochrome path under the influence of the drug is probably may be the result of increasing the enzymatic activity of the ETC complexes as well as the result of formation of super complexes, the existence of which is shown in [54].

12.3.2 SALICYLIC ACID

SA or 2-hydroxybenzoic acid was first isolated from the bark of willow.

Salicylic acid (SA)

This compound attracts the attention of researchers because of its ability to induce systemic-acquired resistance of plants to various nature pathogens. This phytohormone has an important regulatory effect on many physiological and biochemical processes in plants, including photosynthesis, growth and development of thermogenesis, the formation of flowers and seed ripening, ageing, and cell death. To date, numerous results convince evidence in favor of the involvement of SA in the induction and development of plant resistance to the action of stress factors such as drought, cooling, heavy metal toxicity, thermal, and osmotic stress. It is shown that the SA activates the generation of ROS, accumulation of defensive PR proteins, the synthesis of phenolic compounds of the phenylpropanoid pathway and is a modulator of hypersensitivity reactions [55]. Its content in the tissues of different plant species varies from 10 μg/g (potato) to 0.25 μg/g (in *Arabidopsis thaliana*) [56]. In addition, SA helps to maintain the redox state of the cell by regulating the activity of antioxidant enzymes [57] and the induction of alternative pathways in the mitochondrial respiratory chain [58]. Therefore, the introduction of 2–20 μM SA to the suspension cell culture of tobacco (*N. tabacum*) supports cyanide-resistant respiration within 2 h, which is accompanied by a 60% increase in thermogenesis. In addition, the hormone induces concentration-dependent gene expression *NtAOX1*. The number of transcripts of this gene increases 2–6 times later, 4 h after treatment with SA, decreases almost to control values after 24 h after the treatment [59]. These authors showed that the introduction of 0.1 mM of SA in a suspension of cells or in the incubation medium of mito- chondria of tobacco leads to stimulation of respiration in connection with uncoupling of oxidation and phosphorylation. At higher concentrations, SA inhibits respiration through inhibition of electron transport in complex I of the respiratory chain of mitochondria, probably by inhibiting the flow of electrons from dehydrogenases to the pool UQ. It has been suggested that the phenolic nature of SA in millimolar concentrations can act as an analog of quinone, preventing the interaction between dehydrogenases and a pool of UQ. Moreover, the mitochondria of animals have shown that SA induces changes in the permeability of the inner membrane of mitochon- dria and opening of PTP. The induction of PTP was accompanied by a drop in membrane potential ($\Delta\Psi$), as well as swelling of these organelles, and by the releases of cytochrome *C* and the apoptotic proteins, triggering PCD (programed cell death) of animal cells [60]. In Laboratory Shugaev [61] on mitochondria, which were isolated from etiolated cotyledons lupine (*Lupinus angustifolius* L, grade "Dikaf 14"), was demonstrated that the addition into the suspension breathable organelles with high $\Delta\Psi$ 0,5 mM SA induced a dramatic increase in the permeability the inner membrane of mitochondria after a certain lag period. The presence of ATP in the incubation medium did not affect the process of mitochondrial membrane permeabilization. Increasing the concentration of SA, 1.0–3.0 mM was promoted to accelerate the total collapse $\Delta\Psi$, which is consistent with the data, which were obtained in mitochondria of animals. It can be assumed that the inhibition of electron transport in the mitochondrial respiratory chain should increase ROS generation and may be the cause of oxidative stress, induction of PTP, and PCD.

Note that, the SA has an impact on the development of pigment appa- ratus of plants. Therefore, the treatment of seedlings of rye sowing (*Secale cereal* L.) 100 μm fractions of the SA contributed to a twofold increase in the pool of chlorophyll *a*. In this case, the pool of chlorophyll *b* was increased by 2.5 times, the pool of carotenoids—in two times, and the pool of anthocyanin—2.75 times that indicates the ability of SA to stimulate the development of pigment apparatus [62].

12.3.3 BRASSINOSTEROIDS

No less interesting is the use in agricultural production of epibrassinolide (EB)—a synthetic brassinosteroid, analog of natural plant hormone, the regulator of water metabolism of plants [63].

Brassinolide—the first dedicated brassinosteroid

To date from various plant sources, 40 allocated brassinosteroids. Brassinosteroids (BS) are a new class of plant hormones which are key components of many signaling pathways and involved in the regulation of cell proliferation and differentiation [64], providing a change of cell metabolism, increasing the resistance of plants to biotic [65, 66] and abiotic stress [67, 68]. BS are effective immunomodulators that increase resistance of plants to stress and pathogens. BS at concentrations of 10^{-6}–10^{-11} M increase the drought resistance of cucumber plants, increasing resistance to dehydration and overheating. In the plant cells, BS increase the content of free amino acids and amides, which contributes to better retention of water in the cells [69]. Recently, the induction of BS synthesis of proteins with chaperone functions has been shown [70]. Under the influence of BS, there is an increase in resistance of seedlings and adult plants of wheat to salt stress. The growth of the productivity of crops increases by

15–25% [71]. BS at a concentration of 10^{-7}–10^{-9} M contribute to the activation of synthetic processes in tissues of germinating seeds, increasing their content of protein and nucleic acids by 1.2–1.4 times. At the same time, there is a temporary inhibition of proteases and nucleases—DNase and RNase [70]. Improved growth and seed productivity of plants in the processing of EB were shown. Thus, the respiration intensity is increased by 120–130% after 1 h after soaking seeds of haricot or tomatoes in the solution of BS. Against this background, 36% increase in the growth of the embryonic root was observed. An increase in the intensity of breathing is manifested in the early stages of germination of seeds at different phases of ontogenesis leaves. BS work as antistress adaptogens: they did not regulate a separate stage of growth but activated own phytohormones of plants, participating in the synthesis of antistress proteins. When drought spraying of BS enhances the ability of the root to absorb moisture under excess of soil water, it increases the evaporative capacity of the leaves, with a lack of light—speeds up and increases the synthesis of chlorophyll. Under the influence of the BS occurs an increase in the level of growth- stimulating hormones (cytokinins, IAA) and a decrease of ABA in leaves of tomatoes or beans. The ratio of cytokinins + IAA/ABA and cytokinins/ABA under the influence of EB increases. The leaves of plants were treated with EB, amid higher content of zeatin characterized by a higher content of chlorophyll and increased photosynthetic process. On the basis of changes, the hormonal balance observed increase in the intensity energetic processes (photosynthesis, respiration), which increased the rate of growth, vegetative and generative organs, and the productivity of tomato plants and beans [72]. Protective effect of BS under stress conditions is likely associated with decreased ROS generation by mitochondria by activation of electron transport in an alternate path with AOX, as evidenced by a 50% inhibition of the alternative electron transport in the processing plant synthesis inhibitor BS–brasinozol [73]. In addition, the reduction of the pool of ROS in the cell after the processing plant with BS is achieved by increasing the activity of antioxidant enzymes. In this way, many studies have shown a positive effect of BS on the activity of superoxide dismutase. Under the action, exogenous BS noted an increase in the activity of enzymes neutralizing hydrogen peroxide, catalase, peroxidases, and glutathione reductase in plants of many species [70]. On the background of the general trend, increase in antioxidant activity was registered, and some features the action of BS on separate enzymes. So far, the rice plants has shown a decrease in SOD activity by the action of 24-EB in a physiologically normal and saline conditions [74]. The authors suggested that these species' BS induce other defense systems. For the wheat plants revealed varietal differences in response to 24-EBL: for the more salt-tolerant varieties noted a significant increase in SOD activity in response to the processing with BS and the action of BS on the background salt stress [75]. In case of plants, *Raphanus sativus* when processing 24-EB revealed a significant increase in the activity of catalase on the background of decreased activity of peroxidase. Identical to the effect of BS on the activity of catalase and peroxidase in conditions of osmotic stress is shown for sorghum plant. The opposite effect (increased activity of peroxidase and decreased catalase activity) was observed under the action of 24-BS on tomato plants [70]. Along with the change in the activity of antioxidant enzymes system under the influence of BS, certain activities have shown an increase in the content of reduced glutathione, ascorbic acid, and proline in plants of different taxonomic groups [73, 76]. Note, however, that a number of phytohormones, and hence their synthetic analogs, regarding their physiological functions, correspond to the hormones of animals and humans, which is confirmed by data regarding their biosynthesis [77], and consequently their safety must be proved. At present, therefore, of partic-ular urgency is the search for ecologically safe growth factors of plants, which have adaptogenic properties under adverse environmental condi-tions in micromolar and ultra-low concentrations.

12.4 REGULATORS OF PLANT GROWTH AND DEVELOPMENT, WHICH HAVE ANTIOXIDANT PROPERTIES

Among the PGRs with antioxidant properties, practical application received ambiol-derived 5-hydroxybenzimidazole (2-methyl-4-dimethyl-aminomethyl-benzyl-imidazol-5-ol dihydrochloride) [78].

Ambiol

Ambiol is a drug with complex action, with radio protective properties, which reduces the content of radionuclides in products [79]. The use of ambiol has leaded to reduce susceptibility to phytopathogens net blotch (*Drechslera teres*), mycosis (*Phynchosporium spaminicola*), and smut (*Ustilago nuda*) [80]. Processing potatoes with ambiol increased the intensity of photosynthesis and photochemical activity of chloroplasts in leaves of potato by increasing the intensity of transpiration and the water holding capacity of the sheet. Established the intensification of growth processes and increasing productiveness on 20% after handling of potato by ambiol compared with the control group [81]. Protective properties of the drug, apparently, determined by action of ambiol on bioenergetics characteristics of mitochondria. Indeed, the formulation according to the concentration had an influence on the functional state of mitochondria of plant and animal origin. The drug increases the rate of oxidation of NAD-dependent substrates in concentrations of 10^{-5}–10^{-6} and 10^{-9} M, wherein the rates of pair malate and glutamate oxidation in the presence of ADP were increased by 1.4–1.5 times and in the presence of FCCP by 1.6–1.7 times. It increases the efficiency of oxidative phosphorylation: the respiratory control rate (RCR) increased from 2.09 ± 0.06 to 2.83 ± 0.07 (10^{-6} M) up to 3.00 ± 0.10 (10^{-5} M). The drug had no effect on the rates of oxidation of succinate by mitochondria of sugar beet and stimulates the rates of oxidation of succinate by mitochondria of rat liver and increases the efficiency of oxidative phosphorylation [82]. This probably underlies the adaptive nature of the action of ambiol as mitochondria of storage organs are characterized by relatively low rates of oxidation of NAD-dependent substrates [83]. The result of maintaining high activity of NAD-dependent dehydrogenases is the activation of the energy processes in cell that promotes the resistance of plant to varying environmental conditions. Furthermore, having antioxidant properties, the drug reduces the generation of ROS under stress conditions (water deficit) [84].

12.5   DERIVATIVE TRIETHANOLAMINE

From among biologically active substances of particular interest for crop, production can present harmless to animal organisms, some derivatives of triethanolamine (TEA). One of the oldest regulators of growth and development of plants of this class is krezatsin. It is an adaptogen broad-spectrum.

Krezatsin-(*tris* (2-hydroxyethyl) ammonium *o*-tolyloxyacetate) (general formula $C_{15}H_{25}NO_6$)

Krezatsin increases the body's resistance to long influence of adverse factors: low and high temperature and a reducing oxygen, drought. The drug intensifies the biosynthesis of proteins and nucleic acids. Krezatsin enhances the resistance of the organisms to diseases what apparent in the increase of crop yields by 15–40% depending on the species and variety of plants [85]. Drug improves the quality of the products (increases the content of sugars and vitamins, extended shelf life, reduced content of nitrates). In addition, krezatsin prevents the shedding of flowers and ovaries of all species of plants, accelerates flowering and increases commercial specifications decorative flowers [48]. Krezatsin is a low-toxic drug. Toxicity index LD_{50} is 3–10 g per 1 kg body weight, depending on the method of administration and species of animal. He does not exhibit carcinogenic, teratogenic, gonadotoxic, embryotoxic, mutagenic properties, and allergenic actions, does not accumulate in the body. Krezatsin reduces the lesion of grain crops root rot by 15–20%, leaf spotting by 25–30%, brown leaf rust of 10–20% [86]. Based on it, the promising containing silicon PGRs–silatranes were created.

12.6   SILATRANES

The positive impact of silatranes in moderate doses on the growth and development of plants is shown for a wide range of crops. Silatrane are tricyclic chelate silicon esters of TEA with the general formula $RSi(OCH_2CH_2)_3N$. The nature of the physiological effect silatranes, mainly associates with their atranov heterocycle. Considerable dipole moment provides at molecules silatranes the high permeability into the cell membranes while the stiffness increases the resistance of latter to various adverse factors. Carbofunctional substituents at the silicon atom and organic radicals alter the physical and chemical properties of silatranes, such as the distribution ratio in the lipid–water system, while the electron-donor effect of atranov core causes a redistribution of the electron density in the radical at the silicon atom. This affects on penetration of silatranes molecules into the cell membrane, alters their interaction with the receptors and possibly also the speed and path of metabolic conversions, which consequently affects the orientation and the effectiveness of the biological action [87]. Some of them possess wide spectrum of biological activity that have found application in medicine, cosmetics, animal husbandry, poultry farming, practical entomology, and plant breeding. Silatrane exhibits immunostimulatory effect, increases biosynthesis of proteins and nucleic acids and possesses bacteriostatic and insecticidal activity [87]. Among these from the class of BAS, silatranes are found methyl chloromethyl and ethoxysilil-silatranes (respectively, MS (methy-silatranes), CHMS (chloromethyl-silatranes), and ES (ethoxylil-silatranes). These organosilicon compounds affect metabolic processes at low concentrations (10^{-7}–10^{-4} M). The effect of CHMS ($ClCH_2Si(OCH_2CH_2)_3N$) on the germination of seeds, growth of roots, and cytogenetic patterns was investigated on *Crepis capillaries*. The germination of seeds treated with 10^{-6} M solution of CHMS was significantly higher than nontreated or subjected to the action of 10^{-4} M solution. The mitotic index was determined in meristematic cells of root tip in their growth dynamics, which indicates that the roots were treated with 4.4 × 10^{-6} M solution of CHMS. After 16, 30, and 100 h of growth, the number of dividing cells was significantly higher compared to untreated roots. The increase in the mitotic index when treated of seeds with CHMS may be due to several factors:

1. Processing with CHMS facilitates the entrance into the fission of inactive part of the of root cells. Indirectly, this is confirmed by the increase of germination of the dry seeds treated with 10^{-6} M CHMS.

2. CHMS speeds up the passage of the mitotic cycle by shortening duration of the individual stages. In this case, it needs to contribute to the activation of metabolic processes, inducing cell division. The increase in mitotic index occurs much earlier than an increase in the length of the roots (30 h). In the basis of the earlier increase of mitotic index of the experienced group of roots and subsequent stimulation of their growth lies stimulate of PGR of a stretching of cages and transition of their bigger number to a phase of stretching

Further research revealed a wide range of cultivated plants, on which silatranes act as biostimulators. The most responsive to the processing of CHMS—this plants intensive type of growth. The increase from application of CHMS for vegetable crops yield was 20–43% and more and has improved product quality. In the productive organs of plants, the content of sugars and vitamins increases, the accumulation of nitrates and heavy metals reduces. Vegetables retain long marketability, well–storage, and transportation. Thus, CHMS can not only increase yield but also can resist to adverse environmental factors. Adaptogenic effect of CHMS on the stability of membrane structures and functions of chloroplasts of plant cells to high temperatures (40–45°C) was studied on wheat by EPR of spin probes. CHMS increases the resistance of plant cells to high temperature. HMS stimulates the synthesis of galactolipids in seedlings of winter wheat, thereby improving the structure of the photosynthetic apparatus of plant cells. However, the stabilizing effect is not accompanied by strengthening of the protein-lipid contacts and accordingly the change of the temperature of the main phase transitions. The study of the rate of photosynthesis of wheat and chlorella at different temperatures confirmed that CHMS prevents negative structural changes in the membrane systems of cells, especially membranes of chloroplasts.

CHMS's contribution to the thermostability of Hill reaction is the most important stage of the photolysis of water, which compensates the losses of electrons in chlorophyll and supports the restoration of NADP and ADP phosphorylation to ATP. At the same time, cryoprotective effect of CHMS and ES accompany a change in ratio of low and high molecular weight proteins and free amino acids in the phloem tissues of the shoots of grapes [87]. CHMS increases structural and functional stability of the plasma membrane of cells of wheat seedlings to the action of detergents. This reduces the loss of K^+ ions and sensitivity of seedling roots to the action of the respiratory poisons.

In addition, CHMS increases the resistance of plants to water deficits. CHMS increases the resistance of crops to the effects of herbicides that suppress the Hill reaction. Thus, preliminary treatment of barley seeds 10^{-7}–10^{-6} M CHMS increases the resistance of this culture to Simazine [87].

Silatranes with different efficiency affect on heat tolerance of plant. MS and CHMS significantly increased (nearly doubled increases a survival pea seedlings at 45°C in concentrations) 10^{-3}, 10^{-7}, and 10^{-13} M. ES enhances heat tolerance only at a concentration of 10^{-3} M.

A marked increase in the survival of seedlings at a temperature of 45°C is indicative of the fact that the MS and CHMS can activate defense mechanisms of the plant organism. According to the data of immunochemical analysis, these mechanisms are different: CHMS increases the concentration of the HSP, stabilizes the respiratory activity and improves the stability of the membrane, whereas the MS only affects the stabilization of breathing [88].

12.7 GERMATRANES

Tricyclic chelate compounds of silicon and germanium, respectively, silatranes and germatranes, generally own nearly the same biological activity [87], probably due to the similarity of the elements Si and Ge regarding their atomic radius and electronegativity. However, germatranes, having the same or higher biological activity, are less toxic than their silicon counterparts [89]. Unlike silatranes, which are widely used in medicine and agriculture, germatranes have been paid much less attention. Less studied properties of these compounds work as regulators of plant growth and development (PGRs). Although it is shown that treatment of strains of tissue culture plants of tropical *Polyscias filicifolia LX-5* and *Polyscias filicifolia*, most studied germatranolom-1 increases the intracellular content nucleic acids and proteins as well as increases the activity of antioxidant enzymes [90]. In addition, this drug exhibits antistress properties, increasing the resistance of plants (wheat germ) at the temperature stress [91].

Research of protective properties of 1-(germatran-1-yl)-1-acetylamine (gross formula—$N(CH_2CH_2O)_3Ge(OCH_2CH_2)NH_2$), hereinafter referred to as germatrane (GM), has shown that 10^{-5} and 10^{-11} M cheated on bioenergetic characteristics of pea sprouts mitochondria.

Maximum rates of oxidation of NAD-dependent substrates increased by 27.7% (10^{-5} M)–42.7% (10^{-11} M). As the germinating seeds are characterized by rather low rates oxidation of NAD-dependent substrates, the stimulating of activity of NAD-dependent dehydrogenases germatrane apparently contributes to the activation of energy processes in the cell and increases the resistance of seedlings to the action stressors [92].

Indeed, soaking pea seeds in a 10^{-5} M solution of GM prevented the activation of LPO and change of morphology of mitochondria of 6-day-old seedlings in conditions of water deficit and reduce the temperature to 14°C [93] (Fig. 12.4).

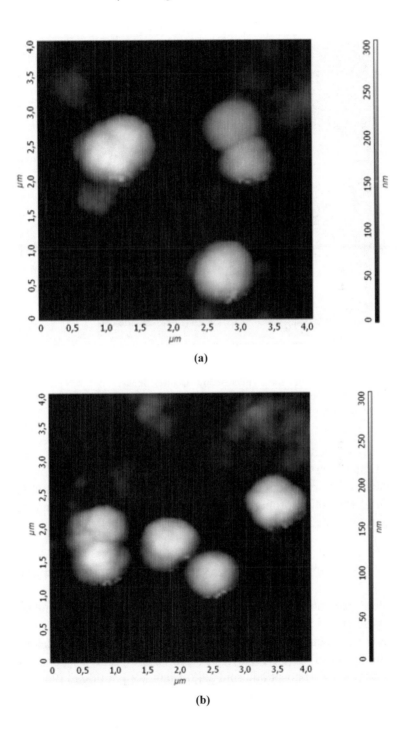

(a)

(b)

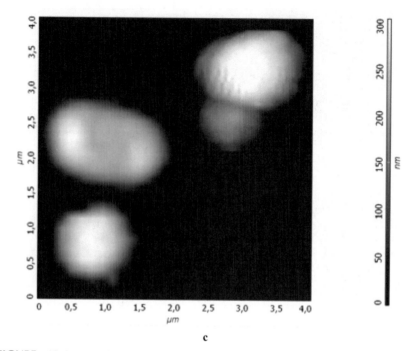

c

FIGURE 12.4 Atomic force microscopy (AFM) images of the pea seedlings mitochondria in the control (a); water deficit + temperature 14°C + GM (b); and water deficit + temperature 14°C (c) [93].

Preventing activation of LPO, with germatrane, apparently contributes to the preservation of the functional state of mitochondria, which is reflected in the preservation of the morphological characteristics of mitochondria. Protective properties of the drug are apparently bound to its ability to activate NAD-dependent dehydrogenases.

12.8   ORGANOPHOSPHATES AS PLANT GROWTH REGULATORS

The new generation of regulators of growth and development plants were synthesized at the Institute of Organic and Physical Chemistry. AE Arbuzov of Kazan Scientific Center RAS. These are drug "Melaphen" (melamine salt of bis(oxymethyl)-phosphinic acid) and "Pyraphen" (salt of bis(oxymethyl)-phosphinic acid 2,4,6-triaminopyrimidine) [94].

Melaphen   Pyraphen

It has been shown that the processing of cereals (rice, sunflower, winter wheat, winter barley, corn, soybeans, etc.) and vegetable crops (vegetable beans, tomato, radish, and red beet) with melaphen in a concentration of 10^{-7}–10^{-8}% enhances the growth and formative processes, significantly increases their yield and quality. So the yield of winter wheat when using the drug has increased by 12.8%, the yield of winter barley by 11.4 and 20.2%. Pretreatment of rice seeds contributed to the yield increased by 20.1%. Processing of sunflower and soybeans increased the yield by 11.9 and 19.3% and oil yield per hectare at 15.5% and 25.9%. The growth of productivity at processing of seeds by melaphen on a vegetable bean was amounted to 11%, the table beets—35.9%, radish—44.4%, and tomatoes—66.2% [95].

The drug prevents the change in the functional state of mitochondria under stress. Thus, a combined effect of insufficient watering and moderate cooling (14°C) (Drought Cooling) resulted in changes in the mitochondria morphology of pea (*P. sativum*) cv. Flora-2, isolated from 5-day etiolated seedlings as compared with the control (22°C), explored by the method of atomic force microscopy (AFM). AFM images of mitochondria of pea seedlings, which were exposed to 2-day impact of insufficient moisture and moderate cooling (*Drought Cooling*), significantly changed and differed from the control samples. In the Drought Cooling group, an increase in the volume of AFM images of some mitochondria was observed, and the number of divisible mitochondria decreased significantly (Fig. 12.5) [96].

A comparison of the published data with the obtained results can be assumed that a combined effect of moderate cooling and insufficient watering of pea seedling mitochondria promotes the increase of generation of ROS and subsequent swelling of mitochondria [97]. Indeed, simultaneous cooling and deficit of moisture have led to LPO activation in the mitochondrial membranes of pea seedlings. In this case, the fluorescence intensity of LPO products increased by 3 and 2.5 times, respectively. Soak the seeds in a 2×10^{-12} M melaphen solution resulted in a decrease of LPO

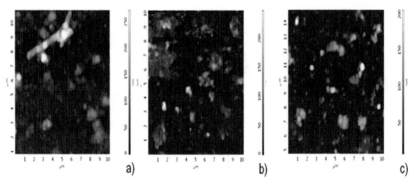

FIGURE 12.5   The two-dimensional AFM image of mitochondria (10×10 mkm^2) isolated from 5-day pea seedlings (a) in the control group, (b) in the Mph drought cooling group (2×10^{-12} M melaphen), and (c) in the Drought Cooling group [96]. (Reprinted from Zhigacheva, I.; Mil', E; Binukov, V.; Generozova, I.; Shugaev, A.; Fatkullina, L. Combined Effect of Insufficient Watering, Moderate Cooling, and Organophosphorous Plant Growth Regulator on the Morphology and Functional Properties of Pea Seedling Mitochondria. Annu. Res. Rev. Biol. 2014, 4(19), 3007–3025. https://creativecommons.org/licenses/by/3.0/)

products in the membranes of the mitochondria: the fluorescence intensity of LPO products decreased almost to the control level [97]. Such treatment prevents the morphology of mitochondria, approaching the size of mitochondria toward control. The effect was accompanied by an increase in the number of divisible mitochondria similar to that in the control [96]. A combined effect of insufficient watering and moderate cooling (14°C) (drought cooling) resulted in decreasing the maximum oxidation rates of NAD-dependent substrates. Rates of oxidation of NAD-dependent substrates in the presence of FCCP (carbonyl cyanide-p-trifluoromethoxy-p-henylhydrazone) decreased by 1.5 times and the RCR decreased by 30%. The soaking of seeds in a 2×10^{-12} M melaphen has prevented by changes in the efficiency of oxidative phosphorylation. In addition, such treatment had prevented in decreasing the oxidation rates of NAD-dependent substrates in the presence of ADP or FCCP. It was found out that the treatment of pea seeds with a pyrimidine analog of melaphen, a 10^{-14} M pyraphen (which has a minor effect on the LPO activity under conditions of insufficient watering and moderate cooling), almost had no effect on the bioenergetic properties of pea seedling mitochondria. The maximum oxidation rates of NAD-dependent substrates, both in the presence of ADP and FCCP, differed little from the maximum rates of oxidation; these substrates by mitochondria of the seedlings were exposed to insufficient watering and moderate cooling. However, the efficiency of oxidative phosphorylation increased: the RCRs increased by 20%. It can be assume that the protective properties of drugs due to their antioxidant properties. Indeed, the rate constant interaction of melaphen with superoxide radical was 1.67×10^4 $(Ms)^{-1}$ and was comparable with nitroblue tetrazolium, and the effective rate constant of the interactions melaphen with peroxyl radicals (k_7) in the oxidation of ethylbenzene $(60°)$ ranged from 1.64×10^6 $(Ms)^{-1}$ [98]. For pyraphen, these values were slightly lower. Melaphen in concentrations affects the fluorescence intensity of LPO; it changes microviscosity of annular lipid membranes of mitochondria isolated from storage parenchyma of sugar beet and rat liver mitochondria [96], which may also have been due to the antioxidant activity of the drug. Changes microviscosity in the area of annular lipids perhaps were reflected in the activity of enzymes of the mitochondrial respiratory chain. Antistress properties of the drugs become apparent on the physiological indices. Moderate cooling and insufficient watering inhibited the growth of roots and shoots of pea seedlings. Soaking of pea seeds with melaphen or pyraphen under conditions of insufficient watering and moderate cooling induces an increase in growth of root at 5 and 1.75 times, respectively. However, pyraphen had almost no effect on the growth of sprouts, whereas melaphen at 3.5-fold stimulated the growth of sprouts. Found stimulation of the growth of roots of seedlings under conditions of water deficiency is of great importance for adaptation to water deficit. The efficiency of melaphen in the protection of plants from water stress is somewhat higher than that of pyraphen, which may be associated with a more efficient protection of membranes with melaphen from LPO [99].

12.8.1 ZIRCONIUM

A promising for agricultural production is the drug Zirconium, which was developed based on the plant *Echinacea purpurea*, which is known in pharmacology as a source of safe and effective biologically active substances for medicines. Hydroxycinnamic acid complex of Echinacea, that performs in this preparation the function of the active principle, determines its effect on the metabolism of the plant cell, involved in the regulation of its enzymatic and hormonal profile. One of the components of the active substance of the drug zirconium is the chlorogenic acid that increases the activity of several antioxidant enzymes such as superoxide dismutase, catalase, and peroxidase. Growth regulating and growth promoting effects of the drug are associated with the activation of phytohormones and inhibition of the activity of auxin oxidase. Processing plant zircon leads to increased synthesis of chlorophyll, to increase the germination energy of seeds and to increase the resistance of plants to stressful environmental conditions. So soaking tomato seed in a solution of zirconium (1 ml/3.5 l of water for 7 h) increases the energy of germination and germination of seeds on average by 2–5%. Seedlings obtained from the treated seed have a large mass. In the control group, weight of one plant is 2.1 g, whereas after seed treatment with zirconium-3.8, processing of cucumber seed with this drug increases germination from 32 to 48%. This improves the quality of seedlings. Plants are more developed with a strong root system. Moreover, the quality of the seedlings was enhanced. The plants were better developed with a strong root age system. The green mass above ground of the plants after the treatment of seeds with zirconium was 40% larger than that of the control plants. Afterwards, the plants grown from the treated seeds had better leafage and a larger leaf area [100].

The treatment with zirconium reduces considerably the degree of affection of plants with various phytopathogens due to inhibiting the viral integrase and realization of the direct antiviral effect [48, 100, 101].

12.9 CONCLUSIONS

Thus, the PGRs of the new generation produce a triple effect on plants: stimulation of the physiological processes, enhancement of the resistance of plants to abiotic and biotic stresses, and improvement of the nonspecific immunity. The result of this effect is the enhancement of the crop productivity and quality of agricultural products.

KEYWORDS

- cytokinin 6-benzylaminopurine
- salicylic acid
- germatranes
- melaphen
- pyraphen
- zirconium

REFERENCES

1. Shakirova, F. M. *Nonspecific Resistance of Plants to Stresses and Regulations Thereof.* Gilem: Ufa, 2001, p. 160 (In Russian).
2. Tailor, N. L.; Day, D. A.; Millar, A. H. Tagrets of Stress-Induced Oxidative Damage in Plant Mitochondria and Their Impact on Cell Carbon/Nitrogen Metabolism. *J. Exp. Botany.* **2003,** *55*(394), 1–10.
3. Sweetlove, L. J.; Heazlwood, J. L; Hearld, V.; Holtzapffel, R.; Day, D. A.; Leaver C. J.; Millar, A. H. The Impact of Oxidative Stress on *Arabidopsis* Mitochondria. *The Plant J.* **2002,** *32,* 891–904.
4. Halliwell, B. *Free Radicals in Biology and Medicine*; Gutteridge, J. M. Claredon Press: Oxford, 1989, p. 215.
5. Chirkova, T. V. Cellular Membranes and Resistance of Plants to Stresses. *Soros Educ. J.* **1977,** *9,* 12–17 (In Russian).
6. Skulachev, V. P. Membrane Bioenergy; Bogachev, A. V; Kasparinskii, F. O. MGU: Moscow, 2012, p. 367 (In Russian).
7. Mokhova E.N.; Khailova, L. S. The Role of Anionic Carriers of the Inner Membrane of Mitochondria in the Dissociative Action of Fatty Acids. *Biokhimiya.* **2005,** *70*(2), 197–202. (In Russian).
8. Vercesi, A.E.; Borecký, J.; Maia, I. G.; Arruda, P. I. M.; Cuccovia, I. M.; Chainovich, H. Plant Uncoupling Mitochondrial Proteins. *Annu. Rev. Plant Biol.* **2006,** *57*, 383–404.
9. Andreev, A. Yu.; Dedukhova, V. I.; Mokhova, E. N. Dissociation of Oxidative Phosphorilation in Liver and Skeletal Muscles Mitochondria by Fatty Acids in Media of Different Ionic Composition. *Biol. Membranes.* **1990,** *7,* 480–486 (In Russian).
10. Petrussa, E.; Casolo, V.; Braidot, E.; Chiandussi, E.; Macri, F.; Vianello, A. Cyclosporine A Induces the Opening of a Potassium Selective Channel in High Plant Mitochondria. *J. Bioenerg. Biomembrane.* **2001,** *33,* 107–117.
11. Grabelnych, O. I. In *Mitochondrial Energy-Dispersive Systems of Plants Under the Action of Low Temperatures. Factors of Resistance of Plants Under the Extreme Natural Conditions and in the Technogenic Medium*, Proceedings of the All-Russia scientific conference, Irkutsk, June 10–13, 2013; Borovik, O. A.; Pobezhimova, T. P.; Voinikov, V. K., Eds.; 2013, 63–65 (In Russian).
12. Grabelnych, O. I.; Pobezhimova, T. P.; Kolesnichenko, A. V.; Sumina, O. N.; Pivovarova, N. Yu.; Voinikov, V. K. Investigation of the Possibility of Free Fatty Acids to Act As an Oxidation Substrate in Winter Wheat Mitochondria. *Bull. Kharkov Nat. Agr. Univ. Ser. Biol.* **2003,** *5*(3), 7–15 (In Russian).
13. Popov, V. N.; Eprintsev, T.; Maltseva, E. V. Activation of Genes Encoding Proteins of the Disjoint Respiration in Tomato Mitochondria Under the Action of Cold and Active Oxygen Species. *Fiziologiya Rastenii.* **2011,** *58*(5), 758–765 (In Russian).
14. Baranova, O. V.; Skarga, Yu. Yu.; Negoda, A. E.; Mironova G.D. Adenine Nucleotide Inhibition of DNF-Induced Transport of Potassium in Mitochondria. *Biokhimiya.* **2000,** *65*(2), 86–91 (In Russian).
15. Pastore, D.; Stoppelli, M. C.; Di. Fonzo N; Passarella, S. J. The Existence of the K+ Channel in Plant Mitochondria. *Biol. Chem.* **1999,** *274*, 26683–26690.

16. Trono, D.; Laus, M. N.; Soccio, M; Alfarano, M.; Pastore, D. Modulation of Potassium Channel Activity in the Balance of ROS and ATP Production by Durum Wheat Mitochondria-An Amazing Defense Tool Against Hyperosmotic Stress. *Front Plant Sci.* **2015**, http://www.ncbi.nlm.nih.gov/pmc/articles/PMC4664611/ doi: 10.3389/fpls.2015.01072.

17. Garlid, K. D.; Paucek, P. Mitochondrial Potassium Transport: The K(+) Cycle. *Biochim. Biophys. Acta.* **2003**, *1606* (1–3). 23–41.

18. Casolo, V.; Braidot, E.; Chiandussi, E.; Vianello, A.; Macri, F. K+ATP Channel Opening Prevents Succinate-dependent H_2O_2 Generation by Plant Mitochondria. *Physiol. Plantarum.* **2003**, *118*, 313–318.

19. Fratianni, A.; Pastore, D.; Pallotta, M. L.; Chiatante, D.; Passarella, S. Increase of Membrane Permeability of Mitochondria Isolated from Water Stress Adapted Potato Cells. *Biosci. Rep.* **2001**, *2*, 181–91.

20. Moller, I. M. Plant Mitochondria and Oxidative Stress: Electron Transport, Nadph Turnover, and Metabolism of Reactive Oxygen Species. *Annu. Rev. Plant Physiol. Plant Mol. Biol.* **2001**, *52*, 561–590.

21. Grabelnych, O. I. Mitochondrial Energy-Dissipative Systems of Plants at the Action of Low Temperatures. Abstract of Doctoral Thesis. Biol. Sci., Irkutsk, 2014, 53 p. (In Russian).

22. Michalecka, A. M.; Agius, S. C.; Moller, I. M.; Rasmusson, A. G. Identification of a Mitochondrial External Nadph Dehydrogenase By over Expression in Transgenicc *Nicotiana sylvestris*. *Plant J.* **2004**, *37*, 415–425.

23. Grabelnych, O. I. Energy Functions of Plant Mitochondria Under Stress Conditions. *J. Stress Physiol. Biochem.* **2005**, *1*(1), 38–54 (In Russian).

24. Atkin, O. K.; Macherel, D. The Erucial Role of Plant Mitochondria in Orchestrating Drought Tolerance. *Ann. Bot.* **2009**, *103*, 581–597.

25. Noguchi, K.; Yoshida, K. Interaction Between Photosynthesis and Respiration in Illuminated Leaves. *Mitochondrion.* **2008**, *8,* 87–99.

26. Fomenko, O. Yu. Free Oxidation. In *Proliferation and Regulation of Expression in Higher Plants*; Popov, V. N., Saarbruken, Germany, Ed; LAP Lambert Academic Publishing, 2010, p. 148 (In Russian).

27. Popov, V. N. In *Reactive oxygen Species in Mitochondria: Formation and Role in the Regulation or Respiratory Processes. Plants Under Conditions of Global and Natural Climatic and Anthropogenic Impacts.* The 8th Congress of the Society of Plant Physiologists, Petrozavodsk, September 21–26, Petrozavodsk, 2015, p. 431 (In Russian).

28. Amirsadeghi, S.; Robson, C. A.; McDonald, A. E.; Vanlerberghe, G. C. Changes in Plant Mitochondrial Electron Transport Alter Cellular Levels of Reactive oxygen Species and Susceptibility to Cell Death Signaling Molecules. *Plant Cell Physiol.* **2006**, *47*, 1509–1519.

29. Grabelnych, O. I.; Sumina, O. N.; Funderat, S. P.; Pobezhimova, T. P.; Voinikov, V. K; Kolesnichenko, A. V. The Distribution of Electron Transport Between the Main Cytochrome and Alternative Pathways in Plant Mitochondria During Short-Term Cold Stress and Cold Hardening. *J. Thermal Biol.* **2004**, 29, 165–175.

30. Generozova, I. P. Respiratory Metabolism of Pea Sprouts Mitochondria After Stresses and Restoration. Plants Under Conditions of Global and Natural Climatic

and Anthropogenic Impacts. The 8th Congress of the Society of Plant Physiologists. Petrozavodsk September 21–26, 2015; Shugaev, A. P. Ed; Petrozavodsk, 2015, p. 126 (In Russian).

31. Maxwell, D. P.; Nickels, R.; McIntosh, L. Evidence of Mitochondrial Involvement in the Transduction of Signals Required for the Induction of Genes Associated with Pathogen Attack and Senescence. *Plant J.* **2002**, *29*, 269–279.

32. Szal, B.; Jolivet, Y.; Hazenfratz-Sauder, M. -P.; Dizengremel, P; Rychter, A. M. Oxygen Concentration Regulates Alternative Oxidase Expression in Barley Roots During Hypoxia and Post-Hypoxia. *Physiol. Plant.* **2003**, *119*, 494–502.

33. Sieger, S. M.; Kristensen, B. K.; Robson, C. A.; Amirsadeghi, S.; Eng, E. W.; Abdel-Mesih, A.; Moller, I. M.; Vanlerberghe, G. C. The Role of Alternative Oxidase in Modulating Carbon use Efficiency and Growth During Macronutrient Stress In Tobacco Cells. *J. Exp. Bot.* **2005**, *416*, 1499–1515.

34. Ezhova, G. P.; Arinbasarova, A. Yu.; Smirnov, V. F.; Guseva, E. The Influence of the Oxidative, Thermal, and Ethanol Stresses on the Survival of *Yarrowia Lipolytica* Yeast. V. *Bull. Lobachevskii Niznii Novgorod Univ. Ser. Biol.* **2010**, *6*, 113–118 (In Russian).

35. Alekhina, N. D.; Balnokin, Yu. V.; Gavrilenko, V. F. Plant Physiology, Moscow, *Academia.* **2005**, 634 pp (In Russian).

36. Clifton, R.; Lister, R.; Parker, K. L.; Sappl, P. G.; Elhafez, D.; Millar, A. H.; Day, D. A. Stress-Induced Co-Expression of Alternative Respiratory Chain Components in *Arabidopsis thaliana. Whelan J. Plant Mol. Biol.* **2005**, *58*. 193–212.

37. Karpova, O. V.; Kuzmin, E. V.; Elthon, T. E.; Newton, K. J. Differential Expression of Alternative Oxidase Genes in Maize Mitochondrial Mutants. *Plant Cell.* **2002**, *14*, 3271–3284.

38. Dojcinovic, D.; Krosting, J.; Harris, A. J.; Wagner, D. J.; Rhoads, D. M. Identification of Regions of the Arabidopsis AtAOX1a Promoter Important for Developmental and Mitochondrial Retrograde Regulation of Expression. *Plant Mol. Biol.* **2005**, *58*, 159–175.

39. Zinchenko V.P.; Goncharov, N. V.; Teplova, V. V.; Petrova, O. I.; Berezhnov, A. V.; Senchenkov, V. E.; Mindukshev, I. V.; Dzhenkins, O.; Radiliv, A. S. Investigation of the Relation Between Intracellular Signal and Metabolic Pathways in Inhibiting Mitochondrial Aconitase With Fuoroacetate. *Cytology.* **2007**, *49*(12), 1023–1031 (In Russian).

40. Gray, G.R.; Maxwell, D. P.; Villarimo, A. R.; McIntosh, L. Mitochondria/Nuclear Signaling of Alternative Oxidase Expression Occurs through Distinct Pathways Involving Organic Acids and Reactive Oxygen Species. *Plant Cell Rep.* **2004**, *23*, 497–503.

41. Fiorani, F.; Umbach, A. L; Siedow, J. N. The Alternative Oxidase of Plant Mitochondria is Involved in the Acclimation of Shoot Growth at Low Temperature: A Study Of Arabidopsis AOX1a Transgenic Plants.*Plant Physiol.* **2005**, *139*, 1795–1805.

42. Dutilleul, C.; Garmier, M.; Noctor, G.; Mathieu, C.; Chetrit, P.; Foyer, C. H.; de Paepe, R. Leaf Mitochondria Modulate Whole Cell Redox Homeostasis, Set Antioxidant Capacity, and Determine Stress Resistance Through Altered Signaling and Diurnal Regulation. *Plant Cell.* **2003**, *15*, 1212–1226.

43. Grabelnych, O. I.; Pivovarova, N. Yu.; Pobezhimova, T. P.; Kolesnichenko, A. V.; Voinikov, V. K. Role of Free Fatty Acids in the Energy Metabolism of Winter Wheat Sprouts Mitochondria. *Fiziologiya Rastenii.* **2009,** *56*(3), 369–381 (In Russian).

44. Laus, M. N.; Soccio, M.; Trono, D.; Liberatore, M. T.; Pastore, D. J. Activation of the Plant Mitochondrial Potassium Chanal by Free Fatty Acids and Acyl-CoA Esters: Positive Defence Mechanism in the Response to Hyperosmotic Stress. *Exp. Bot.* **2011,** *62*, 141–154.

45. Arpagaus, S.; Rawyler, A.; Braendly, R. Occurrence and Characteristics of the Mitochondrial Permeability Transition Pore. *J. Biol.Chem.* **2002,** *277*, 1780–1787.

46. Belosludtsev, K.; Saris, N. E.; Anderson, L. C., Belosludtseva, N.; Agafonov, A.; Sharma, A.; Moshkov, D. A.; Moronova, G. D. On the Mechanism of Palmitic-Acid Induced Apoptosis: The Role of Pore Induced By Palmic Acid and Ca^{2+} in Mitochondria. *J. Bioenerg. Biomembr.* **2006,** *38*, 113–120.

47. Shakirova, F. M.; Gilyazetidinov Sh. Ya.; Kulaeva, O. N. Strategy of Application of Plant Growth Regulators. *Bull. Acad. Sci. Republic Bashkortostan.* **2003,** *8*(1), 14–21 (In Russian).

48. Prusakova, L. D.; Malevannaya, N. N.; Belopukhov, S. L.; Vakulenko, V. V. Plant Growth Regulators Having Antistress and Immunoprotective Properties. *Agrokhimiya.* **2005,** *11*, 76–86 (In Russian).

49. Blokhin, V. G. In *The Effect of Cytokinin (6-BAP) on the Growth and Content of Carotene of Dunaliella Saliva Teod microalgae. Regulators of Plant Growth and Development,* Abstracts of the 4th International Conference; Solovieva, O. E. Ed.; Moscow, 2001, 82–83 (In Russian).

50. Zhizhina, M. N.; Kabulenko, S.N. The Effect of Biologically Active Substances on the Mitotic Activity of the Root Meristem of Corn and Barley Plants Under Conditions of A Saline Stress. Proceedings of the Vernadskii Tavrida National University. *Ser. Biol. Chem.* **2006,** *19*(58), 80–85, no. 4 (In Russian).

51. Kabuzenko, S. N.; Zhizhina, M. N.; Kabuzenko, S. N.; Ponomarenko, S. P.; Rivnya, I.V. The Effect of Synthetic Plant Growth Regulators – Ivin and BAP on Indices of The Water Cycle of Corn and Barley Sprouts Against The Background of A Chloride Salinization. *Physiol. Biochem. Cult. Plants.* **2009,** 41(2), 146–153 (In Russian).

52. Miller, C. O. Cytokinin Inhibition of Respiration in Mitochondria from Six Plant Species. *PNAS.* **1980,** *77* (8), 4731–4735.

53. Belozerova, N. S.; Pozhidaeva, E. S.; Shugaev, A. G.; Kuznetsov, V. V. The Method of Run-On Transcription to Study the Regulation of Expression of the Mitochondrial Genome. *Plant Physiol.* **2011,** *58*, 133–138 (In Russian).

54. Dudkina, N. V.; Sunderhaus, S.; Braun H. P.; Boekema, E. J. Characterization of Dimeric ATP Synthase and Cristae Membrane Ultrastructure from *Saccharomyces* and *Polytomella* Mitochondria. *FEBS Lett.* **2006,** *580*, 3427–3432.

55. Mrtraux, J. P. Systemic Acquired Resistance and Salicylic Acid: Current State of Knowledge. *Eur. J. Plant Pathol.* **2001,** *107,* 13–18.

56. San Vicente, M. R. J.; Plasencia J. Salicylic Acid Beyond Defense: Its Role in Plant Growth And Development. *J. Exp. Bot.* **2011,** *62*(10): 3321–3338.

57. Slaymaker, D. H.; Navarre, D. A.; Clark, D.; del Pozo, O.; Martin, G. B.; Klessig, D. F. The Tobacco Salicylic Acid-Binding Protein 3 (SABP3) is the Chloroplast

Carbonic Anhydrase, Which Exhibits Antioxidant Capacity and Plays A Role in the Hypersensitive Response. *PNAS.* **2002,** *99,* 11640–11645.

58. Moore, A. L.; Albury M. S.; Crichton, P. G.; Affourtit, C. Function of the Alternative Oxidase: Is it Still a Scavenger? *Trends Plant Sci.* **2002,** *7,* 478–481.

59. Norman, C.; Howell, K. A.; Millar, H.; Whelan, J. M.; Day, D. A. Salicylic Acid is an Uncoupler and Inhibitor of Mitochondrial Electron Transport. *Plant Physiol.* **2004,** *134,* 492–501.

60. Battaglia, V.; Salvi, M.; Toninello, A. Oxidative Stress is Responsible for Mitochondrial Permeability Induction by Salicylate in Liver Mitochondria. *J. Biol. Chem.* **2005,** *280,* 33864–33872.

61. Butsanets, P. A. In *Salicylic Acid and Phenyl Oxide Modify the Permeability for Protons of the Inner Membrane of Plant Mitochondria,* Proceedings of the Conference "Basic and Applied Problems of the Modern Experimental Biology of Plants" Dedicated to the 125th Anniversary of the Timiryazev Institute of Physiology of Plants; Shugaeva, N. A.; Shugaev, A. G. Eds.; Russian Academy of Sciences: Moscow, 2015, 131–134 (In Russian).

62. Feduraev P.V. In The Role of Salicylic Acid in The Formation of the Pigment Apparatus of *S. cereale* L, Proceedings of the Conference "Basic and Applied Problems of the Modern Experimental Biology of Plants" Dedicated to the 125th Anniversary of the Timiryazev Institute of Physiology of Plants; Russian Academy of Sciences: Moscow, 2015, 672–675 (In Russian).

63. Prusakova, L. D.; Chizhova, S. I. Application of Brassinosteroids Under Extreme Condition. *Agrokhimiya.* **2005,** *7,* 87–94 (In Russian).

64. Goda, H.; Sawa, S.; Asami, T.; Fujioka, S.; Shimada, Y.; Yoshida, S. Comprehensive Comparison of Auxin-Regulated and Brassinosteroid-Regulated Genes in Arabidopsis. *Plant Physiol.* **2004,** *134*(4), 1555–73.

65. Wang, Z. Y. Brassinosteroids Modulate Plant Immunity at Multiple Levels. *PNAS.* **2012,** *109*(1), 7–8.

66. Ali, S. S.; Kumar, G. B.; Khan, M.; Doohan, F. M. Brassinosteroid Enhances Resistance to Fusarium Diseases of Barley. *Phytopathology.* **2013,** *103*(12), 1260–1267.

67. Kagale, S.; Divi, U. K.; Krochko, J. E.; Keller, W. A.; Krishna, P. Brassinosteroid Confers Tolerance, in *Arabidopsis thaliana* and *Brassica napusto* a Range of Abiotic Stresses. *Planta.* **2007,** *225*(2), 353–64.

68. Ahammed, G. J.; Choudhary, S. P.; Chen, S.; Xia, X.; Shi, K.; Zhou, Y.; Yu, J. Role of Brassinosteroids in Alleviation of Phenanthrene-Cadmium Co-Contamination-Induced Photosynthetic Inhibition and Oxidative Stress in Tomato. *J. Exp Bot.* **2013,** *64*(1), 199–213.

69. Pustovoitova, T. N. In *The Effect of Epibrassinolide on Adaptive Processes in Cucumus sativus L Plants During a Soil Drought. Collected Articles "Plant Growth and Development Regulators in Biotechnologies",* Proceedings of the 6th International Conference; Zhdanova, N. E.; Zholkevich, V. N., Moscow, 2001, 61 (In Russian).

70. Kolupaev, Yu. E.; Vainer, A. A.; Yastreb, T. O.; Oboznyi, A. I.; Khripach, V. A. Reactive Oxygen Species and Calcium K Ions in the Reaction of a Stress-Protecting Action of Brassinosteroids on Plant Cells. *Appl. Biochem. Microbiol.* **2014,** *50*(6), 593–598 (In Russian).

71. Nemchenko, V. V. *Application of Plant Growth Regulators for Increasing the Resistance of Plants to Unfavorable Conditions for Growth. Collected articles "Plant Growth and Development Regulators in Biotechnologies"*, Proceedings of the 6th International Conference, Moscow, 2001, 263 (In Russian).

72. Likhacheva, T. S. The Effect of Epibrassinolide on the Hormonal Balance, Energy-Release Processes, Growth and Productivity of Plants. Abstract of Ph.D. Thesis, Moscow, 2004 (In Russian).

73. Derevyanchuk, M. V.; Grabelnyh, O. I.; Litvinovskaya, R. P., Voinikov, V. K., Sauchuk, A. L.; Khripach, V. A.; Kravets V. S. Influence of Brassinosteroids on Plant Cell Alternative Respiration Pathway and Antioxidant Systems Activity Under Abiotic Stress Conditions. *Biopolym. Cell.* **2014,** *30*(6), 436–442.

74. Ozdemir, F.; Bor, M.; Demiral, T.; Turkan, I. Effects of 24-Epibrassinolide on Seed Germination, Seedling Growth, Lipid Peroxidation, Proline Content and Antioxidative System of Rice (*Oryza sativa* L.) under Salinity Stress. *Plant Growth Regul.* **2004,** *42*, 203–211.

75. Talaat, N. B.; Shawky, B. T. 24-Epibrassinolid Alleviates Salt-Induced Inhibition of Productivity by Increasing Nutrients and Compatible Solutes Accumulation and Enhancing Antioxidant System in Wheat (*Triticum aestivum* L.). *Acta Physiol. Plant.* **2013,** *35*, 729–740.

76. Li, Y. H.; Liu, Y. J.; Xu, X. L.; Jin, M.; An, L. Z.; Zhang, H. Effect of 24-Epibrassinolide on Drought Stress-Induced Changes in *Chorispora bungeana*. *Biol. Plant.* **2012,** *56.* 192–196.

77. Rosati, F.; Danza, G.; Guarna, A.; Cini, N.; Racchi, M. L.; Serio, M. New Evidence of Similarity Between Human and Plant Steroid Metabolism: 5alpha-Reductase Activity in *Solanum malacoxylon*. *Endocrinology.* **2003,** *144*(1), 220–229.

78. Kuznetsov, Yu. V. *Ambiol as Base of New Effective Drugs*. New York, Nova Science Publishers, 2009, p. 125.

79. Kirillova, I. G.; Evsyunina, A. S.; Puzina, T. I.; Korableva, N. P. The Effect of Ambiol and 2-Chloroethylphosphonic Acid on The Content of Hormones in Leaves and Tubers of Potatoes. *Appl. Biochem. Microbiol.* **2003,** *39*(2), 237–241 (In Russian).

80. Filipas, A. S.; Ulyanenko, L. N.; Loi, N. N.; Pimenov, E. P.; Arysheva, S. P.; Dyachenko, I. V.; Stepanchikova, N. S. Resistance of Barley Plants to Phytopathogens Under Technogenic Contamination of Soil. *Agric. Biotechnol.* **2003,** *5*, 74–78 (In Russian).

81. Kirsanova, E. V.; Kirillova I.G. The Effect of the Preparation of Ambiol on the Production Process of Pea and Potatoes. *Proc. Orel Univ.* **2007,** *4,* 7–9 (In Russian).

82. Zhigacheva, I. V. The State of the Electron-Transport Chain of Mitochondria and Physiological Parameters of Animal and Plant Organisms Under the Action of Stresses and Biologically Active Compounds. Abstract of the Doctoral Thesis, Moscow, 2012, 53 p. (In Russian).

83. Shugaev, A. G.; Vyskrebentseva E.I. Ontogenetic Changes in the Functional Activity of Mitochondria of Sugar Beet Roots. *Physiol. Plants.* **1985,** *32,* 259–267 (In Russian).

84. Zhigacheva, I. V. In *Plant Growth and Development Regulators Prevent Dysfunction of Mitochondria of Pea Sprouts Under Conditions of Water Deficiency*, Proceedings of the Conference "Basic and Applied Problems of the Modern Experimental Biology

of Plants" Dedicated to the 125th Anniversary of the Timiryazev Institute of Physiology of Plants; Burlakova, E. B.; Generozova, I. P.; Russian Academy of Sciences: Moscow, 2015, 242–246 (In Russian).

85. Medvedev, G. A.; Kamyshanov, I. G. The Effect of Treatment of Seeds with Bischofite, Mival, And Cresacyne on the Productivity of Spring Varieties of Barley on Chestnut Soils in the Volgograd Region. Adaptive Principles of Stabilization of Ecosystems and the Social Sphere. *Modern Rec.* **2006,** *5,* part 2, 266–250, M. (In Russian).

86. Nemchenko, V. V. Use of Inductors of Stability to Reduce the Affection of Grains with Diseases. *Plant Growth Dev. Regul.* Abstracts of the fourth conference, 1997, 215, M. (In Russian).

87. Voronkov, M. G; Baryshok, V. P. Silatranes in Medicine and Agriculture. Novosibirsk, Nauka, Siberian Branch RAS, 2005, 258 pp. (In Russian).

88. Makarova, L. E.; Sokolova, M. G.; Borovskii, G. B.; Voronkov, M. G.; Kuznetsova, G. A.; Abzaeva, K. A. Temperature Dependence of the Effect of Triethanolamine and Silatranes on the Growth of Pea Sprouts. *Agrokhimiya.* **2009,** *1,* 27–32 (In Russian).

89. Voronkov, M. G; Baryshok, V. P. Development of the Method of Synthesis of Biologically Active Organogermanic Compounds Comprising Chlorine Derivatives of Ethylene. *Vestnik RAN.* **2010,** *80*(11), 985–992 (In Russian).

90. Spasenkov, A. I. Protein Synthesis Ability and Stability of Two Strains of the *Polyscias filicifolia* Tissue Culture Under Stress. Abstract of Ph.D. Thesis, St.Petersburg, 2006, 24 p. (In Russian).

91. Shigarova, A. M. In *Factors of Stability of Plants under Extreme Natural Conditions and Technogenic Environments*, Proceedings of the All-Russia Scientific Conference 10–13 June; Borovskii, G. B.; Zang, T; Le, H; Bartyshok, V. P., Irkutsk, 2013, 294–297 (In Russian).

92. Generozova, I. P.; Shugaev, A. G. Respiratory Metabolism of Mitochondria of Different Age Pea Sprouts Under Conditions of Water Deficiency and Re-Irrigation. *Physiol. Plants.* **2013,** *59,* 262–273 (In Russian).

93. Zhigacheva, I. V.; Binyukov, V. I.; Mil, E. M.; Generozova, I. P.; Rasulov, M. M. Morphological and Bioenergetical Characteristics of Mitochondria Under Stress and the Action of the Organogermanium Compound. *J. Nat. Sci. Sust. Technol.* **2015,** *9*(2), 439–451.

94. Fattakhov, S. G.; Loseva, N. L.; Konovalov, A. I.; Reznik, V. S.; Alyabiev, A. Yu., Gordon, L. Kh.; Tribunskikh, V. I. The Effect of Melaphen on the Growth and Energy Processes in A Plant Cell. *Doklady RAN.* **2004,** *394*(1), 127–129 (In Russian).

95. Barchukova, A. Ya.; Chernyshova, N. V.; Tosunov, Ya; Kazan, K. The Use of the Preparation – Melaphen In Plant Growing. In Melaphen: Mechanism of Action and Fields of Application, Kazan, Press-Service XXI Century, 2014, pp. 177–207 (In Russian).

96. Zhigacheva, I.; Mil', E; Binukov, V.; Generozova, I.; Shugaev, A.; Fatkullina, L. Combined Effect of Insufficient Watering, Moderate Cooling, and Organophosphorous Plant Growth Regulator on the Morphology and Functional Properties of Pea Seedling Mitochondria. *Annu. Res. Rev. Biol.* **2014,** *4*(19), 3007–3025.

97. Zhang, L.; Yinshu, Li; Xing, Da; Gao, Caiji. Characterization of Mitochondrial Dynamics and Subcellular Localization of ROS Reveal that HsfA2 Alleviates Oxidative Damage Caused By Heat Stress in *Arabidopsis. J. Exp. Bot.* **2009,** *60,* 2073–209.

98. Zhigacheva, I. V.; Fatkullina, L. D.; Rusina, I. F.; Shugaev, A. G.; Generozova, I. P.; Fattakhov, S. G.; Konovalov, A. I.. Anti-stress properties of the drug Melaphen. *Doklady RAN.* **2007,** *414* (2), 263–265 (In Russian).

99. Zhigacheva, I.; Generozova, I.; Shugaev, A.; Misharina, T.; Terenina, M.; Krikunova, N.; Generozova, I.; Shugaev, A.; Misharina, T.; Terenina, M.; Krikunova, N. Organophosphorus Plant Growth Regulators Provides High Activity Complex I Mitochondrial Respiratory Chain *Pisum sativum* L Seedlings in Conditions Insufficient Moisture. *Annu. Res. Rev. Biol.* **2015,** *5*(1), 85–96.

100. Dorozhkina, L. A.; Bairambekov, Sh. B.; Korneva, O. G. Plant Growth Regulators— Zirconium, Epin-Extra, and Siliplant for Increasing the Productivity of Vegetables and Melon and Gourds. *Agrokhimiya.* **2011,** *5*, 56–68 (In Russian).

101. Korneva, O. G. The Effect of Plant Growth Regulators and Biologically Active Substances on the Productivity of Potatoes Under Conditions of the Lower Volga region Abstract of Ph.D. Thesis, Rostov-na-Don, 24, 2009 (In Russian).

CHAPTER 13

AMARANTH—BIOINDICATOR OF TOXIC SOILS

SARRA A. BEKUZAROVA[1*], JOHNNY G. KACHMAZOV[2], and
EKATERINA S. AYSKHANOVA[3]

[1]*Gorsky State Agrarian University, 37, Kirov St., Vladikavkaz,
Republic of North Ossetia–Alania 362040, Russia*

[2]*A.A. Tibilov South Ossetian State University, 8, Moscow St.,
Tskhinval 100001, Republic of South Ossetia*

[3]*Chechen State University, 17, Dudayev Blvd, Grozny, Chechen
Republic 364060, Russia*

Corresponding author. E-mail: bekos37@mail.ru

CONTENTS

ABSTRACT

To reduce soil toxicity, culture of the amaranth has a high sorption capacity, especially in the joint sealing of the soil with zeolite clay alanit. The experimental results indicate that the incorporation into the soil of green mass as green manure crops in a mixture with clay alanit reduces the oil pollution from 3.2 thousand to 0.2 mg/kg of dry soil.

13.1   INTRODUCTION

A limited set of crops in agricultural production leads to the outbreak of the different plant diseases, which causes enormous damage to both the material costs and in environmental pollution. Therefore, the introduction of crop rotation of new and innovative plant contributes to the sustainability and stability in plant growing.

Irrational use of chemicals and fertilizers leads to soil pollution with heavy metals and radionuclides, which accumulate in plant foods. The application and use of plants that bind harmful to human chemical elements is a global task of environmental protection.

In recent years, the culture is considered promising amaranth, it is used for food or feed purposes. The uniqueness of this culture is that it in all parts of the plant contains a lot of biologically active substances: essential and nonessential amino acids, microelements, minerals, vitamins, proteins, fatty acids, choline, bile acids, spirin, steroids, and selenium [1–5].

Amaranth has a high tolerance to little fertile soil [6], which barely grows cereals. It grows well in soils with a wide range of content of batteries—from sandy to clay, but prefers well aerated, loose, unsilting of soil with well-developed profile, good drainage, and light texture possibilities [2, 4].

Amaranth has excellent environmental resistance. It can withstand soil waterlogging, drought, salinity, acidity, and alkalinity, a high content of heavy metals. This ability of amaranth to adapt opposite effect demonstrates its adaptive features. Sowing, mowing, and subsequent plowing amaranth allow its use as an excellent green manure—green organic fertilizer [1]. Amaranth has a high stern virtue [2]. Amaranth is a promising crop for the development of low-fertile soils, as he accumulates more biomass even when there is insufficient provision of plants with nitrogen [4].

The undoubted benefits of plants amaranth include good regulation of water metabolism at deficiency of water in terms of atmospheric and soil drought. It is a promising crop for arid and semi-arid zones. The need of water is 42–47% [1, 5]. The demand for water supply of other widespread crops is much higher: winter wheat—51–52% and corn—79%.

Highly ecological significance of an amaranth is not only as a source of dietary and environment-friendly products, but also due to the possibility of cleaning and upgrading with the help of soil [7–10].

Scientific and practical results of many researchers indicate that the application of forage crops as organic fertilizers equivalent siderates [8]. Especially, great role of forage grasses is that it absorbs toxic elements and cleans the soil from heavy metals and radionuclides. In this respect, amaranth has a leading position among all cultivated grasses. Amaranth has acquired this ability through the accumulation of leaves containing high calcium, which varies from 8% to 26% depending on soil and climatic conditions. Most of the silicon content in the leaves of amaranth plants is associated with organic components of plant tissue (proteins, lipids, and fiber), which shows great potential of culture for one season that greatly purify the soil of petroleum hydrocarbons, radionuclides, and heavy metals [7–10]. An important feature of amaranth is the accumulation of silicon compounds (organogenic—soluble and polymer—total), which is capable of absorbing toxic elements (including hydrocarbons oil contaminated lands) and accelerating the metabolism in the plant amaranth. Consequently, silicon is having a high sorption capacity in those compounds that are found in amaranth, and large plants biomass provides up to 60 t/ha.

Especially, high sorption capacity has amaranth when used in mixtures with zeolite clay–alanit [2, 10].

Alanit (zeolite clay deposits North Osetin) comprises (%): silica—51–53, aluminum—16–17, iron—5.6, calcium—30–33, potassium, phosphorus, manganese, sulfur, magnesium (in the range 0.1–0.9), and small amounts of zinc, copper, and other microelements. Due to the high content calcium of the reaction medium alanita alkaline (pH 8.6).

Alanit, like all zeolites, able to retain moisture (water loss does not exceed 3%) and retain heat in the root zone of plants, which is very important in the process of survival for small seeds of amaranth [1, 8].

Having sorption properties, silicon is able to neutralize heavy metals in the soil solution [8].

13.2 MATERIALS AND METHODOLOGY

The study was carried out at the experimental field of Department of Plant Breeding of Gorsky State Agrarian University and on in the Chechen Research Institute of Agriculture. Soil composition: Soil-leached humus, medium loam, pH–5.9.

The objects of the study were the crops of alfalfa, potatoes, corn, soy, and amaranth. In the system of crop rotation under potato and corn used fertilizer for soybean before herbicide treatment. Alfalfa, clover, and amaranth cultivated without the use of chemicals. All cultures sowed after winter wheat. The chemical composition of plants and soil was determined in the laboratory of Gorsky State Agrarian University.

To accelerate purification and remediation of contaminated oil and oil products of lands in the first year, annual culture amaranth in a mixture of alanit was sown at 0.8–1 t/ha. In the early phase of maturation, seeds of the plants were mowed and plowed in the soil.

The following year, ripened by the time of plowing amaranth seeds come up and give the plants. In this year, these carried out sowing of legume–cereal mixtures of perennial grasses, which comprises 50–60% of leguminous plants, with their subsequent plowing into the flowering stage. In this paper, forage legumes were sown in an amount of 4 kg of each component.

The study was carried out jointly by the Centre of Geophysical Research of the Russian Academy of Sciences and Government of North Ossetia–Alania (Vladikavkaz), as well as a comprehensive research by University of the Russian Academy of Sciences (Grozny) on sites contaminated by oil and oil products. Chemical analyses of soils on the content of toxic substances were determined in Gorsky State Agrarian University.

13.3 RESULTS AND DISCUSSION

The studies revealed that the use of amaranth on contaminated soil and plowing the green mass in early maturation of seeds helps clean the soil from toxic substances. At this stage of development, it accumulated a good green mass. Its incorporation into the soil together with partially matured seeds (which will to germinate next year) provides a sufficient amount of organic matter. It will promote rapid decomposition of cultivated plants and will help reduce toxicity.

If reseeding of perennial grasses carry out in the following year, the process of reducing the toxicity of the soil will continue. Bean herbs are also as sorbents of toxic substances. That will contribute to the greening of agricultural production. It was previously shown [6, 8, 9] that soil fertility depends not only on the humus content and reserves the necessary nutrients, but also on agrochemical, physico-chemical, and other properties of the soil. A particularly negative impact on the productivity of most forage crops, including legumes and amaranth, has increased the acidity of the soil. Toxic soils, generally, have higher acidity. Therefore, the introduction of alanit, having an alkaline reaction, together with amaranth seeds, provides more favorable conditions for the development of culture as it is very responsive to soil acidity [11]. The data presented in Table 13.1 show the advantage of the variants of the experiment, where the planting of amaranth in conjunction with the clay alanit was carried out. The following year perennial grasses were sown with the advantage of legume components.

Justification of timing of mowing (early maturity) is explained by biological characteristics of amaranth, which has a high rate of reproduction. At a seeding rate of 1 kg/ha, amaranth is capable of producing more than 3 t of seeds. Therefore, along with the green crop into the soil, 1/8 of the seeds vegetate with perennial grasses in the following year. Due to the fact that amaranth seeds are very small (weight of 1000 pieces—0.5–0.6 g), seeding them with chopped alanit improves flowability, even distribution of the mixture on the treated area provides a nutritious environment in the seedbed.

Of legumes, as evidenced by our long-term data [9, 11–13], the highest sorption ability toward toxic substances have clover, alfalfa, sainfoin, coronilla, and clover. In this work, the total number grass–legume mixture was 35 kg/ha.

Results of chemical analyses showed that after the amaranth preserved, the minimum amount of zinc is much greater than maximum allowable concentration (MAC): 2.37 mg/kg dry soil, at MAC of 23 mg/kg. In clover and alfalfa, the amount of zinc was slightly higher (2.5–2.9). With the use of chemicals level of zinc, corn and potatoes reached 3–4 mg/kg, that is, within acceptable limits. The manganese content under the amaranth ranged from 12 to 18 mg/kg (MAC 20 mg/kg). The maximum number of manganese amount was observed in sowing soybean and corn (7–16 mg/kg exceeding of MAC). The most significant indicators were in the copper

TABLE 13.1 Cleaning The Soil From Chemicals In Experimental Plots Contaminated With Oil and Heavy Metals Pb and Co.

Methods of treatment	The concentration of oil in the soil, %	The content of substances, mg/kg soil		
		Petroleum products	Plumbum, Pb	Cobalt, Co
The first year after soil contamination				
The area contaminated with oil—control	3.2	8600	12.08	5.86
Sowing amaranth	1.4	3200	5.24	2.82
Sowing amaranth with alanit 0.3–0.5 t/ha	0.6	2600	2.62	1.92
Sowing amaranth with alanit 0.8–1.0 t/ha with plowing in the early phase of maturation of seeds	0.3	1500	1.82	1.64
The second year after the soil treatment				
Amaranth plants that were sown in the first year (with alanit 0.8–1.0 t/ha and with plowing in the early phase of maturation of seeds) + sowing of perennial grasses with a predominance of legumes to 50–60% in the following year	0.2	600	0.92	1.26
Sowing of perennial with a predominance of legumes	0.8	2400	2.16	1.47

content in the soil. The use of fertilizers increases the content of copper up to 4 mg/kg exceeding the MAC of 1 mg/kg. Under the amaranth, the amount of copper did not exceed 2 mg/kg, with 1.2–1.9 mg/kg observed in crops of legumes.

Consequently, amaranth as a sorbent can be used for crop rotation, accumulate organic matter, and maintains a clean environment for subsequent culture.

The results of the data obtained (see Table 13.1) suggest that oil content decreases from 3.2 thousand to 0.2 mg/kg dry soil, and heavy metals such as lead from 12.08 to 0.92 mg/kg, and reduced maintenance cobalt from 5.86 to 1.26 mg/kg. The concentration of oil in the soil falls from 8.6 to 0.6%.

Working jointly with Kabardino-Balkarian State Agrarian University, they conducted an experiment to reduce the toxicity of the soil in the cultivation of corn [9–11]. The essence of the new technological solutions was the fact that after treatment of corn, a mixture of herbicides (Milagro— 1.5 g/ha and Harmony—1.5 g/ha) was carried out sowing of amaranth (*Amarantus caudatus*) in the aisle of corn (in the phase of development of 3–5 leaves). In the phase of branching, amaranth was engaged in treatment of crops with a mixture of biologics Baikal EM-1 and Baikal EM-5; then, the green mass was patched into the soil as a green manure with simultaneous treatments of aisle.

The research results are summarized in Table 13.2, from which it can be derived that while incorporating the amaranth as green manure crops at sowing in the aisle of corn in the phase of branching, and at treatment of crops with a mixture of biologics Baikal EM-1 and EM-5 provide increase of nitrogen content in soil for up to 224 kg/ha and the reduction of toxic substances to the maximum permissible concentrations.

The data in Table 13.2 showed that amaranth and biologics significantly reduce the content of heavy metals on the background of increasing nitrogen content.

On the basis of the conducted research, it can be concluded that the annual sowing of amaranth, perennial legume grasses (clover, alfalfa, sainfoin, coronilla, and sweet clover), having high absorbing capacity can provide full rehabilitation of oil-contaminated lands for two years, or 1.5–2 times faster than at known methods and reduce the content of heavy metals in soil contaminated by them.

TABLE 13.2   The Effect of Green Fertilizer of Amaranth and Biologics Containing Microorganisms on Soil Toxicity in the Cultivation of Corn.

Method of sowing	The content of nitrogen in soil, kg/ha	The content of heavy metals, mg/kg		
		Pb	Cd	Zn
The control (without herbicides and cover crops)	172	4.2	2.6	15.8
Treatment of crops with herbicides + seeding of amaranth in the aisle	180	6.9	3.1	22.6
Treatment of corn + biologics	198	3.8	0.76	13.2
Herbicide in the aisle + inter-row treatments with incorporation of green manure crops of amaranth and treatment of crops with biologics	224	2.0	0.32	8.2
Maximum allowable concentration	120	5.0	1.0	20.0

The great role of amaranth is to reduce erosion on sloping lands along with perennial grasses, the value of which is determined by many factors, first of all, entering into the soil large amounts of harvest residues, enrichment of soils with biological nitrogen due to the impact of legumes, protection, and environmental protection [6]. Particularly, acute problem in the North Caucasus region is the loss of humus at the cultivation of row crops, which reaches 60–67% of the total losses.

In biological agriculture, an important direction is the use of forage crops as green manure, which is equivalent to organic fertilizers, although the cost of green manure is much lower.

Especially, the great role of legumes plays an important role in the ability to absorb toxic elements and clean the soil from heavy metals and radionuclides. In recent years, these properties are open to many cultures and, in particular amaranth, contains significant amount of silicon (50 kg per 1 t), phosphorus (164 kg/t), potassium (156 kg/t), calcium (58 kg/t), magnesium (77 kg/t), as well as many trace elements [2, 6, 10].

In addition, the culture of the amaranth has high allelopathic properties. Assessing the amaranth in joint crops with legumes, we came to the conclusion that in the vicinity of coronilla motley greatly apparent competitive ability and therefore amaranth–coronilla mixture is not very effective.

Given the high sorption capacity of each culture and their allelo-pathic properties, we developed a method of placement of crops on soils contaminated with heavy metals. Method lay in the fact that coronilla and amaranth were sown in isolated strips of widths drills (3T-3.6 A). Separate sown of strips each planting of crops on green fertilizer, due to their low competitive ability. So amaranth contained flavonoids and a number of acids produced by these plants that inhibit the nitrogen-fixing ability of legumes—coronilla.

Given the biological characteristics of each species of the mixture, the culture of coronilla was sown early in the spring as more resistant to frosts. After 2–3 weeks in available, bands were carried out by sowing amaranth. Seeding rate of coronilla—15–20 kg/ha and amaranth—0.5–1.0 kg/ha.

Aboveground mass of coronilla in the year of sowing is poorly developed. In the following year, the root system is able to accumulate organic matter and biological nitrogen up to 200 kg/ha.

Developing in the individual sowing, the seeds of annual crops of amaranth accumulate significant amounts of macro- and trace elements (vanadium, manganese, molybdenum, cobalt, copper, and other elements) [1].

Mowing of aboveground mass of amaranth was carried out in the year of sowing in the phase of milky-wax ripeness so that a portion of mature amaranth seeds remained in the soil and germinates in the next year, together with coronilla nearby in the form of strips. Getting into the ground, amaranth seeds together with the top mass function sorbing substances. In contact with infected soil, chemical reactions occur that neutralize heavy metals and radionuclides.

By the time of ploughing, green manure (grown coronilla and sprouted from seeds of amaranth) accumulates a sufficient amount of organic substances that can reduce the toxicity of the soil. As evidenced from the data of Table 13.3, jointly planting amaranth and coronilla in alternating strips reduces the concentration of heavy metals in the soil.

These data show that the lead content in the mixed sowing of amaranth and coronilla is 30.6 mg/kg. On a site where sown these crops alternating strips, the lead content decreased to 26.4 mg/kg. The advantage of across band sowing can be seen for other elements (nickel, copper, and zinc).

TABLE 13.3 The Content of Nitrogen and Heavy Metals in the Soil Depending on Method of Sowing Coronilla and Amaranth.

Method of sowing	The content of nitrogen, kg/ha	Content, mg/kg			
		Ni	Pb	Cu	Zn
Joint sowing amaranth and coronilla	123	26.4	30.6	4.2	28.0
Sowing amaranth	148	19.8	28.4	3.0	24.0
Sowing coronilla	162	15.4	32.0	3.8	32.0
Sowing amaranth and coronilla, sown separate strips	206	13.2	26.4	2.2	23.2
Maximum allowable concentration		20.0	32.0	6.8	35.0

Amaranth and coronilla, sown separate strips, give the highest content of nitrogen in the soil, almost twice the joint inoculation and a significant decrease of heavy metals content. At joint cultivation, green manure crops are necessary to conduct a preliminary assessment of their compatibility. In the case of incompatibility of cultures there should be sowed separate strips alternately for each culture, as is in our experience with coronilla.

13.4 CONCLUSIONS

The results of the research revealed that by sowing grasses with high sorption capacity, amaranth and perennial forage legumes provide the rehabilitation of oil-contaminated lands, reducing the content of heavy metals, which allows to obtain ecologically pure products.

By incorporation of amaranth in the aisles of sowing corn and simultaneous treatment of the biologics reduces the toxicity of the soil, increases crop yields.

Together with coronilla, amaranth should be sown strips, without mixing the seeds of both crops in the crop, indicating the need to assess the compatibility of cultures when sown mixtures.

KEYWORDS

- alanit
- crude oil
- heavy metals
- coronilla
- clover
- alfalfa

REFERENCES

1. Bekuzarova, S. A.; Kuznetsov, I. Yu.; Gasiev, V. I. *Amaranth is a Universal Culture.* Publishing House Colibri: Vladikavkaz, 2014; p 145 (In Russian).
2. Burykina, S. I.; Moskaliuk, P. J.; Archipenko, Z. P.; Altukhova, E. I. Yield and Fodder Value of Amaranth in the South of Ukraine. *Forage Prod.* **2000,** *9,* 22–25 (In Russian).
3. Zheleznov, A. V. Amaranth–Bread, the Sight and the Cure. *Chem. Life.* **2005,** *6,* 56–61 (In Russian).
4. Zelenkov, V. N.; Gulshina, V. A.; Tereshkina, L. B. Amaranth. Agrobiological Portrait. The Publication of the Russian Academy of Natural Sciences: Moscow, 2008; p 101 (In Russian).
5. Kononkov, P. F.; Gins, V. K.; Gins, M. S. Amaranth is a Promising Culture of XXI Century. Publishing house Evgeny Fedorov: Moscow, 1998; p 310 (In Russian).
6. Danilov, K. N. Sowing Amaranth. *Forage Prod.* **1997,** *10,* 20–21 (In Russian).
7. Bekuzarova, S. A.; Alexandrov, E. N.; Weisfeld, L. I.; et al. Patent No. 2555595 "Method of reproduction of oil contaminated lands". Published on 10.07.2015. Bulletin No. 19; 2015 (In Russian).
8. Bekuzarova, S. A.; Weisfeld, L. I.; Alexandrov, E. N. Bioremediation of Oil Contaminated Soils. In *Heavy Metals and Other Pollutants in the Environment. Biological Aspects*; Zaikov, G. E., Weisfeld, L. I., Lisitsyn, E. M., Bekuzarova S.A. Eds.; Apple Academic Press, 2016. In Press AAP.
9. Zaalishvili, V. B.; Bekuzarova, S. A.; Bataev. D. -K. S.; Budziewa, O. R.; Mazhiev, K. H.; Mazhieva, A. H. Patent RF No. 2481162 "Method of reclamation of oil polluted lands". Published on 10.05.2013. Bulletin No. 13, 2013 (In Russian).
10. Bekuzarova, S. A.; Shabanova, I. A. Patent RF No. 2222930 "Way to use legumes in toxic soils". Published 10.02.2004. Bulletin No. 4, 2004 (In Russian).
11. Zaalishvili, V. B.; Bekuzarova, S. A.; Mazhiev, H. N.; Bataev, D. -K. S.; Budziewa, O. R.; Mazhiev, K. H.; Mazhieva, A. H. Patent RF No. 2396133 "Method of rehabilitation of oil contaminated lands ". Published on 10.08.2010, 2010 (In Russian).

12. Zherukov, B. H.; Khaniyeva, I. M.; Bekuzarova, S. A. Patent RF No. 2444879 "Method for reducing the toxicity of soil in the cultivation of corn". Published on 20.03.2012. Bulletin No. 8, 2012 (In Russian).
13. Bekuzarova, S. A.; Efimova, V. A. Patent No. 2190315 "Method of sowing of green manure crops on toxic soil". Published 10.10.2002, 2012 (In Russian).

CHAPTER 14

ANTIRADICAL PROPERTIES OF ESSENTIAL OILS AND EXTRACTS FROM SPICES

TAMARA A. MISHARINA* and EKATERINA S. ALINKINA

Emanuel Institute of Biochemical Physics of Russian Academy of Sciences, 4, Kosygina St., Moscow 119334, Russia

Corresponding author. E-mail: tmish@rambler.ru

CONTENTS

ABSTRACT

The antiradical properties of essential oils and extracts from the clove bud (*Eugenia caryophyllata* Thumb.) and berries of tree (*Pimenta dioica* (L.) Meriff.), extracts from berries of white and black pepper *Piper nigrum* L., from fruits of red cayenne *Capsicum annuum* L., and green chili pepper *Capsicum frutescens* L. were studied in model reactions with the stable-free 2,2-diphenyl-1-picrylhydrazyl radical. The essential oils of clove bud and pimento had qualitatively close composition of the main components but differed by their quantitative content. In the studied samples, eugenol was the main compound with high antiradical activity. The values of antiradical efficiency were close for essential oils and were twice that for extracts and ionol. Spice extracts contained flavonoids, di- and triterpenoids, phenolic acids, alkaloids, and carotenoids. These substances possessed the antiradical activities. The values of antiradical efficiency were the highest for pimento and clove bud essential oils and decreased in the follow order: pimento essential oil > clove bud essential oil > pimento extract > clove bud extract > black pepper extract > cayenne pepper extract > chili jalapeno pepper extract > white pepper extract.

14.1 INTRODUCTION

Natural spices and their derivative products—essential oils and extracts—during many centuries are widely used in the all countries of the world as really natural flavorings for foodstuffs [1, 2]. The addition of spices to food makes it possible to improve significantly the organoleptic characteristics of the products—taste and flavor, diversify their assortment. The spices enrich the food of the natural preservatives and antioxidants that can content in many herbs [3–5]. The use of spices increases the shelf life of products due to the decrease in microbiological activity and intensity of lipid oxidation. As it is recently known, the search for harmless and effective natural antioxidants is promising challenge in the chemistry of food substances [1–3].

Extracts are obtained by treating milled spices with solvents, followed by evaporation of the solvents, and they contain up to 40% of essential oils and various nonvolatile compounds that possess the taste of the spices. The main components of such extracts are biologically active flavonoids, di- and triterpenoids, phytosterols, tocopherols, phenolic acids, alkaloids, carotenoids, vitamins, and other compounds. Spices, essential oils, and extracts are shown to have various biological activities. Ability of spices possess antiviral, anticancer, antiinflammatory, and geroprotective activity depending upon their antioxidant and antiradical properties [4–7]. Daily spice use is thought to benefit health and prevent many diseases [1–3]. The findings point to the prospects for further studies on spices, essential oils, and extracts as natural flavorings and antioxidants for food industry and as preventive and therapeutic agents for the treatment of oxidative stresses of the people [7–12].

The goal of this work is to study the reaction of free stable 2,2-diphenyl-1-picrylhydrazyl (DPPH) radical with components of essential oils and extracts obtained from popular and frequently used spices: the clove bud *Eugenia caryophyllata* Thumb., berries of tree *Pimenta dioica* (L.) Meriff., berries of white and black pepper *Piper nigrum* L., from fruits of red cayenne *Capsicum annuum* L., and green chili pepper *Capsicum frutescens* L.

14.2 MATERIALS AND METHODOLOGY

DPPH radical (Sigma-Aldrich, Germany) was used in the present work. Essential oils and extracts from the clove bud *E. caryophyllata* Thumb., pimento—berries of tree *P. dioica* (L.) Meriff., fruits of pepper *P. nigrum* L., pods of red cayenne *C. annuum* L., and green chili jalapeno pepper *C. frutescens* L. as industrial products were made by the company Plant Lipids Ltd. (India). These products are used by many Russian companies for production of multifunctional flavorings and additives for sausages, meat products, and other foodstuffs.

14.2.1 CHROMATOGRAPHY–MASS SPECTROMETRY

The composition of essential oils was analyzed with a HP 5890/5980 device (Hewlett Packard, United States) equipped with silica-fused capillary column HP-1 (25 m × 0.30 mm with the phase layer 0.25 μm) with the temperature programed from 50°C to 250°C at a rate of 4°/min. The temperature of the injector and mass-detector was 250°C. The mass spectra were obtained in the electronic strike regime at an ionizing power of 70 eV. Components were identified by the comparison of the values of the retention indexes, and the mass spectra obtained as a result of the analysis of samples with indexes and spectra of standards were estimated with the same column or taken from the mass-spectrum libraries (NBS, NIST, Wiley 275).

Quantitative analysis of the component content was performed by measurement of the peak areas obtained during the chromatographic analysis under conditions similar to that of chromatographic-mass spectrometry with a flame-ionizing detector by the method of simple normalization.

14.2.2 DETERMINATION OF ANTIRADICAL ACTIVITY OF ESSENTIAL OILS AND EXTRACTS

Solutions of essential oils and extracts of spices were added to a solution of DPPH radical in methanol (1 mL, 200 μM) up to predetermined concentration and added a volume of 2 mL methanol. The initial DPPH radical concentration was 10^{-4} M (100 nmol/mL or 39.4 mg/L) in all of the reaction mixtures; the optical density of the initial solutions was about 1. A series of model reactions was studied for each essential oil and extract, with varying of the concentration of substrates in the range of 10–2000 mg/L. To obtain kinetic curves of DPPH radical reduction by antioxidants, the reaction mixtures were placed into a quartz cell (10 mL) with tightly closing lids, and the optical density was registered on a spectrophotometer SP-2000 (SDB Spectrum, Russia) at a wavelength of 515 nm and room temperature in the dark in every 5 min during 120 min.

A graph of the linear dependence of the optical density on the DPPH radical concentration was built for solutions of DPPH radical in methanol. The value of molar extinction coefficient, which is equal to 10,010 mol/ (L cm), was determined by the graph. The residual concentration of DPPH radical in model reactions was calculated based on optical density value.

Every series of kinetic measurements was performed three times; mathematical treatment was carried out in Microsoft Excel 2007 and Sigma Plot 10 Software. The standard deviation of average values from 3 to 4 measurements did not exceed 3% (relative).

14.3   RESULTS AND DISCUSSION

To estimate the antiradical activity, we used a simple and sensitive method based on the ability of an antioxidant to release protons and reduce DPPH radical. In the process of reduction, the color of DPPH radical solution changes from intensive violet to straw-yellow, which is registered by the spectrophotometer at a wave length of 515–517 nm. In all reaction mixtures under study, the optical density of the initial DPPH radical solutions was about 1. The remaining radical concentration in model reactions was calculated in accordance with the optical density. The treatment of obtained kinetic curves for the reaction of DPPH radical with extract components at different concentrations in model systems and the calculation of parameters that reflect the antiradical activity of extracts are given in detail in this paper [13].

For each series of the studied essential oils and extracts, kinetic curves of radical reduction during 2 h were obtained. On the basis of the data obtained, dependence curves of the radical reduction level in 30 min on the extract concentration were composed. The linear questions of dependences of extent of radical reduction (y) from concentrations for each oil or extract (x) for 30 min were obtained:

$$y = A - B \times x \qquad (14.1)$$

Coefficients of the questions 1 (A and B) and coefficients of linear correlation R^2 are given in Table 14.1.

TABLE 14.1   Coefficients of the Linear Correlation of Dependence of Extent of Radical Reduction from Essential Oil or Extract Concentrations and EC_{50} Values for Different Spices.

Spice	A	B	R^2	EC_{50}, mg/L
Clove bud essential oil	61.70	0.967	0.9994	12
Pimento essential oil	69.06	0.953	0.9971	20
Pimento extract	71.23	0.901	0.9217	24
Clove bud extract	88.84	0.938	0.9682	41
Black pepper extract	98.57	0.119	0.9918	408
Cayenne pepper extract	90.86	0.100	0.9665	406
Chili jalapeno pepper	99.25	0.029	0.9780	1670
White pepper extract	100.13	0.032	0.9870	1590

These questions were used for the calculation of parameter of EC_{50}, which is equivalent to the quantity of an extract containing components with antiradical activity that is required to reduce 50% of radical concentration for 30 min. The obtained data are shown in Table 14.1, EC_{50} values were expressed as mg/L. It is suggested that this parameter allows us to compare the antiradical activity of complex antioxidant mixtures. It is obvious that EC_{50} values were the lowest for two essential oils, clove bud and pimento extracts. These preparations had the highest antiradical activity. EC_{50} values were more and similar for extracts of black and cayenne peppers, and much higher for extracts of white and chili peppers (Table 14.1). However, the shortcoming of this test is the absence of its connection with the reaction period. Therefore, it was suggested to use an alternative test, which would combine the time required for the reduction of 50% of radical concentration (TEC_{50}) at the required concentration of the extract (EC_{50}). This is the parameter of antiradical efficiency (AE, L/g · sec), which is calculated by question (14.2):

$$AE = \frac{1}{\left(EC_{50} \cdot TEC_{50}\right)}$$

(14.2)

As a result of these calculations, the AE values were estimated (Table 14.2). The AE parameter is the main antiradical characteristic for essential oils and extracts, and it is sufficient for comparing the antiradical properties of different plant preparations.

TABLE 14.2 Antiradical Efficiency Characteristics for Extracts of Different Peppers.

Pepper extract	AE, L/(g s)	Concentration of active components, eqv. radical, mmol/100 g	Antioxidant content, mmol/100 g of dried spice [3]
Clove bud essential oil	5.22×10^{-2}	58.5	–
Pimento essential oil	6.67×10^{-2}	39.9	–
Clove bud extract	2.97×10^{-2}	21.8	170.1
Pimento extract	3.47×10^{-2}	47.2	75.2
Black pepper extract	2.02×10^{-3}	11.9	8.7
Cayenne pepper extract	1.86×10^{-3}	11.7	8.0
Chili jalapeno pepper	3.31×10^{-4}	3.0	7.5
White pepper extract	2.50×10^{-4}	2.8	2.1

As can be seen from Table 14.2, studied preparation may be separated on three groups on the basis of AE values. The highest AE values were for essential oils and extracts of pimento and clove bud. AE values of extracts of black and cayenne peppers were 10 times less than for pimento and clove bud. Extracts of white and chili pepper had the lowest AE values (Table 14.2). Thus, AE values of studied preparations decreased as follows: essential oil of pimento > essential oil of clove > extract of pimento > extract of clove > extract of black pepper > extract of cayenne pepper > extract of chili jalapeno pepper > extract of white pepper. Therefore, the antiradical activity of the studied preparations was higher or the same as that of ionol (AE = 2.97×10^{-2}), the synthetic antioxidant used as the standard [13]. It is noteworthy that concentrations of $10^{-3}\%$ were optimal for the organoleptic properties of the essential oil and extracts of clove bud and pimento in food.

As it was shown earlier for essential oils oregano, savory and thyme [13], the reaction of a radical with components of each extracts of different peppers proceeded in two phases. The first phase was fast and finished, when the most active antiradical components were fully involved in the reaction. The second phase was slow. The gradual decrease in the concentration of the free radical in the second phase suggests the presence in the extracts of components with low antiradical activity. If the kinetic curve shows saturation in the presence of free radicals in the system, it indicates the total disappearance of antiradical substances and the end of the reaction. In this case, the concentration of active substances in the studied sample can be calculated. For other curves, which do not show saturation, the concentration of antiradical substances can be calculated for both fast and slow phases on the basis of the kinetic curve at the intersection of two tangents [13]. So, it was found that the rate of the second stage of the reaction ionol with radical was 10 times lower than the rate of the first stage [13]. For the studied essential oils of clove bud and pimento, the reaction rates first and second stages differed by 57 and 30 times, while for the extracts, it differed by 23 and 15 times, respectively. The second slow stage of the reaction for the essential oils and extracts lasted up to 12 days, whereas for ionol, it lasted only for 5 h [13]. The long duration of the antiradical effect, which was characteristic for a variety of natural extracts with complex composition containing substances with different antioxidative activity, is the advantage of multicomponent preparations.

Theoretically, this reaction is stoichiometric and proportional to the number of hydrogen atoms involved in the reaction with the DPPH radical. In other words, it may be expected that the quantity of reduced radicals would be proportional to the concentration of the antiradical component (or components if a mixture of antioxidants is used). In real systems, the situation is more complicated. A reaction may involve several antioxidants or, alternatively, one antioxidant molecule may have more than one hydrogen atom that undergoes elimination. Anyway, several, rather than one, reactions occur, in which reaction products may also participate. The calculated concentrations of antiradical components in the extract samples (unit equivalent to the radical concentration) are shown in Table 14.2. However, one should bear in mind that the obtained values of the content of antiradical components are conditional and can be different from the real ones. Therefore, the chemical nature of substances that are included in substrates must be taken into account. For comparison, the data on the antioxidant content in dried milled spices [3] are also shown in Table 14.2.

Chromatogram of the studied essential oil of clove bud is given in Figure 14.1 [14]. The composition of components in this oil is shown in Table 14.3. Dry clove buds contained 16–19% of essential oil; the main

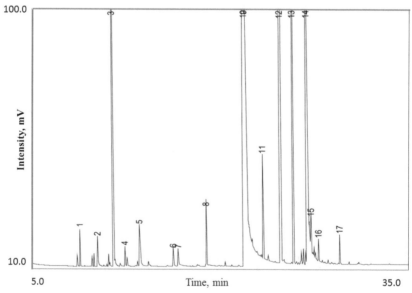

FIGURE 14.1 Chromatogram of essential oil from clove bud [14]. Peak numbers correspond compounds in Table 14.3.

component of this essential oil was eugenol (73%). The content of caryo-phyllene was 10.4%, of humulene 6.1%, eugenyl acetate—5.3%, and 1.8-cineol was about 3.1% (Table 14.3). The amount of other mono- and sesquiterpenes did not exceed 0.5%. The clove bud extract contained about 30% of essential oil of the same composition as the individual essential oil. Clove buds are known to contain about 3 g/100 g of gallic acid, its derivatives and catechins, and about 0.5 g of flavonoids [2]. All of these compounds were characterized by high antioxidative activity and were present in the clove extract [2]. Clove buds have actively been used as a spice, drug, and antimicrobial preparation since ancient times.

TABLE 14.3 Composition of Essential Oil of Clove Bud *Eugenia caryophyllata* Thumb.

Number of peak, Figure 14.1	Compound	Content, %
1	α-Pinene	0.16
2	β-Myrcene	0.26
3	1,8-Cineol	3.06
4	γ-Terpinene	0.10
5	Linalool	0.45
6	Terpinene-4-ol	0.14
7	α-Terpineol	0.15
8	Chavicol	0.45
10	Eugenol	73.08
11	Methyleugenol	0.54
12	β-Caryophyllene	10.39
13	α-Humulene	3.07
14	Eugenyl acetate	5.30
15	Bicyclogermacrene	0.10
16	δ-Cadinene	0.10
17	Caryophyllene oxide	0.15

The berries of pimento have the name allspice in the Europe because it has the flavors of cinnamon, nutmeg, and clove combined in it. The pimento berries content is about 2–5% of essential oil, the composition of studied pimento essential oil is given in Table 14.4, chromatogram is given in Figure 14.2 [14].

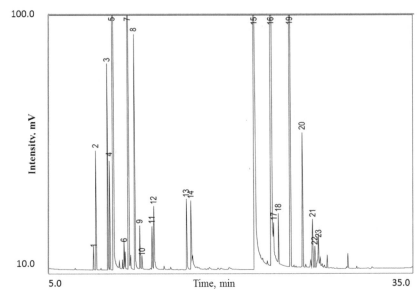

FIGURE 14.2 Chromatogram of essential oil from pimento berries [14]. Peak numbers correspond compounds in Table 14.4.

TABLE 14.4 Composition of Essential Oil of Pimento, Berries of Tree *Pimenta dioica* (L.) Meriff.

Number of peak, Figure 14.2	Compound	Content, %
1	α-Thujene	0.17
2	α-Pinene	0.80
3	Sabinene	1.50
4	β-Pinene	0.81
5	β-Myrcene	8.55
6	*p*-Cymene	0.47
7	1,8-Cineol	5.62
8	Limonene	2.12
9	γ-Terpinene	0.32
10	Sabinene hydrate	0.13
11	α-Terpinene	0.41
12	Linalool	0.65
13	Terpinene-4-ol	0.77
14	α-Terpineol	0.12

TABLE 14.4   *(Continued)*

Number of peak, Figure 14.2	Compound	Content, %
14	Cinnamal	0.59
15	Eugenol	35.42
16	Methyleugenol	28.02
18	β-Elemene	1.00
19	β-Caryophyllene	8.66
20	α-Humulene	1.34
21	Methylisoeugenol	0.48
22	Eugenyl acetate	0.89
23	Bicyclogermacrene	0.25

The major constituents of the essential oil were eugenol (35.4%), methyl eugenol (28.0%), β-myrcene (8.6%), caryophyllene (8.7%), and 1,8-cineol (5.7%) (Table 14.4). The extract from pimento berries content was 23% of essential oil, more than 25 substances with high antioxidant activity were found in the extract, including phenolic acids, flavonoids, catechins, and some phenyl propanoids [1]. The compounds possessing anticancer activity are found in the leaves of pimento. High antioxidant and antiradical activities of the extract from pimento berries are connected with eugenol [1, 2]. The highest content of eugenol was in clove bud essential oil (73%), less in pimento essential oil (35.4%). About 10% of eugenol was in the extract from pimento berries, and 21% was in clove bud extract, which were studied by us; due to these, extracts had the highest antiradical activity in comparison to other peppers extracts (Table 14.2). The effective content of antiradically active substances in the studied eugenol containing essential oils and extracts, which was calculated from the equivalent concentration of the radical reduced in the first stage of the reaction, decreased as follows: essential oil of clove bud > pimento extract > essential oil of pimento > clove bud extract. This order is different from the order showing the decrease in the level of eugenol, the main active substance: clove bud essential oil > pimento essential oil > clove bud extract > pimento extract. In pimento essential oil, the level of eugenol was almost two times lower than in clove essential oil (Tables 14.3 and 14.4). The effective concentration of substances with high antiradical activity was also lower, but the value of AE of this oil was higher than that of clove essential oil (Table 14.2). Apparently, the synergetic effect of eugenol and other components of pimento essential oil, which affect one another, probably by changing the reaction of certain substances with the added DPPH radical and newly formed intermediate reaction products, increased the total value of AE in the whole oil. This demonstrates that it is not correct to identify the quantitative amount of antioxidative components in the sample as the total antiradical activity. To provide the antioxidative and antiradical properties of multicomponent preparations, it is apparently important to take into account not only the level of certain components with high activity but also their combination and the presence of other compositions. Some components of essential oils, for example, monoterpenes, such as terpinene, α- and γ-terpinenes, demonstrate a low reactive capacity with DPPH radical but high antioxidative activity in reactions with active oxygen- and nitrogen-containing radicals, the main oxidants in cells or in the environment [13]. Therefore, many essential oils, which do not contain phenols, are effective inhibitors of the autooxidation of lipids and other compositions. In such complex multicomponent systems as essential oils or extracts of spice or other aromatic plants, both synergetic and antagonistic interactions of certain components with one another are possible. This may not change, increase, or decrease the total antiradical activity of the studied preparation. Indeed, the pimento extract demonstrated a higher level of active antiradical compositions than essential oils, though the value of AE of the extract was lower than that of essential oils. Clove extract contained less antiradical substances than essential oil, and its antiradical efficacy was also lower (Table 14.2).

One of the most popular spices in the world is black pepper. The pepper, its essential oil, and extracts possess a wide spectrum of biological activity; they are involved in digestion processes, and the presence of tocopherols and polyphenols is responsible for the antioxidant properties of pepper [1, 2]. The treatment of black pepper or piperine with food in rats reduced the oxidation level of lipids in different organs of animals, induced antioxidant enzyme activity, and increased their antioxidant status [4, 6]. The alkaloids piperine, chavicine, isopiperine, isochavicine, sarmentine, sarmentosine, and tricholein, as well as flavonoids quercetin, iso-quercetin, isorhamnetin, kaempferol, and others are biologically active components of pepper and its extracts [1–4]. These substances, as well as volatiles with low molecular weight (essential oil), are presented in black pepper extract under study. The major volatile components of black pepper were α- and β-pinenes, limonene, sabinene, 3-carene, and caryophyllene;

these compounds give a specific flavor with a pine sent to the pepper, but antioxidant activity of pepper essential oil was low [15, 16]. Total antiradical activity of black pepper extract was 10 times lower than activity of pimento extract (Table 14.2).

White pepper is obtained from ripened grains of black pepper after soaking for 7–9 days, removal of the swollen peel, and drying. The studied extracts of black and white pepper contained about 20% and 12% piperine, respectively. The AE value for black pepper extract was eight times higher than white pepper extract (Table 14.2). Black pepper extract also contained compounds with antiradical properties four times higher than in white pepper extract (Table 14.2). The piperine content in both extracts differed by two times; therefore, the total antiradical effect of black pepper extract was provided by the total action of alkaloids, phenolic acids, and flavonoids. It is known that these compounds are mainly contained in the peel of vegetables, fruits, and spices. Removal of the black pepper peel resulted in the loss of a considerable part of the active compounds; thus, white pepper extract had a much lower EA value (Table 14.2).

Various types of pod peppers are worldwide popular food supplements, such as red cayenne C. annuum L. and green chili pepper C. frutescens L. Seven natural capsaicinoids, which give a pungent taste, are found in pod pepper. Its total content depends on many factors. It was found that the capsaicin content, the main component, and dihydrocapsaicin composes 80–90% in ripe fruits of red cayenne pepper. Among them, norcapsaicin, nordihydrocapsaicin, nornordihydrocapsaicin, homocapsaicin, and homo-dihydrocapsaicin are found; their content is much lower than the capsaicin content. A person can determine the taste of capsaicin at a concentration of 10 parts per million. Depending on the degree of pungency, which is caused by the presence of capsaicinoids, there are several types of red pepper. Sweet pepper is notable for large fruits, which contains 0.02–0.1% capsaicin and now is cultivated in many countries. Moderately, hot pepper has a common name "chili;" the capsaicinoids content reaches 2.5% in chili fruits. Depending on the form, color, and size, there are several types of this pepper: ancho has red pods; jalapeno has dark green (hot); mulato has brown; pasilla has long, slim, and brown pods with rich and burning taste; and serrano has small, green, and very hot pods. The red cayenne pepper with little pods has hot and pungent taste and color from yellow to dark-red [1, 2]. The capsaicinoid content was six times greater in the extract of red pepper compared with extract of green jalapeno pepper.

The second group of main components for pod pepper is carotenoids. Altogether 34 carotenoids were found and identified in pepper fruits, the content of four main carotenoids (capsanthin, capsorubin, zeaxanthin, and cryptoxanthin) was about 90% of all the carotenoids; they give red color to pepper. Carotenoids are stable in fresh rods, but they are exposed to autoxidation in dried and milled rods. The yellow-orange color of chili is from β-carotene and violaxanthin. The content of capsanthin, the major carotenoid in ripe fruits, being the more stable, increases proportionally with advanced stages of ripeness [1, 2]. Pod pepper also contains flavonoids, vitamins, and about 50 volatile compounds that cause the flavor. However, the total content of volatiles is too small and composes less than 0.01%. The key substance for the flavor of sweet and chili peppers is 2-isobutyl-3-methoxypyrazine; its threshold concentration in water is 0.001 part per billion [17].

A number of studies showed that red pepper extracts displayed antioxidant properties in different model systems, not only carotenoids and flavonoids but also capsaicinoids possessed activity [18–20]. Pepper extracts inhibited the oxidation of cholesterol, linolenic, and docosahexaenoic acids [20]. It was found that the content of ascorbic acid, carotene, capsanthin, quercetin, and luteolin, as well as the total activity for red pepper, is higher than for green pepper [21]. These data are in good agreement with the AE values determined for these two types of pepper.

We studied on the antiradical properties of extracts from chili jalapeno and red cayenne pepper; the capsaicin content in these peppers was 1.0 and 6.0%, respectively. Indeed, the AE value was also six times higher for red cayenne pepper extract, which contained six times more capsaicin than chili pepper. Moreover, the carotenoid content was higher, as cayenne pepper is red, whereas chili is green. The content of compounds with high antiradical activity was four times higher for red cayenne pepper than for chili pepper (Table 14.2). It should be mentioned that the antiradical characteristics were the same for black pepper extract and cayenne pepper extract (Table 14.2). Both pepper species differed in the content, structure, and chemical properties of the main alkaloids, polyphenols, including flavonoids and carotenoids, but they effectively reduced free radicals in model system. Thus, for comparison, the AE value for black and red peppers was much higher than for many flavonoids and was equal to $1.84–2.02 \times 10^{-3}$ (Table 14.2), whereas the AE value was 5.2×10^{-4} for α-tocopherol, 1.9×10^{4} for quercetin, and 0.5×10^{-4} for resveratrol [22].

Apparently, the study of all possible reactions of even one radical with multicomponent essential oils and extracts is a complicated task. However, this should not be considered as an obstacle to the application of these plant preparations as antiradical agents. Moreover, experiments with animals showed that clove bud essential oil demonstrated properties of biological antioxidants in vivo when administered in low doses mice for 6 months [6]. Study of the biological activity of essential oils of oregano, savory, clove, and the extract of ginger showed a promising effect of the uptake of small doses of these and other essential oils for the prophylaxis of diseases connected with oxidative stress and an increase of longevity [10, 11]. Therefore, the present study showed that industrial samples of essential oils and extracts of clove bud and pimento demonstrated a high antiradical efficacy that exceeded that of the synthetic phenolic antioxidant ionol. The antiradical properties of essential oils and extracts considerably depended on their composition. However, the quantitative level of antioxidant components in the samples did not always correspond to their antiradical activity. The ratio of these components was especially important, as the synergetic effects provided higher antiradical activity of the multicomponent mixtures in comparison with individual substances. Study of the connections between the composition and properties of plant preparations is expected to help predict the properties and regulate different types of biological activities via the preparation of particular compositions.

14.4 CONCLUSIONS

Thus, our study showed that essential oils and extracts from clove bud, pimento, black, white, red cayenne, and green chili peppers possess antiradical properties in model systems with diphenylpicrylhydrazyl radical. Pimento and clove bud preparations content eugenol had the highest activity. The AE values for extracts from black and red cayenne peppers were almost the same; it was 10 times lowest than activity of clove bud and pimento and was 10 times higher than the effectiveness of white and chili pepper extracts. The differences in the antiradical properties of studied substances are determined by the lower content of polyphenols and piperine in white pepper compared with black pepper. The content of capsaicinoids and carotenoids was lower in green pepper compared with red pepper.

The synergistic effects are probably of great importance for the anti-radical properties of extracts, which are connected with the mutual effect of polyphenols, vitamins, and alkaloids of capsaicinoid and piperine types.

ACKNOWLEDGMENT

This work was supported by the Russian Science Foundation, project No. 14-16-00102.

KEYWORDS

- pimento
- peppers
- 2,2-diphenyl-1-picrylhydrazyl radical
- clove bud
- berries

REFERENCES

1. Charles, D. J. *Antioxidant Properties of Spices, Herbs and Other Sources*. Springer: New York, 2013; p 610.
2. Parthasarathy, V. A., Chempakam, B., Zachariah, T. J., Eds.; In *Chemistry of Spices*. CAB Int.: Oxfordshire, 2008; p 445.
3. Crozier, A.; Jaganath, I.B.; Clifford, M. N. Dietary Phenolics: Chemistry, Bioavailability and Effects on Health. *Nat. Prod. Rep.* **2009**, *26*, 1001–1043.
4. Miguel M. G. Antioxidant and Anti-Inflammatory Activities of Essential Oils: A Short Review. *Molecules.* **2010**, *15*, 9252–9287.
5. Berger R.G., Eds.; In *Flavours and Fragrances. Chemistry, Bioprocessing and Sustainability*. Springer Verlag: New York, 2007; p 635.
6. Migue, M. G. Antioxidant Activity of Medicinal and Aromatic Plants. *Flavour Fragr. J.* **2010**, *25*, 291–312.
7. Edris, A. E. Pharmaceutical and Therapeutic Potentials of Essential Oils and Their Individual Volatile Constituents: A Reviews. *Phytother. Res.* 2007, *21*, 308–323.
8. Burlakova, E. B.; Misharina, T. A.; Vorobyeva, A. K.; Alinkina, E. S.; Fatkullina, L. D.; Terenina, M. B.; Krikunova, N. I. Inhibition of Mouse Aging by Using an Essential Oil Composition. *Doklady Biochem. Biophys.* **2012**, *444*, 167–170.

9. Misharina, T. A.; Burlakova, E. B.; Fatkullina, L. D.; Alinkina, E. S.; Vorobyeva, A. K.; Medvedeva, I. B.; Erokhin, V. N.; Semenov, V. A.; Nagler, L. G.; Kozachenko, A. I. Effect of Oregano Essential Oil on the Engraftment and Development of Lewis Carcinoma in F1 DBA C57 Black Hybrid Mice. *Appl. Biochem. Microbiol.* **2013,** *49,* 432–436.

10. Misharina, T. A.; Fatkullina, L. D.; Alinkina, E. S.; Kozachenko, A. I.; Nagler, L. G.; Medvedeva, I. B.; Goloshchapov, A. N.; Burlakova, E. B. Effects of Low Doses of Essential Oil on the Antioxidant State of the Erythrocytes, Liver, and the Brains of Mice. *Appl. Biochem. Microbiol.* **2014,** *50,* 88–93.

11. Maffei, M. E.; Gertsh, J.; Appendino, G. Plant Volatiles: Production, Function and Pharmacology. *Nat. Prod. Rep.* **2011,** *28,* 1359–1380.

12. Misharina, T. A. Antiradical Properties of Essential Oils and Extracts from Coriander, Cardamom, White, Red and Black Peppers. *Appl. Biochem. Microbiol.* **2016,** *52,* 79–86.

13. Alinkina, E. S.; Misharina, T. A.; Fatkullina, L. D. Antiradical Properties of Oregano, Thyme, and Savory Essential Oils. *Appl. Biochem. Microbiol,* **2013,** *49,* 73–78.

14. Misharina, T. A., Alinkina, E. S., Medvedeva, I. B. Antiradical Properties of Essential Oils and Extracts from Clove Bud and Pimento. *Appl. Biochem. Microbiol.,* **2015,** *51,* 119–124.

15. Misharina, T. A.; Terenina, M. B.; Krikunova, N. I. Antioxidant Properties of Essential Oils. *Appl. Biochem. Microbiol.* **2009,** *45,* 642–647.

16. Frankel, E. N.; Meyer, A. S. The Problems of Using One-Dimensional Methods to Evaluate Multifunctional Food and Biological Antioxidants. *J. Sci. Food Agric.* **2000,** *80,* 1925–1941.

17. Bauer, K.; Garbe, D.; Surburg, H. *Common Fragrance and Flavor Materials.* VCH. Verlag: Weinheim, 1990; p 218.

18. Rosa, A.; Deiana, M.; Casu, V.; Paccagnini, S.; Appendino, G.; Ballero, M.; Dessi, M. A. Antioxidant Activity of Capsaicinoids. *J. Agric. Food Chem.* **2002,** *50,* 7396–7401.

19. Ochi, T.; Takaishi, Y.; Kogure, K.; Yamauti, I. Antioxidant Activity of a New Capsaicin Derivative From *Capsicum annuum. J. Nat. Prod.* **2003,** *66,* 1094–1096.

20. Kim, J. S.; Ahn, J.; Lee, S. J.; Moon, B.; Ha, T. Y.; Kim, S. Phytochemicals and Antioxidant Activity of Fruits and Leaves of Paprika (*Capsicum annuum* L., var. Special) Cultivated in Korea. *J. Food Sci.* **2011,** *76,* 193–198.

21. Materska, M.; Perucka, I. Antioxidant Activity of the Main Phenolic Compounds Isolated from Hot Pepper Fruit (*Capsicum annuum* L.). *J. Agric. Food Chem.* **2005,** *53,* 1750–1756.

22. Miller, A. L. Antioxidant Flavonoids: Structure, Function and Clinical Usage. *Alt. Med. Rev.* **1996,** *1,* 103–110.

CHAPTER 15

THE CHEMICAL COMPOSITION OF ESSENTIAL OILS FROM WILD-GROWING AND INTRODUCED PLANTS OF THE ASTRAKHAN REGION

ANATOLY V. VELIKORODOV, VYACHESLAV B. KOVALEV, SVYATOSLAV B. NOSACHEV, ALEXEY G. TYRKOV, MARIA V. PITELINA*, and EKATERINA V. SHCHEPETOVA

Astrakhan State University, 20a, Tatishchev St., Astrakhan 414056, Russia

Corresponding author. E-mail: avelikorodov@mail.ru

CONTENTS

ABSTRACT

This chapter submits the data on studying the chemical composition of essential oils isolated from land parts of wild-growing plants of the Astrakhan region—European Bugleweed (*Lycopus europaeus* L.), Bugleweed High (*Lycopus exaltatus* L.); four endemic types of wormwood of the Astrakhan region such as *Artemisia lerchiana, Artemisia santonica, Artemisia arenaria, Artemisia austriaca,* as well as two species of the plants are introduced in the Astrakhan region—fennel gianthisson (*Lophantus anisatum* Benth.) and common hyssop (*Hyssopus officinalis* L.). The essential oils were extracted by the use of steam distillation during the vegetation and blossoming periods of the plants. The dependence of the yield of essential oil on plant species, habitat, and vegetation period are studied, the appearance and change range of reflective index are characterized. The quantitative analysis of components in essential oils was carried out by the method of gas–liquid chromatography. The yield of essential oil from *L. exaltatus* was 0.7–0.9%, *L. europaeus* gave 0.5–0.7%, four types of wormwood including *A. lerchiana, A. santonica, A. arenaria,* and *A. austriaca* gave 0.24–0.7% fennel gianthisson; 0.25–0.55%, common hyssop; and 0.1–0.8% in equivalent to air-dry raw materials, respectively, during vegetation and blossoming of plants. The chromatography–mass spectrometry was used to study the chemical composition of essential oils. In the sample of essential oil from European lycopus, 31 components were identified, the essentials ones are α-terpineol, caryophyllene oxide, and isoeugenol methyl ether. In the essential oil from high bugleweed, 12 components are most abundant, of which are 2,4-decadienal, 2,4-hexadien-1-ol, and α-limonene diepoxide are identified. The predominant components of essential oil from four studied types of wormwood are 1,8-cyneol, camphor, isoborneol, terpine-4-ol, α-bisabolol, β-pinene, α-pinene, limonene, *trans*-pinocarvyl acetate, germacrene D, and γ-elemene. The essential oil from fennel gianthisson differs for the high concentration of methyl chavicol, methyl eugenol, caryophyllene, and D-limonene. The main components of the essential oil from common hyssop are isopinocamphone and pinanediol. It is found that the essential oil from fennel gianthisson has a high antifungal activity versus *Microsporum canis, Trichophyton rubrum,* and *Candida albicans.*

15.1 INTRODUCTION

The unique character of the biota in the Astrakhan region is explained by the specifics of the geographical location and climate. The natural features make the flora of the Astrakhan region diverse; there are both widespread species and those with a very limited habitat. Here is an area of a big variety of wormwoods (*Artemisia* L.).

The Caspian deserts are the land of suffrutescent sagebrushes among which white wormwood, black wormwood, sand wormwood, and common wormwood are the most widespread. The family of wormwoods is presented in the Astrakhan region by 10 species, many of them have a large habitat and form the considerable phytomass that can explain the prospects of their practical use [1].

As a result of evolution, desert plants developed a number of features that help them to survive the lack of water and the salinity of soil. The leaves of many species have changed—the surface area of a leaf became much smaller. The sprouts of some have strengthened.

As a rule, the underground part of desert plants surpasses their elevated parts in the power of development by 19–20 times. Such salt-loving species of plants as *Salicornia, Halocnemum strobilaceum, Tamarix ramosissima,* and *Limonium gmelinii* grow here. *Ephedra distachya* L., *Koeleria,* matgrass, *Nitraria schoberi, Ceratoides papposa, Leymus racemosus, Festuca valesiaca,* and *Agropyron desertorum* are the typical plants for the desert flora of our region. The vegetative cover of the desert is characterized with a high dynamic range that is caused by dramatic changes in conditions of habitat including soil deformation. In general, the flora of the desert totals 160–200 species, the leading families are composites (*Compositae*), pigweeds (*Chenopodiaceae*), and gramineous (*Gramineae*).

The family of wormwood—*Artemisia* L. (*Asteraceae family*)—unites over 400 species mainly widespread in the moderate zone of the northern hemisphere, 174 species grow in the CIS (Commonwealth of Independent States). The wormwood species are often met in steppes, others grow in semi-deserts and deserts, and some are weeds in all zones.

The interest to wormwoods is explained by pharmacological active agents, sesquiterpene lactones that were found in the studied species. As a result of the comprehensive study, some preparations from wormwood are offered for medical application.

The study of the dependence of chemical composition of essential oils on ecological factors shows the allelopathic phenomena in phytocenoses and has an important practical value.

This chapter represents the data on studying the chemical composition of essential oils from four species of wormwood, on the influence of ecological factors on accumulation, the contents and composition of essential oil from plants of the *Artemisia* L. growing in the climatic conditions of the Astrakhan region.

The components in the composition of essential oil are often identical, but there are also differences [2].

Austrian wormwood (*Artemisia austriaca*) is a perennial gray-white root-sucker herbaceous plant [3]. Stems are upright, branching, and leafy. Leaves are twice pinnatisect into small linearly mucronated parts. Flower baskets are broadly ovate, small, drooping, and are collected in paniculate inflorescence. Involucral leaflets are linear and pilary. All flowers and floral calathidia are tubular. The blossoming period is July–August.

Wormwood sand (*Artemisia arenaria*) is a subshrub with a height of 20–100 cm. At the bottom, the stems are the ligneous, and vegetative sprouts are truncated. Leaves are green, slightly succulent, almost naked, dissected into narrow segments, and linearly lanceolate final segments; the lower leaves are macropodus, the others are assidenous. Anthodes are ovoid, assidenous, or on the truncated pedicles, collected in a sprawling whisk. *A. arenaria* grows on sand in steppes and on the sea coasts of the Balkan Peninsula, on the coast of the Azov and Black seas, in western Ciscaucasia, in the Caspian Sea region, and in the Aral Sea region. Thanks to the fast vegetative reproduction, *A. arenaria* fixes friable sand easily forming unproductive pastures–sandy wormwood areas.

Wormwood santonian (*Artemisia santonica*) is a desert-steppe species of the Caspian region. It is a perennial subshrub with a height of 70 cm. Leaves are alternate, twice plumose-dissected, and the lower stem leaves are petiolar, let down, and gray. Flowers are small, 2–3 mm long, telianthus, without flower-cups; corollas are tubular, quinquedentate with oil droplets. Inflorescences are oblong-ovoid, assidenous anthodes collected in narrow contracted panicles. Halophilous-meadow-steppe Black Sea-Kazakhstan species grows on damp saline soils, on salt-marsh edge, in the areas where salted ground waters come out.

Wormwood Lerch (*Artemisia lerchiana*) or white wormwood is a perennial subshrub with a height of 16–50 cm. At the beginning, all the plant is covered with grayish dense fluffy hairs; later, it is partially naked. The bush is built up of perennial ligneous, strongly truncated stems and short leaf-bearing one-year sprouts. Fruit-bearing sprouts are numerous; in the top half, they are branchy. Leaves in unfertile sprouts and in bottom stems are petiolate, 2–3 pinnatisected; middle—sessile, 2 pinnatifid, at the basis with pinnatisected ears; overhead leaves are simple, linear. Inflorescences are assidenous anthodes collected in contracted panicles. Corollas are yellow or pink. Fruit is achene. Blossom period is from August–September. It grows in meadows, pastures, in steppes with strongly saline, black humus, and brown soils. It is the indicator of soil alkalinity and a fodder plant. Because of soil salination, *A. lerchiana* shows properties that are specific to succulents. This fact singles *A. lerchiana* out of mesophytes as the cells of the latter lose water in the conditions of water shortage [4].

Nowadays, when preventing and treating many diseases, pharmaceutical preparations are demanded to possess specificity, maximum efficiency, and lack of by-effects. Multicomponent forms that contain biologically active agents from medicinal vegetative raw materials meet this demand. Studying phytochemical properties of wild-growing plants and preparing medicines on their basis to meet the necessary requirements are of some practical interest.

Lycopus high (*Lycopus exaltatus* L.) and European lycopus (*Lycopus europaeus* L.) are wild-growing perennial grassy plants of *Lamiaceae* family, they grow in a temperate climate of many countries of Europe, in the European part of Russia, Central Asia, Caucasus, and in Western and Eastern Siberia. In the Astrakhan region, large populations of these plant species dominate in the humid meadows of the Delta and the floodplain of the Volga River, and these also can be met as weed plants in vegetable gardens.

The folk-medicine advises to use European lycopus tea as antiinflammatory, restorative, and analgesic agent to treat Basedow's disease, to reduce high blood pressure and tachycardia. The extracts of lycopus high are used for the treatment of paludism, diarrhea, gastric distresses, metrorrhagia, and neurosis.

The scientific interest to vegetative raw materials of lycopus high and European lycopus is growing. Shelukhina *et al.* [5] studied the chemical composition of European lycopus. Through the methods of ^1H, ^{13}C NMR, UV-spectroscopy, and HPLC–MS methods, they have isolated phenolic compounds and identified caffeic acid ethyl ester,

3,4-dimethoxybenzaldehyde,   5,3',4'-trihydroxy-6,7-dimethoxyflavone,
apigenin, apigenin-7-glucuronide ethyl ester, luteolin, luteolin-7-gluc-
uronide methyl ester, luteolin-7-glucuronide ethyl ester, caffeic acid,
rosmarinic acid, and rosmarinic acid methyl ester. The sum of phenolic
compounds in terms of rosmarinic acid and absolutely dry European
lycopus extract made 3.5%. After studying a number of researches on the
influence of extractions from European lycopus on thyroid body and tissue
metabolism of iodine in experimental animals (guinea pigs, etc.), Alefirov
et al. [6] proved that extracts from grassy European lycopus possess
antihypothyroid activity on the experimental model of thyrotoxicosis in
rats that led to the normalization of the state and behavior of animals,
the thyroid hormones level in blood serum. It was found out that water
extraction from grassy lycopus showed the most evident medicinal effect
similar to the action of thyrozol. The study of pharmacological activity of
vegetative raw materials of European lycopus by Alefirov *et al.* showed
antihypothyroid action of extracts from the plant makes possible to use
the preparations on the basis of European lycopus in the treatment of
Basedow's disease as alternatives in case of intolerance to hormonal antithyroid agents. The study of pharmacological activity of vegetative raw
materials of European lycopus by Alefirov *et al.* and contributors showed
antihypothyroid action of extracts from the plant that makes possible to
use the preparations on the basis of European lycopus in the treatment of
Basedow''s disease as alternatives at while the intolerance to hormonal
antithyroid agents. Earlier, during the phytochemical researches of aboveground parts of European lycopus and lycopus high [7], at the department
of organic, inorganic, and pharmaceutical chemistry of the Astrakhan state
university, in water extracts of vegetative raw materials of these plants,
when carrying out qualitative reactions, hydrolyzable and condensable
tannins were found. The content of tannin in leaves of *L. exaltatus* L. and
L. europaeus L. made 0.01% and 1.6%, respectively. The evaluation of
total flavonoids in equivalent to luteolin-7-glucozide in aqueous–alcoholic
(60%) extracts of lycopus high showed in stems was 7.2 mg, in leaves—
11.7 mg; the European lycopus extracts showed in stems were 5.3 mg, in
leaves—8.5 mg per 100 g of dry raw materials. The sum of flavonoids in
equivalent to rutin in stems and leaves made: 6.8 and 12.2 mg per 100 g of
dry raw materials respectively for lycopus high; 5.3 and 6.25 mg per 100
g of dry raw materials respectively for European lycopus. In water extraction from grassy lycopus high, the sum of triterpene saponins made 3.8%
in equivalent to oleanolic acid.

European lycopus and lycopus high are the plants with similar pleasant aromatic smells that become stronger when grinding both fresh and dried-up raw materials. The smells are kept during a storage time (2 years) that provides the evidence for the content of essential oil in different parts of these plants. According to the present data, the yield of essential oil from raw materials of European lycopus that grows in the conditions of Uzbekistan made 0.2%. In essential oil, limonene, terpinene, linalool acetate, linalool, bornyl acetate, geranyl acetate, nerol, geraniol, *p*-cymene, γ-terpinene, α-pinene, camphene, terpinolene, etc were identified. The study of European lycopus that grows in the northern part of Serbia [8] showed that the yield of essential oil is 0.5%. Moreover, in the composition of the oil, the following components were identified: copaene, geranyl acetate, selinene, cadiene, ledol, hexadienol, borneol, terpineol, decanal, geraniol, furfural, hexanol, benzaldehyde, nonadienal, isocitral, lavandulol, nonalol, etc. While further studying the sample of European lycopus from the northern part of Serbia, antimicrobial activity of essential oil components against *Escherichia coli* and *Klebsiella pneumoniae* was found. The data concerning the composition of essential oil from lycopus high are not submitted in the scientific literature, the pharmacological activity of essential oil from this plant is not studied.

In recent years, the interest to essential oil plants of *Lamiaceae* family to which hyssop belongs has increased significantly. Probably, this plant originally comes from Southwest Asia and Southern Europe. This subshrub is cultivated in Eastern and Central Europe, in France, Italy, the Balkans, Crimea, and in Asia [9, 10].

The essential oil is the main physiologically active component of *Hyssopus officinalis*. The content of oil in leaves makes 0.3–1.5%, in inflorescences 0.9–2.0%, and in stems, only trace amounts are found. Above-land parts of the plant (leaves, inflorescences, and softwood stems) are used as raw materials to receive essential oil which is consumed by food, cosmetic, and pharmaceutical industry. The yields of essential oil from dried and fresh plant raw materials received by steam distillation make 0.15–0.3% and 0.3–0.8%, respectively. In large volumes, this oil is produced in France, Italy, and in the countries of Former Yugoslavia.

The essential oil from *H. officinalis* is a light green or light yellow liquid with a characteristic camphor smell.

It possesses antibacterial, antiviral, antifungal, and expectorative activity [11–13]. The recent researches figured that the essential oil produced from a hyssop shows antiplatelet activity [14]. In addition, spasmolytic activity of essential oil from a hyssop medicinal is revealed [15].

The yield and chemical composition of essential oil from *H. officinalis* depend on many external factors (climatic conditions, soil type, plant origin, time of raw materials preparation, etc.) [16, 17].

According to the literary data, the main components of essential oil from *H. officinalis* are isomeric pinocamphones, β-pinene, pinocarvone, limonene, linalol, β-caryophyllene, germacrene D, tujones, and myrtenol [18–20].

The chemotype of hyssop growing in Turkey differs from the chemotype of hyssop cultivated in Poland by the content of its dominating ingredient—pinocarvone [21, 22]. The essential oil of hyssop medicinal from Spain is characterized by high concentration of 1,8-cyneol (52.89%) [23]. Hyssop medicinal cultivated in France differs by the domination of linalool in its essential oil (49.6%). This oil is also characterized by the low content of monoterpene ketones [24].

The study of the chemical composition of essential oil (yield of 0.34%), received by the method of gas chromatography–chromato-mass-spectrometry, from hyssop leaves picked near Khandiza (former Uzbek SSR), showed that its main components are pinocamphone (71%), β-pinene (8.6%), and 1,8-cyneol (6.4%), but limonene and isopinocamphone were not found [25].

The chemical composition of essential oil from hyssop medicinal, cultivated in the Astrakhan region, has not been studied before. The need to detailed studying the components of essential oil composition is caused by the potential content of toxic compounds. According to the scientific data [26], methyl eugenol, for example, possesses a cancerogenic activity, and unterpene ketones show an obvious epileptogenic activity [27].

In recent years, the interest to fennel gianthisson (*Lophantus anisatum* Benth.) has significantly increased in Russia. It was improved by Ukrainian plant selection breeders [28], then gardeners began to grow it up in their private grounds more often, and it is wildly cultivated by beekeepers as this plant is an excellent honey herb [29].

Fennel gianthisson belongs to Labiate family (Lamiaceae), it is a perennial, winter-hardy plant, and a semi-grassy bush with the height not exceeding a meter. Stems are tetrahedral; leaves are petiolar oval, and seldom have rough edges, 7–10 cm long and 4–5 cm wide. Root is fibrous. Flowers are hermaphrodite with a long stoma. Inflorescences are spiciform, white, or violet, sometimes different colored, up to 20 cm long or more, with anisic smell. The vegetative period lasts till steady frosts. In the first year of crops, seeds ripe at the end of September, and in the next years, it happens 2–3 weeks earlier.

In folk medicine, *L. anisatum* Benth. is applied as antiinflammatory and bactericide agent.

L. anisatum Benth. is believed to increases body resistance and facilitates the adaptation to adverse environment as well as has a sedative effect on the central nervous system.

Water extracts from leaves of this plant are used to treat inflammatory processes in gastrointestinal tract, liver, and urinary tracts diseases, as a medicine for acute respiratory diseases, bronchitis, pneumonia, and bronchial asthma, to remove radionuclides and to reduce cholesterol content in blood. Gel produced from leaves of *L. anisatum* Benth. cures the skin diseases caused by fungi successfully.

From this point of view, it is important to know what substances including biologically active compounds are contained in this plant.

Despite the wide range of pharmacological properties of *L. anisatum* Benth., its chemical composition is nearly unknown.

The present work is purposed to study the chemical composition of essential oil samples from four endemic species of wormwood (*A. lerchiana, A. santonica, A. arenaria,* and *A. austriaca*), two species of *Lycopus*—Bugleweed high (*L. exaltatus* L.) and European bugleweed (*L. europaeus* L.), and also *H. officinalis* and *L. anisatum* Benth., growing in the Astrakhan region, depending on type of above-land parts and vegetation period, as well as quantitative determination of essential oil main components [30–33].

15.2 MATERIALS AND METHODOLOGY

15.2.1 RAW MATERIALS

Above-land parts of *A. santonica, A. lerchiana* Web., *A. arenaria,* and *A. austriaca* Jack. were gathered and collected not far from the populated areas of the Astrakhan region (Dzhakuevka, Privolzhe, Kucherganovka, Yaksatovo, Streletskoye, Kamyzyak, and Enotaevka). The raw materials were picked in a blossoming phase in July and analyzed in a dry state.

To receive essential oils samples from two species of bugleweed, the vegetative raw materials were gathered in natural habitats: the vegetative and blossoming stems of lycopus high (*L. exaltatus* L.) were picked in Il'inka village; European bugleweed (*L. europaeus* L.) were gathered from the banks of the Volga river in the Astrakhan surroundings during the period from the end of June to September.

H. officinalis and *L. anisatum* Benth. (above-land parts) were provided by All-Russian Research Institute of the irrigated melon growing (a pharmaceutical kitchen garden, Kamyzyak, Russia). The raw materials were analyzed in fresh and dry states.

The dry raw materials were received according to the rules of picking and drying pharmaceutical herbs [34]. To avoid the destruction of biologically active agents and to remove excessive moisture, all raw materials were dried up right after gathering by the most widespread method—the air drying based on a free access of air to the plant material which is spread out in the darkened place.

The essential oil was extracted out from air-dry raw materials weighing 5 kg that consisted of beated land parts (leaves, stems, and inflorescences) through the method of hydrodistillation at atmospheric pressure in the device made of stainless steel, the distillate was collected throughout 5 h. The oil was dried by waterless sodium sulfate; then, it was separated from the drying agent by decantation. The duration of hydrodistillation process is fixed experimentally on the basis of studying the dynamics of change in essential oil yields in time. The yield of essential oil was estimated in percentage (%) terms of the weight of absolutely dry raw materials. Physical and chemical properties of essential oils were determined by the standard techniques [35].

Qualitative and quantitative compositions of essential oil samples were carried out by the chromatograph with a mass-selective detector Shimadzu QP 2010. To identify components, the mass spectra libraries of NIST 02, 05, 11 were used.

The sample of essential oil was dissolved in benzene to the concentration of 0.1% by volume. The column MDN-1 (methyl silicone, firmly bound) is 30 m, the diameter is 0.25 mm. The chromatography mode is the following: the injector is 180°C; the detector is 200°C; the interface is 210°C; the carrier gas is helium (99.99999%), 1 ml/min at the division of the stream 1:10; the thermostat is 60°C in 1 min, 2°/min to 70°C, 5°/min to 90°C, 10°/min to 180°C, 20°/min. to 280°C, further the isotherm is

1 min. The mode of spectra mass registration is 39–350 m/z. To determine linear indices, the samples of essential oil and normal paraffins (nonane, undecane, tridecane, and pentadecane) were dissolved in benzene. n-Paraffins were diluted to the concentration of 0.007% by volume, the essential oil is 1:30,000 by volume. The quantitative content of essential oil components was calculated over the areas of gas-chromotographic of peaks without using correcting indices. The qualitative analysis was carried out by the comparison of linear retention indices [36] and full mass-spectra of components with the relevant data of pure compounds.

Linear retention indices (RI) were calculated by the following formula:

$$RI_x = 100n + 100k \left(\frac{t_{Rx} - t_{Rn}}{tR_{(n+k)} - t_{Rn}} \right),$$

where n is the number of carbon atoms of n-paraffin, k is the difference of the number of carbon atoms in two n-paraffins, t_{Rx} is the time of substance retention, t_{Rn} is the retention time of n-paraffin with n carbon atoms, and $tR_{(n+k)}$ is the retention time of n-paraffin with $n + k$ carbon atoms.

15.2.2 THE STUDY OF ANTIFUNGAL ACTIVITY OF LOPHANTUS ANISATUM BENTH. ESSENTIAL OIL

Studying of antifungal activity was carried out according to the M27 standard by the method of serial dilution of NCCLS [37, 38] in Saburo solid and liquid medium [38].

In a test tube, a microorganism suspension was added to the preparation serially diluted in dimexidum, and the minimum concentration of substance capable to detain growth of test culture was determined. Microorganisms of *Microsporum canis, Trichophyton rubrum*, and *Candida albicans* were used as test cultures.

Test tubes were thermostated at $24 \pm 3°C$ for 7 days (*C. albicans*) and for 30 days (*M. canis, T. rubrum*). To determine the nature of the preparation activity (fungistatic—FS) or (fungicide—FTs), the wort-agar from all the test tubes was plated in the Petri-dish. The cups were placed into the thermostat at $24 \pm 3°C$ for 7 days (*C. albicans*) and for 30 days (*M. canis, T. rubrum*). Econazole was used as a preparation of comparison. The results were statistically processed with the use of Student's t-test.

15.3 RESULTS AND DISCUSSION

The samples of essential oil from four species of wormwood were submitted to the determination of color and refraction index, and the results of these indicators study are given in Table 15.1.

TABLE 15.1 The Area of Picking Raw Material, Yield, Appearance, Refraction Index of Essential Oil from Four Species of Wormwood.

Wormwood species	Area of picking raw material	Appearance	Refraction index n_D^{20}	Yield of essential oil, %
Artemisia austriaca	Kucherganovka	The jellylike mass of light yellow color with a strong smell of camphor	–	0.42
	Kamyzyak		–	0.33
	Privolzhye		–	0.30
	Streletskoye		–	0.37
Artemisia santonica	Dzhakuyevka	The oil of yellowish color with a strong smell of camphor	1.5044	0.70
	Kucherganovka		1.4984	0.65
	Streletskoye		1.5031	0.70
	Yaksatovo		1.5033	0.54
Artemisia lerchiana	Kucherganovka	Colorless oil	1.4822	0.24
	Dzhakuyevka		1.4820	0.32
	Privolzhye		1.4821	0.27
	Streletskoye		1.4822	0.26
Artemisia arenaria	Privolzhye	The oil of yellowish-green color	1.5049	0.48
	Dzhakuyevka		1.5170	0.68
	Streletskoye		1.5055	0.64
	Enotayevka		1.5048	0.67

In Table 15.2, the compounds identified in essential oil from four species of wormwood are given as well as their quantitative contents as percentage of the whole oil.

The evidence presented in Table 15.2 suggests that the essential oils from the four studied species of wormwood differ from each other in their chemical compositions considerably. The main components of essential oil from wormwood of *A. lerchiana* are 1,8-cyneol (31.0–32.3%), camphor (48.5–49.4%), isoborneol (5.6–6.5%), and terpine-4-ol (3.4–3.5%).

TABLE 15.2 The Composition of Essential Oil Samples From *Artemisia lerchiana* (Samples No. 1–4), *Artemisia santonica* (Sample Nos. 5–8), *Artemisia arenaria* (Sample Nos. 9–12), *Artemisia austriaca* (Sample Nos. 13–16)

The name of a sample	RI	Sample number															
		1	2	3	4	5	6	7	8	9	10	11	12	13	14	15	16
α-Pinene	931	0.3	0.3	0.2	0.3	–	–	–	–	8.4	7.9	8.2	7.9	0.1	0.2	0.2	0.3
Camphene	946	2.0	1.9	1.8	2.1	0.6	–	–	–	0.5	0.4	0.6	0.5	–	–	–	–
Amyl vinyl carbinol	951	0.3	0.3	0.4	0.2	0.5	0.4	0.2	0.5	–	–	–	–	0.3	0.4	0.2	0.4
Sabinen	972	–	–	–	–	–	–	–	–	1.3	1.1	0.9	1.4	–	–	–	–
β-Pinene	975	0.5	0.4	0.6	0.5	–	–	–	0.4	23.3	24.3	23.1	23.6	–	–	–	–
Oct-1-en-3-ol	978	0.2	0.2	0.3	0.1	–	–	–	–	–	–	–	–	–	–	–	–
Δ²-Carene	984	0.1	–	0.1	0.1	–	–	–	–	–	–	–	–	–	–	–	–
β-Myrcene	991	0.1	–	0.2	0.1	–	–	0.3	–	1.5	1.1	1.3	1.5	–	–	–	–
p-Cymol	1023	0.8	0.6	0.5	0.7	–	0.3	0.2	–	1.4	1.5	1.4	1.4	–	–	–	–
Limonene	1028	–	–	–	–	–	–	–	–	7.3	7.2	7.1	6.9	–	–	–	–
1,8-Cyneol	1033	31.0	32.3	31.5	31.0	15.4	15.0	15.2	14.0	1.3	0.9	1.2	1.4	1.6	1.6	1.8	1.6
Santolina alcohol	1037	–	–	–	–	0.4	0.4	0.3	0.4	–	–	–	–	–	–	–	–
γ-Terpinen	1058	0.4	0.2	0.5	0.5	0.2	–	0.1	0.1	0.3	0.3	0.2	0.5	–	–	0.2	–
M = 152	1078	0.5	0.6	0.5	0.5	–	–	0.2	–	–	–	–	–	–	–	–	–
Camphenilone	1085	–	–	–	–	–	–	–	–	–	–	–	–	0.1	–	0.4	0.1
cis- Sabinen hydrate	1094	–	–	–	–	0.6	0.7	0.5	0.7	–	–	–	–	–	–	–	–
Linalool	1100	–	–	–	–	–	0.1	–	–	0.1	–	0.1	0.3	–	–	–	–
M = 154	1103	0.3	0.3	0.1	0.4	0.3	0.4	0.4	0.3	–	–	–	–	1.0	0.7	1.1	0.9
isopentyl-3-methylbutanoate	1105	0.3	0.3	0.2	0.2	0.3	0.4	0.4	0.5	–	–	–	–	1.0	0.9	0.9	0.9

TABLE 15.2 *(Continued)*

The name of a sample	RI	Sample number															
		1	2	3	4	5	6	7	8	9	10	11	12	13	14	15	16
3-Tujone	1107	–	–	–	–	–	–	–	0.1	–	–	–	–	0.1	0.2	0.3	0.1
α-Tujone	1108	–	–	–	–	–	–	0.2	–	–	0.2	–	0.1	1.5	1.3	1.5	1.6
M = 152	1118	–	–	–	–	–	–	–	–	–	–	–	–	0.8	0.8	0.9	0.7
M = 140	1134	0.5	0.4	0.4	0.4	–	–	–	–	–	–	–	–	–	–	–	–
trans-Pinocarveol	1139	1.0	0.5	0.8	1.0	–	0.1	–	0.2	0.5	0.6	0.5	0.6	0.3	0.3	–	0.4
Camphor	1146	48.8	49.0	48.5	49.4	60.6	59.8	60.2	59.7	5.4	5.2	5.3	5.4	74.1	73.9	73.8	74.0
Isoborneol	1148	5.6	6.0	6.5	5.7	9.8	9.6	9.5	9.8	1.3	1.0	1.3	1.4	3.4	3.6	3.5	3.1
trans-Verbenol	1150	–	–	–	–	–	0.2	–	0.1	–	–	–	–	0.6	0.5	0.5	0.6
Pinocarvone	1161	0.5	0.4	0.4	0.4	0.8	0.8	0.7	1.0	0.2	0.4	0.1	0.3	0.7	1.0	0.8	0.8
Estragol	1172	0.3	0.1	0.3	0.2	–	0.3	0.1	0.1	0.2	0.1	0.4	0.4	–	–	–	–
Terpine-4-ol	1177	3.4	3.5	3.4	3.4	4.9	4.7	4.8	5.1	1.0	0.9	1.3	0.9	2.1	1.7	1.9	2.3
Myrtenal	1180	0.6	0.3	0.4	0.5	0.6	0.6	0.6	0.5	–	–	–	–	0.7	0.8	0.6	0.4
α-Terpineol	1189	1.1	1.0	0.9	1.2	2.1	1.9	2.4	3.1	0.4	–	0.6	0.5	0.7	0.6	0.7	0.5
Myrtenol	1196	0.5	0.4	0.5	0.2	0.6	0.6	0.4	0.5	0.3	0.3	0.5	0.4	0.8	0.9	0.8	0.8
Cytronellol	1215	–	–	–	–	–	–	–	0.2	0.2	0.3	0.3	0.2	–	–	–	–
trans-Carveol	1219	0.2	0.1	0.2	0.1	0.4	0.5	0.4	0.4	–	–	–	–	0.2	0.5	0.2	0.4
3Z-Hexenyl-2-methylbutanoate	1233	–	–	–	–	–	–	–	–	–	–	–	–	0.7	0.7	0.6	0.7
trans-Pinocaryl acetate	1237	–	–	–	–	0.1	0.4	0.4	0.5	–	–	–	–	3.0	2.9	3.0	3.1
Carvone	1243	0.2	0.3	0.1	0.3	0.4	0.6	0.3	0.3	–	–	–	–	0.8	0.6	0.6	0.8
Phelladral	1271	–	–	–	–	0.4	0.5	0.4	0.4	–	–	–	–	0.5	0.5	0.4	0.6

TABLE 15.2 (Continued)

The name of a sample	RI	Sample number															
		1	2	3	4	5	6	7	8	9	10	11	12	13	14	15	16
Mertenyl acetate	1325	–	–	–	–	–	–	–	–	–	0.1	–	–	0.1	0.1	0.3	0.3
Nerol acetate	1346	–	–	–	–	–	–	0.1	–	0.2	0.3	0.2	0.2	–	–	–	–
α-Terpinyl acetate	1350	–	–	–	–	0.1	0.2	0.1	0.1	–	–	–	–	–	–	–	–
Octahydro-1,4,9,9-tetramethyl-1H-3a,7-methanoazulene	1390	–	–	–	–	–	–	–	–	0.2	0.3	0.1	0.2	–	–	–	–
Eugenol methyl ether	1406	0.5	0.6	0.7	0.5	0.4	0.5	0.4	0.2	1.8	2.2	2.4	1.9	0.4	0.5	0.5	0.4
β-Cedrene	1413	–	–	–	–	–	–	0.2	–	0.4	0.3	0.5	0.4	–	–	–	–
β-Farnesene	1456	–	–	–	–	–	–	–	–	0.2	0.2	0.1	0.1	–	–	–	–
Germacrene D	1483	–	–	–	–	–	0.5	0.2	0.2	–	–	–	–	2.9	3.0	3.0	2.7
γ-Elemene	1500	–	–	–	–	–	–	–	0.1	–	0.2	–	0.1	0.2	0.2	0.2	0.3
Nerolidol	1540	–	–	–	–	–	–	0.2	0.1	1.2	1.3	1.4	1.1	–	–	–	–
Spathulenol	1578	–	–	–	–	–	–	–	–	1.9	1.9	1.7	1.8	–	–	–	–
1,7,7-trimethylbicyclo[2.2.1]hept-2-yl acetate	1584	–	–	–	–	0.5	0.5	0.5	0.4	2.8	2.9	2.6	2.7	0.2	0.4	0.1	0.1
M=196	1615	–	–	–	–	–	–	–	–	–	–	–	–	0.5	0.6	0.5	0.7
Bisabolol oxide II	1655	–	–	–	–	–	–	–	–	1.6	1.6	1.7	1.3	–	–	–	–
α-Bisabolol	1688	–	–	–	–	–	–	–	–	34.7	34.9	34.8	34.5	0.5	0.6	0.5	0.4

Note: M—unidentified compounds and RI—retention index.

At the same time, *A. santonica* essential oil is characterized by higher concentration of camphor (59.7–60.6%), isoborneol (9.5–9.8%), terpine-4-ol (3.9–4.9%), and much smaller contents of 1,8-cyneol (14.0–15.4%). The distinctive characteristic of the chemical composition of *A. arenaria* essential oil is a higher concentration of α-bisabolol (34.5–34.9%), β-pinene (23.1–24.3%), α-pinene (7.9–8.4%), the presence of limonene (6.9–7.3%) unlike other types of wormwood, and the lower contents 1.8-cyneol of (0.9–1.4%) and camphor (5.2–5.4%). In *A. austriaca* essential oil as the main components, there are camphor (73.8–74.1%), isoborneol (3.1–3.6%), *trans*-pinocarvil acetate (3.9–3.1%). Unlike other types of wormwood, in *A. austriaca* essential oil gemacrene D and γ-elemene are identified.

The received samples of essential oil from two species of bugleweed are mobile liquids of yellow color with a characteristic pleasant smell. The essential oil from European bugleweed (*L. europaeus* L.) possesses an evident flower smell with a light aroma of menthol and bergamot. The essential oil of bugleweed high (*L. exaltatus* L.) has a light flower smell. The content of essential oil in vegetative raw materials of *L. exaltatus* L. and *L. europaeus* L. made 0.9% and 0.7%, respectively, during the blossoming period of plants; 0.7% and 0.5% per 100 g of air-dry raw materials, respectively, during the vegetative period. While determining the chemical composition of samples in the essential oil from *L. europaeus* L., the percentage of 31 components was calculated and 12 components were identified. The results are given in Table 15.3.

TABLE 15.3 The Blend Composition of the Essential Oil from *Lycopus europaeus* L. and *Lycopus exaltatus* L.

Component	Content, %	
	Lycopus europaeus L.	*Lycopus exaltatus* L.
Amyl vinyl carbinol	1.59	–
Benzyl alcohol	1.52	–
Phenylethyl alcohol	2.62	–
trans-Pinocarveol	1.14	–
cis-Verbenol	2.61	–
Myrcenol	0.34	–
4-Terpineol	3.65	0.52
α-Terpineol	9.06	0.48

TABLE 15.3 *(Continued)*

Component	Content, %	
	Lycopus europaeus L.	*Lycopus exaltatus* L.
Benihinal	2.72	–
Carvone	2.13	–
Pelargonic acid	1.08	–
Isoeugenol	3.00	–
Damascenone	0.44	–
Jasmone	0.83	–
Copaene	1.95	–
Isoeugenol methyl ether	9.43	–
Caryophyllene	2.51	–
Geranyl acetone	0.56	–
α-Caryophyllene	0.48	–
β-Farnesene	0.78	–
β-Ionone epoxide	0.39	–
Isocyclocytral	1.39	–
α-Selinene	3.48	–
Viridiflorol	1.32	–
Z-α-*trans*-Bergamotol	0.98	–
Z-α-*trans*-Bisabolene epoxide	0.62	–
Caryophyllene oxide	10.30	–
Ledene oxide	0.60	–
2,6,10,14-Hexadecatetraen-1-ol, 3,7,11,15-tetramethyl-, acetate	1.56	–
Ledol	2.24	1.05
Patchulane	0.75	0.56
Paraffines	–	14.11
2,4-Hexadien-1-ol	–	5.05
2-Decenal	–	1.08
2,4-Decadienal	–	51.72
A-Limonene, diepoxide	–	1.61
2-Hepten-1-ol	–	0.35
3,4-Dimethyl-2-hexanone	–	0.35
6,10-Dimethyl-5,9-dodecadien-2-one	–	0.33
Unidentified compounds	27.93	22.79

According to the obtained data, the main components of essential oil from *L. europaeus* L. are terpenoids such as *cis*-verbenol, amyl vinyl carbinol, phenylethyl alcohol, benzyl alcohol, *trans*-pinocarveol, myrcenol, α-terpineol, 4-erpineol, benihinal, carvone, isoeugenol, damascenone, jasmone, isoeugenol methyl ether, geranyl acetone, β-ionone epoxide, and isocyclocytral. In addition, sesquiterpenes are met: α-caryophyllene, caryophyllene, copaene, α-selinene, patchulane, and β-farnesene. Sesquiterpenoids of essential oil from *L. europaeus* L. are viridiflorol, Z-α-*trans*-bisabolene epoxide, Z-α-*trans*-bergamotol, caryophyllene oxide, ledene oxide, and ledol.

Similar to the oil from *L. europaeus* L., the main components of essential oil from *L. exaltatus* L. are terpenoids. They are α-terpineol, 4 terpineol, 2,4-hexadien-1-ol, 2-decenal, 2,4-decadienal, α-limonene diepoxide, 2-hepten-1-ol, 3,4-dimethyl-2-hexanone, 6,10-dimethyl-5,9-dodecadien-2-one. The sesquiterpene of essential oil from *L. exaltatus* L. is patchulane, and the sesquiterpenoid is ledol which is also found in the essential oil from *L. europaeus* L. In the essential oil of from *L. exaltatus* L., there are also paraffins.

On the basis of the total contents of different terpenoids groups in essential oil from *L. europaeus* L., in particular, alcohols (not less than 22.53%), phenols (not less than 12.43%), ketones (not less than 3.96%), aldehydes (not less than 1.39%), etc., it is possible to assume that it shows antiseptic activity (it can stop the growth of bacteria, viruses, fungi, or kill the latter), as well as anesthetizing and antiinflammatory activities, stimulates immune system.

In the essential oil from *L. exaltatus* L., the component structure is less various. However, it has the high concentration of terpene derivatives: aldehydes (not less than 52.8%) and alcohols (not less than 19.16%). So, it is supposed to have much higher antimicrobic, febrifugal, and antiinflammatory activity than the essential oil from *L. europaeus* L. In general, the determination of both pharmacological and potential toxic effects of essential oils from *L. europaeus* L. and *L. exaltatus* L. needs further researches.

The study of the dependence of essential oil yield in terms of vegetation and types of above-land parts of *H. officinalis* showed that the greatest oil yield is received from inflorescences (Table 15.4).

In Table 15.5, the compounds identified in the essential oil from *H. officinalis* (above-land parts in the blossoming phase) are given, as well as their quantitative contents.

TABLE 15.4 The Essential Oil Yield from Different Above-Land Vegetative Parts of *Hyssopus officinalis* and in Different Vegetation Terms.

An above-land vegetative part of *Hyssopus officinalis*	Vegetation terms	Essential oil yield, %[*]
Leaves	May–the beginning of June	$\dfrac{0.3}{0.2}$
Stems	May–the beginning of June	$\dfrac{0.1}{0.1}$
Leaves	Mid-June–the beginning of July (the blossoming period)	$\dfrac{0.4}{0.3}$
Stems	The blossoming period	$\dfrac{0.2}{0.1}$
Inflorescences	–	$\dfrac{0.8}{0.6}$

[*]The numerator and the denominator show the essential oil yield from fresh and dry vegetative raw materials, respectively.

TABLE 15.5 The Quantitative Composition of the Essential Oil from *Hyssopus officinalis*.

The name of a component	RI	Whole oil, %
Sabinen	951	0.19
β-Pinene	954	1.58
Mol. mass = 112*	959	0.22
Eucalyptol	1002	0.34
Dihydrocarveole	1081	0.60
Nopinone	1105	0.48
trans-Pinocarveol	1118	1.41
Verbenol	1124	0.22
Mol. mass = 152*	1131	2.52
Isopinocamphone	1143	63.55
Myrtenal	1161	1.58
α-Terpineol	1166	0.21
Myrtenol	1171	1.39
trans-2-Pinalol	1181	0.72

TABLE 15.5 *(Continued)*

The name of a component	*RI*	Whole oil, %
Pinanediol	1212	9.45
Myrtanal	1292	0.86
α-Bourbonene	1372	0.99
β-Caryophyllene	1404	0.37
Aromadendrene	1443	0.21
Limonen-6-ol, pivalate	1452	0.61
α-Caryophyllene	1457	0.32
Germacrene D	1462	0.95
γ-Elemene	1477	0.18
Mol. mass = 220*	1487	4.64
ε-Muurolene	1494	0.19
Elemol	1524	0.71
Spathulenol	1549	1.08
Caryophyllene, oxide	1553	1.42
Mol. mass = 182*	1577	2.31
τ-Cadinol	1614	0.18
Cubenol	1636	0.33
Mol. mass = 268*	1818	0.18

Note: *RI*—Retention Index.

*Unidentified compound.

In Table 15.6, the content of terpenes, terpenoid, seskviterpen, and seskviterpenoid in the essential oil from *H. officinalis* is given.

The evidence presented in Table 15.2 suggests the main components of essential oil from hyssop medicinal are oxygenated monoterpenes: isopinocamphone (63.55%) and pinanediol (9.45%).

The comparative analysis of the obtained experimental results and literary data on the component composition of essential oils from hyssop growing in other countries (Serbia, Poland) [12, 23] shows a significant difference in the chemical composition of essential oil from hyssop medicinal cultivated in the Astrakhan region. So, the content of β-pinene in the essential oil is much low (1.58%) than in the essential oil from hyssop which is grown up in Poland (6.14%) or in India (18.4%).

TABLE 15.6   The Content of Main Components of Essential Oil.

Components of essential oil	Content, %
Monoterpene hydrocarbons (β-pinene, sabinen)	1.77
Sesquiterpene hydrocarbons (β-caryophyllene, germacrene D, aromadendrene, α-bourbonene, ε-muurolene, γ-elemene, α-caryophyllene)	3.21
Oxygenated monoterpenes (isopinocamphone, nopinone, α-terpineol, trans-pinocarveol, myrtanal, pinanediol, myrtenol, limonen-6-ol pivalate, trans-2-pinalol, verbenol, dihydrocarveole, eucalyptol, myrtenal)	81.42
Oxygenated sesquiterpenes (cubenol, τ-cadinol, caryophyllene oxide, spathulenol, elemol)	3.72
Unidentified compounds	9.87
Total	100

Studying the dependence of essential oil yield on vegetation terms and a type of above-land parts of *L. anisatum* Benth. showed that the maximum yield is received from inflorescences and leaves of the plant in the blossoming phase (Table 15.7).

TABLE 15.7   The Essential Oil Yield from Different Above-Land Vegetative Parts of *Lophantus anisatum* Benth. and in Different Vegetation Terms

An above-land vegetative part of *Lophantus anisatum* Benth.	Vegetation terms	The essential oil yield, %[*]
Leaves	May–the beginning of June	0.45
		0.43
Stems	May–the beginning of June	0.32
		0.27
Leaves	Mid-June–the beginning of July (blossoming period)	0.50
		0.48
Stems	Blossoming period	0.32
		0.30
Inflorescences	–	0.55
		0.54
Seeds	End-April–the beginning of August	0.25

[*]The numerator and the denominator show the essential oil yield from fresh and dry vegetative raw materials, respectively.

The samples of essential oil with a characteristic pleasant smell of anise were submitted to the determination of color, specific weight at 20°C, and the index of refraction, and the results of these indicators determination are presented in Table 15.8.

TABLE 15.8   The Index of Refraction and the Specific Weight of Essential Oil Samples from *Lophantus anisatum* Benth.

Above-land parts of the plant	Color	d, g/sm³	n_D^{20}
Leaves in the vegetation phase	Slightly yellowish	0.9360	1.4700
Leaves in the blossoming phase	Slightly yellowish	0.9365	1.4780
Stems in the vegetation phase	Light yellow	0.9370	1.4782
Stems in the blossoming phase	Light yellow	0.9372	1.4782
Inflorescence	Yellowish-green	1.0070	1.5200
Seeds	Yellowish	0.9532	1.4932

In Table 15.9, the compounds identified in the essential oil from *L. anisatum* Benth. as well as their quantitative contents are given.

TABLE 15.9   The Quantitative Composition of the Essential Oil from *Lophantus anisatum* Benth.

The name of a component	Retention index, *RI*	Percentage (%) of the whole oil
Amyl vinyl carbinol	957	0.32
β-Myrcene	990	0.06
D-Limonene	1014	8.14
Linalool	1086	0.07
1-Octenyl acetate	1094	0.50
Methyl chavicol	1172	62.08
Chavicol	1215	0.12
Mol. mass = 162*	1217	1.19
Eugenol	1330	0.09
β-Elemene	1394	0.59
Methyl eugenol	1453	24.01
Caryophyllene	1403	1.28
Germacrene D	1480	0.80

TABLE 15.9   *(Continued)*

The name of a component	Retention index, *RI*	Percentage (%) of the whole oil
δ-Cadinene	1516	0.15
Germacrene D-4-ol	1536	0.12
τ-Muurolol	1564	0.24
α-Cadinol	1637	0.24

*Unidentified compound.

The evidence presented in Table 15.9 suggests the main components of essential oil from *L. anisatum* Benth. are methyl chavicol (62.08%), eugenol methyl ether (24.01%) and D-limonene (8.14%).

The results of studying antifungal effect of the essential oil from *L. ophantus anisatum* Benth. are presented in Table 15.10.

TABLE 15.10   The Fungistatic and Fungicide Activities of *Lophantus anisatum* Benth. Essential Oil.

The studied sample	Concentration, mcg/ml*		
	Microsporum canis	*Trichophyton rubrum*	*Candida albicans*
Lophantus anisatum Benth. essential oil	$\dfrac{80}{100^{**}}$	$\dfrac{80^{**}}{100}$	$\dfrac{100}{200^{**}}$
Econazole	$\dfrac{40}{80}$	$\dfrac{40^{**}}{80}$	$\dfrac{40}{80}$

*The numerator is a fungistatic activity; the denominator is a fungicide activity.
**The distinctions between repeated patterns are reliable at $p = 0.95$.

The obtained experimental data show that essential oil from *L. anisatum* Benth. has a fungistatic and fungicide activity against the studied test cultures.

15.4   CONCLUSIONS

The obtained data enlarge and supplement the knowledge of the chemical composition of wormwood growing on the Caspian Plain significantly, as well as they allow to expand the raw-material base of essential-oil plants due to the plant species of the *Artemisia* L. family. Some compounds that are not typical for the *Artemisia* L. family such as Amyl vinyl carbinol, octahydro-1,4,9,9-tetramethyl-1H-3a, 7-methanoazulene, and neryl acetate were identified. It is supposed to be the result of the complex of soil climatic conditions and vegetative biocenoses in the Caspian Sea region.

It is found out that the content of essential oil in *L. europaeus* L. and *L. exaltatus* L. is higher when picking raw materials in a blossoming phase. Through the method of chromatography–mass-spectrometry, 31 and 12 components were identified in grassy European bugleweed (*L. europaeus* L.) and in grassy bugleweed high (*L. exaltatus* L.), respectively. European bugleweed and bugleweed high raw materials are of interest for the pharmaceutical, perfumery, and cosmetic industries as well as for aromatherapy.

The qualitative and quantitative chemical composition of *H. officinalis* L. and *L. anisatum* Benth. essential oils cultivated in the Astrakhan region is determined. The main components of *H. officinalis* L. are isopinocamphone and pinanediol. *L. anisatum* Benth. essential oil that contains methyl chavicol and methyl eugenol as the main components. The study of antifungal activity showed that the essential oil from *L. anisatum* Benth. possesses a rather high antifungal activity against *M. canis*, *T. rubrum*, and *C. albicans*.

KEYWORDS

- **wormwood**
- **lycopus**
- **hyssop**
- **giant hyssop**
- **hydrodistillation**
- **chromatography**
- **isopinocamphone**
- **pinanediol**

REFERENCES

1. Pilipenko, V. N.; Teplyi, D. L.; Vasil'eva, L. A. Medicinal Plants of the Astrakhan Region. Astrakhan, Astrakhan State Pedagogical University, 1996; p 181 (in Russian).
2. Abubakirov, N.K.; Belenovskaya, L.M.; Grushevskii, I.M.; Kozhina, I.S.; Kuzhetsova, G.A.; Kuz'mina, L.V.; Medvedeva, L.I.; Pimenov, M.G.; Sokolov, P.D. (Editor); Cherepanov, S.K.; Shukhobodskii, B.A.; Yunusov, C.Yu. Vegetable Recourses of USSR. Flowering Plants, Their Chemical Composition, Application. The Family *Asteraceae*: Saint Petersburg, 1993; 352p. (in Russian).
3. Cherepanov, S. K. The Tracheophytes of Russia and the Neighbouring Countries, Saint-Petersburg, Peace and Family, 1995; p 992 (in Russian).
4. Orlova, Yu. V. Ecophysiological Characteristic of *Artemisia lerchiana Web* in the Volga Delta Region. Author's Abstract of dis. PhD of Biology Science (03.00.12), Orlova, Yu.V., Volgograd State Pedagogical University, Moscow, 24p (in Russian).
5. Shelukhina, N. A.; Savina, A. A.; Sheichenko, V. I.; Sokol'skaya, T. A.; Bykov, V. A.. Evaluation of Chemical Composition of *Lycopus europaeus L. Prob. Biol. Clin. Pharm. Chem.* **2010**, *11*, 7–11 (in Russian).
6. Alefirov, A. N.; Sivak, K. V. Antithyroid Effect of *Lycopus europaeus* L. (*Lamiaceae*) Extracts in Rats with Experimental Thyrotoxicosis. *Plant Resour.* **2009**, *45(2)*, 117–122 (in Russian).
7. Ionova, V. A.; Gavrilova, T. L.; Shchepetova, E. V.; Imasheva, N. M. Content of Some Biological Active Substances in Herbage of *Lycopus europaeus* L., *Lycopus exaltatus* L. *Nat. Sci.* **2013**, *1*, 93–99 (in Russian).
8. Radulovic, N.; Denic, M.; Stojanovic-Radic, Z.; Skropeta, D. Fatty and Volatile Oils of the Gypsywort *Lycopus europaeus* L. and the Gaussian-Like Distribution of its Wax Alkanes. *JAOCS.* **2012**, *89*(12), 2165–2185.
9. Hoppe, H. A. Drogenkunde. Walter de Gruyter: Berlin, New York, 1975.
10. Baj, T.; Kowalski, R.; Świątek, Ł.; Modzelewska, M.; Wolski, T. Chemical Composition and Antioxidant Activity of the Essentials Oil of Hyssop (*Hyssopus officinalis L. ssp officinalis*). *Ann. Univ. Mariae Curie-Skłodowska.* **2010**, *23*(3), sect. DDD, 55–62.
11. Fraternale, D.; Ricci, D.; Epifano, F.; Curini, M. Composition and Antifungal Activity of Two Essential Oils of Hyssop (*Hyssopus officinalis* L.). *J. Essent Oil Res.* **2004**, *16*(6), 617–622.
12. Garg, S. N.; Naqvi, A. A.; Singh, A.; Ram, G.; Kumar, S. Composition of Essential Oil from an Annual Crop of *H. officinalis* Grown in Indian Plains. *Flav. Fragr. J.* **1999**, *14*(3), 170–172.
13. Renzini, G.; Scazzocchio, F.; Lu, M.; Mazzanti, G.; Salvatore, G. Antibacterial and Cytotoxic Activity of *Hyssopus officinalis* L. Oil. *J. Essent. Oil Res.* **1999**, *11*(5), 649–654.
14. Tognolini, M.; Barocelli, E.; Ballabeni, V.; Bruni, R.; Bianchi, A.; Chiavarini, M.; Impicciatore, M. Comparative Screening of Plant Essential Oils: Phenylpropanoid Moiety as Basic Core for Antiplatelet Activity. *Life Sci.* **2006**, *78*, 1419–1432.
15. Lu, M.; Battinelli, L.; Daniele, C.; Melchioni, C.; Salvatore, G.; Mazzanti, G. Muscle Relaxing Activity of *Hyssopus officinalis* Essential Oil on Isolated Intestinal Preparations. *Planta Med.* **2002**, *68*(3), 213–216.

16. Wolski, T.; Baj, T.; Kwiatkowski, S. Hyzop Lekarski (*Hyssopus officinalis* L.) Zapomniana Roślina Lecznicza, Przyprawowa Oraz Miododajna. *Ann. Univ. Mariae Curie-Skłodowska.* **2006,** *41*(1), sect. DD, 1–10.

17. Wolski, T.; Baj, T. Hyzop Lekarski (*Hyssopus officinalis* L.) Aromatyczna Roślina Lecznicza. *Aromaterapia.* **2006,** *4*(46), 10–18.

18. Gorunović, M.; Bogavac, P.; Chulchat, J.; Chabardi, J. Essential Oil of *Hyssopus officinalis* L. Lamiaceae of Montenegro Origin. *J. Essent. Oil Res.* **1995,** *7,* 39–43.

19. Mazzanti, G.; Battinelli, L.; Salvatore, G. Antimicrobial Properties of the Linalool-Rich Essential Oil of *Hyssopus officinalis* L. Var Decumbens (Lamiaceae). *Flavour Fragr. J.* **1998,** *13,* 289–294.

20. Özer, H.; Şahin, F.; Kiliç, H.; Güllüce, M. Essential Oil Composition of *Hyssopus officinalis* L. subsp. *angustifolius* (Bieb.) Arcangeli from Turkey. *Flavour Fragr. J.* **2005,** *20*(1), 42–44.

21. Mitić, V.; Đorđević, S. Essential Oil Composition of *Hyssopus oficinalis* L. Cultivated in Serbia, Facta univ. *Ser. Phys., Chem. Technol.* **2000,** *2(2),* 105–108.

22. Joulain, D.; König, W. A. *The Atlas of Spectral Data of Sesquiterpene Hydrocarbons.* E.B.-Verlag: Hamburg, 1998.

23. Vallejo, M.; Herraiz, J.; Perez-Alonso, M.; Velasco-Negueruela, A. Volatile Oil of *Hyssopus officinalis* L. from Spain. *J. Essent. Oil Res.* **1995,** *7,* 567–568.

24. Salvatore, G.; D'Andrea, A.; Nicoletti, M. A Pinocamphone Poor Oil of *Hyssopus officinalis* L. var. Decumbens from France (Banon). *J. Essent. Oil Res.* **1998,** *10,* 563–567.

25. Dzhumaev, Kh. K.; Zenkevich, I. G.; Tkachenko, K. G.; Tsibul'skaya, I. A. Essential Oil of the Leaves of *Hyssopus seravschanicus* from South Uzbekistan. *Chem. Nat. Comp.* **1990,** *26*(1), 101–102.

26. Vincenzi, M.; Silano, M.; Stacchini, P.; Scazzocchio, B. Constituents of Aromatic Plants: I. Methyleugenol. *De Vincenzi, M., Silano, M., Stacchini, P., Scazzocchio, B., Fitoterapia,.* **2000,** *71*(2), 216–221.

27. Burfield, T. Safety of Essential Oils. *Int. J. Aromather.* **2000,** *10*(1–2), 16–29.

28. Proshakov, Yu. I. Lophant Anisic is Counteract of Ginseng. *Potato Veg.* **2002,** *1,* 16–17 (in Russian).

29. Pustovalova, N. Aromatic Giant-Hyssop. *Garden Veg. Garden.* **2004,** *5,* 13–16.

30. Velikorodov, A. V.; Morozova, L. V.; Pilipenko, V. N.; Kovalev, V. B. Chemical Composition of Essential Oil of Four Endemic Species Artemisia of Astrakhan region: *Artemisia lerchiana, Artemisia santonica, Artemisia arenaria, Artemisia austriaca. Chem. Plant Raw Mater.* **2011,** *4,* 115–120 (in Russian).

31. Gavrilova, T. L.; Shchepetova, E. V.; Abdurakhmanova, N. M.; Kovalev, V. B. Studding of Chemical Composition of Essential Oil of *Lycopus,* Growing in Astrakhan Region Modern Problems of Science & Education, **2015,** *4,* http://www.science-education.ru/ru/article/view?id=20991 (in Russian).

32. Velikorodov, A. V.; Kovalev, V. B.; Kurbanova F. Kh.; Shchepetova E. V. Chemical Composition of Essential Oil of *Hyssopus officinalis* L., Cultivated in Astrakhan region. *Chem. Plant Raw Mater.* **2015,** *12,* 71–76 (in Russian).

33. Velikorodov, A. V.; Kovalev, V. B.; Tyrkov, A. G.; Degtyarev, O. V. Studding of the Chemical Composition and Antifungal Activity of *Lophantus anisatum* Benth. Essential Oil. *Chem. Plant Raw Mater.* **2010,** *2,* 143–146 (in Russian).

34. Kuznetsova M.A. *The Rules for Harvesting & Drying of Drug Plants.* Moscow, 1985; p 321 (in Russian).

35. Goryaev, M. I.; Pliva, I. *Methods of Study of Essential Oils.* Alma-Ata, 1962; p 751.

36. Tkachev, A. V. The Study of Plant Volatiles. Novosibirsk, Offset, 2008; 969p. (in Russian)

37. Espinel-Ingroff, F.; Boyle, K.; Sheehan, D.J. In Vitro Antifungal Activities of Voriconazole and Reference Agents as Determined by NCCLS Methods: Review of the Literature. *Mycopathologia.* **2001,** *150,* 101−115.

38. Rex, J. H.; Pfaller, M. A.; Galgiani, J. N.; Bartlett, M. S.; Espinel-Ingroff, A.; Ghannoum, M. A.; Lancaster, M.; Odds, F. C.; Rinaldi, M. G.; Walsh, T. J.; Barry, A. L. Development of Interpretive Breakpoints for Antifungal Susceptibility Testing: Conceptual Framework and Analysis of In Vitro−In Vivo Correlation Data for Fluconazole, Itraconazole, and Candida Infections. *Clin. Infect. Dis.* **1997,** *24*(2), 248−249.

GLOSSARY

Air deodorization Removing malodorous gaseous substances from air.

Alanit The zeolite clay of North Ossetia deposits contain 51–53% silicon, aluminum 16–17%, 5–6% iron, 30–33% calcium, potassium—0.07%, phosphorus—0.38%, manganese—0.04%, sulfur—0.98% magnesium—1.6%, and small amounts of zinc, copper, cobalt, and other trace elements. The reaction is medium (pH) 8.64 due to the high calcium content.

α,β-Alkenenitriles In organic chemistry, an alkene is an unsaturated hydrocarbon that contains at least one carbon–carbon double bond. The words alkene, olefin, and olefin are used often interchangeably (see nomenclature section below). The α,β-alkene nitriles have a double bond next to the nitrile group.

Amaranth *Amaranthus* is a widespread genus of mainly annual herbaceous plants with small flowers, gathered in dense spike-paniculate inflorescences.

Amide An amide, also known as an acid amide, is a compound with the functional group $R_nE(O)_xNR'_2$ (R and R' refer to H or organic groups). The simplest amides are derivatives of ammonia wherein one hydrogen atom has been replaced by an acyl group. The ensemble is generally represented as $RC(O)NH_2$. Amides are usually regarded as derivatives of carboxylic acids in which the hydroxyl group has been replaced by an amine or ammonia.

Annelation In organic chemistry, it is a chemical reaction in which a new ring is constructed on a molecule.

Antiangiogenic activity The idea of antiangiogenic therapy was the brainchild of Dr. Judah Folkman in the early 1970s. He proposed that by cutting off the blood supply, cancer cells would be deprived of nutrients and, hence, treated. His efforts paid off when bevacizumab, a monoclonal antibody targeting vascular endothelial growth factor, was approved as antiangiogenic therapy in 2004 for the treatment of colon cancer. Since then, an array of antiangiogenic inhibitors, either as monotherapy or in combination with other cytotoxic and chemotherapy drugs, have been developed, used in clinical trials and approved for the treatment of cancer.

Antiinflammatory The property of a substance (drug) to reduce inflammation or swelling.

Antioxidants The molecules inhibit the oxidation of other molecules. Oxidation is a chemical reaction that can produce free radicals, leading to chain reactions that may damage cells (lipid peroxidation). Antioxidants such as thiols or ascorbic acid (vitamin C) terminate these chain reactions. To balance the oxidative state, plants and animals maintain complex systems of overlapping antioxidants, such as glutathione and enzymes (e.g., catalase and superoxide dismutase) produced internally or the dietary antioxidants, vitamin A, vitamin C, and vitamin E.

Antiviral activity Compounds with antiviral activity (antiviral drugs) are a class of medication used specifically for treating viral infections. Like antibiotics and broad-spectrum antibiotics for bacteria, most antivirals are used for specific viral infections, whereas a broad-spectrum antiviral is effective against a wide range of viruses. Unlike most antibiotics, antiviral drugs do not destroy their target pathogen; instead, they inhibit their development.

Apoptosis A regulated process of programed cell death that occurs in multicellular organisms. Biochemical events lead to characteristic cell changes (morphology) and death.

Autophagy A regulated process for degradation of unnecessary or dysfunctional cellular components.

Azoles Azoles are a class of five-membered heterocyclic compounds containing an nitrogen atom and at least one other non-carbon atom (i.e., nitrogen, sulfur, or oxygen) as part of the ring.

Beckmann fragmentation The rearrangement of an oxime to its corresponding amide in the presence of an acid and is generally known as the Beckmann rearrangement. The Beckmann rearrangement of oxime derivatives proceeds stereospecifically, and the stereo-configuration of the migrating group is retained. Certain oximes, those having a quarternary carbon *anti* to the hydroxyl, have been reported to undergo the rearrangement to form nitriles instead of amides. This type of transformation is called the Beckmann fragmentation, which has been used in modification of steroids. Other oximes, such as the bridged bicyclic ketoximes and oximes with an electron-donating substituent at the α-carbon, also undergo fragmentation.

Bioavailability A measurement of the rate and extent to which a substance (drug) reaches at the site of action. In pharmacology, it is a subcategory of absorption and is used to describe the fraction of an administered dose of unchanged drug that reaches the systemic circulation, one of the principal pharmacokinetic properties of drugs.

Bitumen Peat organic matter compound soluble in organic solvents.

Bulk density A property of "divided" solids, which is defined as the mass of many particles of the material divided by the total volume they occupy.

Chemotype A chemotype (sometimes chemovar) is a chemically distinct entity in a plant or microorganism, with differences in the composition of the secondary metabolites. Minor genetic and epigenetic changes with little or no effect on morphology or anatomy may produce large changes in the chemical phenotype. Chemotypes are often defined by the most abundant chemical produced by that individual, and the concept has been useful in work done by chemical ecologists and natural product chemists.

Conjugate In chemistry, a conjugate refers to a compound formed by the joining of two or more chemical compounds or the term conjugate refers to an acid and base that differ from each other by a proton.

Cytotoxic activity Cytotoxicity is the quality of being toxic to cells. Examples of toxic agents are an immune cell or some types of venom.
Die A specialized tool used in manufacturing industries to cut or shape material mostly using a press.

Dieckmann condensation The Dieckmann condensation is the intramolecular chemical reaction of diesters with base to give β-ketoesters. It is named after the German chemist Walter Dieckmann (1869–1925).

Donor–acceptor bond A term denoting one of the ways in which a chemical covalent bond is formed. The ordinary covalent bond between two atoms is due to the interaction of two electrons, one from each atom.

Doping carboxylic group Carboxylic groups with positive or negative electricity.

Easily hydrolized substances Peat organic matter compound, soluble in weak acids.

Endemic species Species that is found only restricted area is known as endemic species.

EPR spectroscopy A technique electron paramagnetic resonance spectroscopy for studying materials with unpaired electrons.

Eriophorum peat A type of grass peat composed from 40% to 100% *Eriophorum* plants remains.

Erythrocytes Red blood cells are the most common type of blood cell, and the vertebrate organism's principal means of delivering oxygen (O_2) to the body tissues via blood flow through the circulatory system.

Essential oil An essential oil is a concentrated hydrophobic liquid containing volatile aroma compounds from plants. Essential oils are also known as volatile oils, ethereal oils, aetherolea, or simply as the oil of the plant from which they were extracted, such as oil of clove. An oil is "essential" in the sense that it contains the "essence of" the plant's fragrance—the characteristic fragrance of the plant from which it is derived.

Esters In chemistry, esters are chemical compounds derived from an acid (organic or inorganic) in which at least one –OH (hydroxyl) group is replaced by an –O–alkyl (alkoxy) group. Usually, esters are derived from a carboxylic acid and an alcohol.

Extra equivalent adsorption Called sorbent nonexchangeable adsorption, proceeding along with ion exchange or after its finishing without opposite-charged ion displacement from sorbent phase as a result of specific interactions.

Extrusion A process used to create objects of a fixed cross-sectional profile. A material is pushed through a spinneret of the desired cross-section.

***g*-Factor** A factor that connects a gyromagnetic ratio of the particles with the classical value of the gyromagnetic ratio. For classical particles, *g*-factor is 1; for free quantum particles with non-zero spin, these value is 2; for real particles experimentally certain value, *g*-factor may be different from both 1 and 2 and is one of the characteristics of the particles.

Gluconeogenesis The synthesis of glucose from noncarbohydrate carbon substrates.

Grignard reaction The Grignard reaction is an organometallic chemical reaction in which alkyl, vinyl, or aryl-magnesium halides (Grignard reagents) add to a carbonyl group in an aldehyde or ketone. This reaction is an important tool for the formation of carbon–carbon bonds. The reaction of an organic halide with magnesium is not a Grignard reaction but provides a Grignard reagent. Grignard reactions and reagents were discovered by and are named after the French chemist François Auguste Victor Grignard (University of Nancy, France), who was awarded the 1912 Nobel Prize in Chemistry for this work.

Hemolytic activity It is ability to cause hemolysis with hemoglobin releasing.

Heterocyclic A heterocyclic compound or ring structure is a cyclic compound that has atoms of at least two different elements as members of its ring(s). Heterocyclic chemistry is the branch of chemistry dealing with the synthesis, properties, and applications of these heterocycles.

High-moor peat The peat formed from oligotrophic vegetation.

Homeostasis The property of an open system to maintain its internal state and remain stable through a set of coordinated reactions.

Humic acids A principal component of humic substances, which are the major organic constituents of soil (humus), peat, coal, many upland streams, etc.

Humic substances Complex and heterogeneous mixtures of polydispersed materials formed by biochemical and chemical reactions during the decay and transformation of plant and microbial remains (a process called humification).

Humin Peat organic matter compound, nonsoluble in acids and organic solvents.

Hydrate layer A thin layer formed by water-oriented molecules between the phase boundaries.

Hydrazides Hydrazides in organic chemistry are a class of organic compounds sharing a common functional group characterized by a nitrogen-to-nitrogen covalent bond with four substituents with at least one of them being an acyl group. The general structure for a hydrazide is $E(=O)–NR–NR_2$, where the Rs are frequently hydrogens. Hydrazides can be further classified by atom attached to the oxygen: carbohydrazides $(R–C(=O)–NH–NH_2)$, sulfonohydrazides $(R–S(=O)_2–NH–NH_2)$, and phosphonic dihydrazides $(R–P(=O)(–NH–NH_2)_2$.

Hydrazones Hydrazones are a class of organic compounds with the structure $R_1R_2C=NNH_2$. They are related to ketones and aldehydes by the replacement of the oxygen with the NNH_2 functional group. They are formed usually by the action of hydrazine on ketones or aldehydes.

Hydro distillation Hydro distillation is a process where the plant material is boiled with the resultant steam being captured and condensed. The oil and water are then separated; the water, referred to as a 'hydrosol,' can be retained as it will have some of the plant essence. Rose hydrosol, for example, is commonly used for it's mild antiseptic and soothing properties, as well as it's pleasing floral aroma.

Hydrogen bond A type of bond formed when a hydrogen atom bonded to atom A in one molecule makes an additional bond to atom B either in the same or another molecule.

Hyperfine coupling The spacing between the EPR spectral lines indicates the degree of interaction between the unpaired electron and the perturbing nuclei.

Hyperglycemia It is caused by high levels of glucose in the blood.

I The concentration of free radicals, which characterizes the intensity of the EPR-signal; the I value is measured in relative units (spin/g).

IFN The abbreviation of several related proteins that are produced by the body's cells as a defensive response to viruses. They are important modulators of the immune response. Interferon was named for its ability to interfere with viral proliferation. The various forms of interferon are the body's most rapidly produced and important defense against viruses. Interferons can also combat bacterial and parasitic infections, inhibit cell division, and promote or impede the differentiation of cells.

IFN-α Produced by monocytes/macrophages, lymphoblastoid cells, fibroblasts, and a number of various cell types following induction by viruses, nucleic acids, glucocorticoid hormones, and low-molecular weight substances.

IFN-γ Interferon gamma is a dimerized soluble cytokine that is the only member of the type II lass of interferons.

IL The abbreviation of a group of naturally occurring proteins that mediate communication between cells. Interleukins regulate cell growth, differentiation, and motility. They are particularly important in stimulating immune responses, such as inflammation. Interleukins are a subset of a larger group of cellular messenger molecules called cytokines, which are modulators of cellular behavior. Like other cytokines, interleukins are not stored within cells but are instead secreted rapidly, and briefly, in response to a stimulus, such as an infectious agent.

IL-2 Interleukin-2, known as T-cell growth factor, was the name given to the lymphocyte product that stimulated T-cell proliferation.

IL-4 Interleukin-4 is a cytokine that induces differentiation of naive helper T cells (Th0 cells) to Th2 cells.

Intermolecular cross-linking bond An intermolecular bond, when two highly electronegative atoms joined through hydrogen atom.

Intensity of resonance line (I) The concentration of free radicals, which characterizes the intensity of the EPR-signal. The value is measured in relative units (spin/g).

Ion-exchange groups Acidic or a basic group capable of ion-exchange process.

IR-spectroscopy (Infrared spectroscopy) The spectroscopy that deals with the infrared region of the electromagnetic spectrum, that is light with a longer wavelength and lower frequency than visible light.

Isobornylphenols The terpenephenols containing isobornyl groups as alkyl substituents and synthesized with the use of natural monoterpenes.

Linear retention indices In gas chromatography, Kovats retention index (shorter Kovats index, retention index; plural retention indices) is used to convert retention times into system-independent constants. The index is named after the Hungarian-born Swiss chemist Ervin Kováts, who outlined this concept during the 1950s while performing research into the composition of the essential oils.

Low-moor peat The peat formed from eutrophic vegetation.

Membrane-protective properties The ability to protect cell membranes from damage of various kinds, including the effects of free radicals.

Mesophites Mesophytes are terrestrial plants which are adapted to neither a particularly dry nor particularly wet environment. Mesophytes make up the largest ecological group of terrestrial plants, and usually grow under moderate to hot and humid climatic regions.

Metabolic syndrome A disorder, characterized by several pathological conditions, such as obesity, elevated blood pressure, elevated fasting plasma glucose, high serum triglycerides, and low high-density lipoprotein levels.

Metabolism The set of chemical transformations aimed to maintain the living state of the cells and the organism.

Mitochondria A double membrane-bound organelle found in all eukaryotic organisms. The most prominent roles of mitochondria are to produce the energy currency of the cell, ATP (i.e., phosphorylation of ADP), through respiration, and to regulate cellular metabolism. The central set of reactions involved in ATP production is collectively known as the citric acid cycle, or the Krebs cycle. However, the mitochondrion has many other functions in addition to the production of ATP.

Monosaccharides They are simple sugars with the general formula $C_nH_{2n}O_n$ that are not able to be hydrolyzed to simpler compounds, in which each carbon atom bounding with a hydroxyl group is chiral, giving rise to a number of isomeric forms, all with the same chemical formula.

Monoterpenoids These are derivatives of monoterpenes that consists of two HYPERLINK "https://en.wikipedia.org/wiki/Isoprene" \o "Isoprene" isoprene units and bounded with O or S atom.

Oxidative hemolysis The free radical oxidation of lipids and proteins of the plasma membrane of erythrocytes. The result is an increase in the erythrocyte membrane permeability, which further leads to the realization of the mechanism osmotic hemolysis.

Paramagnetism A form of magnetism whereby certain materials are attracted by an externally applied magnetic field, and form internally induced magnetic fields in the direction of the applied magnetic field.

Peat A heterogeneous mixture of more or less decomposed plant (humus) material that has accumulated in a water-saturated environment and in the absence of oxygen.

Peat decomposition degree A relative proportion of humification products of the entire peat substance.

Plant growth regulators The natural or synthetic substances used for controlling or modifying plant growth processes, such as formation of leaves and flowers, elongation of stems, development and ripening of fruit, and for increase in resistance of stress factors.

Polyconjugated systems The systems formed by multiple conjugated double bonds in polymers.

Polypropylene-grafted-acrylic acid Polypropylene functionalized with acrylic acid as a comonomer.

Pro-apoptotic effect The ability of compounds to induce apoptosis in cancer cells. Apoptosis is a process of programed cell death that occurs in multicellular organisms. Biochemical events lead to characteristic cell changes (morphology) and death.

Protein kinase An enzyme that performs the chemical modification of proteins (substrates) by addition of phosphate groups to them (phosphorylation).

Protein phosphatase An enzyme that removes a phosphate group from the phosphorylated amino acid residue of the protein (substrate).

Replication of viruses Viral replication is the formation of biological viruses during the infection process in the target host cells. Viruses must first get into the cell before viral replication can occur. From the perspective of the virus, the purpose of viral replication is to allow production and survival of its kind. By generating abundant copies of its genome and packaging these copies into viruses, the virus is able to continue infecting new hosts. Replication between viruses is greatly varied and depends on the type of genes involved in them. Most DNA viruses assemble in the nucleus while most RNA viruses develop solely in cytoplasm.

Seco- (prefix) Fission of a ring, with addition of a hydrogen atom at each terminal group thus created, is indicated by the prefix **seco-**, the original steroid numbering being retained. (If more than one ring is opened, general systematic nomenclature may be preferable.)

Sedge peat A type of grass peat composed from more than 65% sedge plants remains.

Semiquinone anion radical's production Process when quinone is reduced to a semiquinone radical anion by the addition of a single H atom to a quinone.

Sesquiterpenes Sesquiterpenes are a class of terpenes that consist of three isoprene units and have the empirical formula $C_{15}H_{24}$. Like monoterpenes, sesquiterpenes may be acyclic or contain rings, including many unique combinations.

Sesquiterpenoids Sesquiterpenoids are sesquiterpenes with additional elements besides carbon and hydrogen. Biochemical modifications of sesquiterpenes such as oxidation or rearrangement produce the related sesquiterpenoids.

SHF-power Super high frequency power. The power in the range between 3 GHz and 30 GHz.

Signaling pathway The entire set of cell changes induced by receptor activation.

Sorption properties The ratio of the mass of sorbate to the unit mass of sorbent.

Spin density The electron density applied to free radicals. It is defined as the total electron density of electrons of one spin minus the total electron density of the electrons of the other spin.

Supramolecular associates A well-defined complex of molecules held together by noncovalent bonds.

Synergetic effect The combination of two or more things that creates an effect which is greater than the sum of both separately.

Terpenoids Terpenoids are the hydrocarbons of plant origin of the general formula $(C_5H_8)_n$ as well as their oxygenated, hydrogenated, and dehydrogenated derivatives. The terpenoids, sometimes called isoprenoids, are a large and diverse class of naturally occurring organic chemicals similar to terpenes, derived from five-carbon isoprene units assembled and modified in thousands of ways. Most are multicyclic structures that differ from one another not only in functional groups but also in their basic carbon skeletons. About 60% of known natural products are terpenoids. The basic molecular formulae of terpenes are thus multiples of C_5H_8. Most terpenes are classified by the number of C_5 isoprene units that they contain. Given the many ways the basic C_5 units can be combined, it is not surprising to observe the amount and diversity of the structures. The classes are: hemiterpenes consisting of a single C_5 isoprene unit, monoterpenes (C_{10}), sesquiterpenes (C_{15}), diterpenes (C_{20}), sesterterpenes (C_{25}), triterpenes (C_{30}), carotenoids (C_{40}), and polyterpenes consisting of long chains of many isoprene units.

Thorpe–Ziegler intramolecular cyclizations The Thorpe–Ziegler cyclization is a base-catalyzed intramolecular condensation of α,ω-dinitriles to form cyclic ketones and has been employed especially for the preparation of macrocyclic ketones by a high dilution technique. Occasionally, this reaction is also known as the Thorpe–Ziegler addition. This reaction has been extended to the intramolecular condensation of ω-cyano sulfones. At present, this reaction has been used for the preparation of heterocycles.

Water absorption capacity The ability of peat to keep a certain amount of water after excessive moisture.

Water-soluble substances Peat organic matter compound soluble in water.

Width of resonance line (ΔH) A value that characterizes the width of the EPR-signal. This value is measured in units of magnetic field gauss (G).

X-ray crystallography A tool used for identifying the atomic and molecular structure of a crystal, in which the crystalline atoms cause a beam of incident X-rays to diffract into many specific directions. By measuring the angles and intensities of these diffracted beams, a crystallographer can produce a three-dimensional picture of the density of electrons within the crystal. From this electron density, the mean positions of the atoms in the crystal can be determined, as well as their chemical bonds, their disorder, and various other information.

INDEX

For Product Safety Concerns and Information please contact our
EU representative GPSR@taylorandfrancis.com Taylor & Francis
Verlag GmbH, Kaufingerstraße 24, 80331 München, Germany